IEE TELECOMMUNICATIONS SERIES 41

Series Editors: Professor Charles J. Hughes
Professor David Parsons
Professor Gerry White

WORLD TELECOMMUNICATIONS ECONOMICS

Other volumes in this series:

WORLD TELECOMMUNICATIONS ECONOMICS

Jeffery J. Wheatley

The Institution of Electrical Engineers

Published by: The Institution of Electrical Engineers, London,
United Kingdom

© 1999: The Institution of Electrical Engineers

The Institution of Electrical Engineers,
Michael Faraday House,
Six Hills Way, Stevenage,
Herts. SG1 2AY, United Kingdom

British Library Cataloguing in Publication Data

A CIP catalogue record for this book
is available from the British Library

ISBN 0 85296 936 8

Printed in England by Short Run Press Ltd., Exeter

Contents

Preface

The aim of this book is to describe the economics of telecommunications in a way that will be of use to policy analysts, engineers and managers within the industry and to economists and others wanting to know more about telecommunications. Readers with an unspecialised background in business, research or teaching should find it to be a helpful introduction.

The substance of most economic theories can be grasped without the need for extensive mathematical analysis, making the subject readily accessible. Economic relationships do not follow exact or unchanging rules. Quite complex statistical methods may be needed to estimate current relationships, but the underlying propositions are behavioral and more intuitive. The book takes a world perspective because of the globalisation of the communications media, and the ambitions and competitive reach of many of the individuals, companies and state corporations working within them.

Over the past twenty years, there has been a worldwide policy shift from demand management to a trust in markets, paralleled by striking political changes which have dissolved the command economies and moved parties of all persuasions closer to market solutions. The great state monopolies are losing their power. Private ownership is an essential precondition of competition. The new private owners are fighting off an increasing number of competitors. Some in telecommunications would say that this is an inevitable consequence of a new technology so powerful that, as I had it once put to me by someone holding a piece of five-strand optical-fibre tape, all human knowledge could be transmitted between one's fingers in three seconds.

Marvellous though this may be, if true (even then I recall someone mentioning a switching problem), I think it takes too narrow a view. Markets move faster than governments, and it is this, as much as anything in the technology itself, that has made private ownership a preferred form of management. While studying at the London School of Economics I attended a course of lectures in logic and scientific method, given by the late Professor Karl Popper. It takes time for ideas to flow through to policy. Popper's emphasis on the limits of our knowledge, and the creative role that he gave to intellectual speculation, were to surface later in the economic policies which began to sweep the world in the 1980s. This is no coincidence. Popper, an admirer of Socrates, knew the Austrian economist Friedrich Hayek and there is much in common in their ideas. Both men had good reason to distrust over-powerful governments. Hayek's view that it is our ignorance of what is going on in the market process that prevents us from intervening to make it work better is Socratic. The new liberalisation may succeed or fail. Only by letting it run will we find out.

Forecasting was once a major activity of business and government economists. It worked tolerably well in the conditions of the time but, during the last two decades of the twentieth century, world economics and politics have proved more unstable. Forecasts are less reliable. Economists find more of their time deployed on policy analysis, business planning and the management of uncertainty. The shift has not been confined to the activities of economists. Telecommunications network planning is now conducted in an environment where broadcasting and the telephone spread news and information of every kind around the world with unprecedented speed.

Telecommunications privatisation has generated a large quantity of academic literature but, aside from that, the number of British books on the economics of telecommunications is not large. Morgan's 1956 *Telecommunications Economics* was a valuable contribution to the analysis of the supply side of the industry but had little to say about demand. The immediate antecedent to the present work was Littlechild's *Elements of Telecommunications Economics,* published by the IEE in 1979, which looked forward to a more competitive world about to emerge. It is a book to which many industry analysts, including myself, owe a debt.

I would like to express my thanks to Professor John Flood for his helpful criticism and suggestions during the preparation of the text. David Cracknell, John Harper, Bruce Laidlaw and Claire Milne have provided comments and suggestions on parts of it and their expertise has been extremely useful. The wise words and kind thoughts of Dr Robert Olley, from his vantage point at the University of Saskatchewan and at our meetings here and there, have helped me a lot over the years and he too deserves my thanks. Some of the material in Chapter 19 is from a paper of mine that is in *Telecommunications and socio-economic development,* published by Elvesier Press and I am grateful for their permission to use it here.

Abbreviations

Definition and text references are given in the index.

$: US dollar unless otherwise indicated
ACD: automated call distribution
ACE: average cost elasticity
ADSL: asymmetrical digital subscriber loop
APT: arbitrage pricing theory
AT&T: American Telephone and Telegraph Corporation (company)
ATM: asynchronous transfer mode
AVE: asset-volume elasticity
BABT: British Approvals Board for Telecommunications
BAT: a British company
BCE: Bell Canada Enterprises (Canadian company)
B-ISDN: broadband integrated services digital network
BOT: build/operate/transfer
BP: British Petroleum (company)
BSI: British Standards Institution
BT: British Telecommunications plc (British Company)
C&W: Cable & Wireless (British company)
CAD: computer-aided design
CAGR: compound annual growth rate
CAI: common air interface
CAM: computer-aided manufacture
CAPM: capital asset pricing model
CAS: competitive access supplier
CATV: cable television
CCA: current cost accounting
CCITT: Comité Consultatif International de Télégraphie et Téléphonie
         (International Telegraph and Telephone Consultative Committee of the
         ITU, now called ITU-T)
CFA: Communauté Financière Africaine
CLC: customer line charge (US)
CLEC: competitive local exchange carrier (US)
CLI: calling line identification
COCOM: Coordination Committee for Multilateral Export Controls (US)
CPI: consumer price index
CSO: Central Statistical Office (UK agency)
CT1: (analogue) cordless telephony (UK system)

CT2: (digital) cordless telephony (UK system)
CTI: computer telephony integration
CVE: cost-volume elasticity
CUG: closed user group
DBS: direct broadcasting by satellite
DCF: discounted cash flow
DCME: digital compression and multiplexing equipment
DEA: data envelope analysis
DLC: digital loop concentrator
DQ: directory enquiry service
DSL: digital subscriber loop
DSN: derived services network
E-Mail: electronic mail
EC: exchange connection
ECO: equivalent competitive opportunities test
ECU: european currency unit
EDI: electronic document interchange
EFT: electronic funds transfer
EU: European Union
FAC: fully allocated cost
FCC: Federal Communications Commission (US)
FDM: frequency-division multiplexing
FRBS: financial results by service
GATT: General Agreement on Trade and Tariffs (Treaty)
GDFCF: gross domestic fixed capital formation
GDP: gross domestic product
GEO: see GSO
GHz: gigahertz; billions of cycles per second
GNP: gross national product
GRC: gross replacement cost
GSM: originally Groupe Speciale Mobile, later global system for mobile
    communications
GSO: geostationary orbit (also called geosynchronous)
HCA: historic cost accounting
HDSL: high-rate digital subscriber loop
HEO: highly elliptical orbit (satellite)
HHI: Herfinahl-Hirshman index
HMSO: Her Majesty's Stationery Office
Hz: hertz
IBCN: integrated broadband communications network
IBM: International Business Machines (US corporation)
ICC: Interstate Commerce Commission (US)
ICI: Imperial Chemical Industries (company)
ICO: intermediate circular orbit
ICOR: incremental capital/output ratio
IDATE: Institut de l'Audiovisual et des Télécomunications en Europe (France)
IDD: international direct dialling
IDQ: international directory enquiry service

IEA: Institute of Economic Affairs (UK)
IN: intelligent network
ILEC: incumbent local exchange carrier (US)
INTUG: International Telephone Users Group
IP: information provider
IPLC: international private leased circuit
IPR: intellectual property rights
IRR: internal rate of return
IRU: indefeasible rights of use
ISD: international subscriber dialling
ISDN: integrated services digital network
ISO: International Satellite Organisation
ISR: international simple resale
IT: information technology
ITAP: Information Technology Advisory Panel (UK)
ITC: Independent Television Commission (UK)
ITS: International Telecommunications Society
ITU: International Telecommunication Union
IVANS: international value added network services
IVPN: international virtual private network
IVR: interactive voice response
kbit/s: kilobits per second
KDD: a Japanese PTO
kHz: kilohertz
LEO: low earth orbit (satellite)
LATA: Local Access and Transport Area (US)
LRMC: long run marginal cost
MA: Massachusetts
MAN: metropolitan area network
Mbit/s: megabits per second
MCI: Microwave Communications Incorporated (US operator)
MEA: modern equivalent asset
MEO: mid earth orbit (satellite)
MFN: most favoured nation
MHz: megahertz
MMC: Monopolies and Mergers Commission (UK)
MR: marginal revenue
MRS: marginal rate of substitution
NAP: network access point (Internet)
NP: number portability
NPV: net present value
NSP: network service provider (Internet)
NTT: Nippon Telephone and Telegraph (Japanese operator)
OCP: optional calling plan
OECD: Organisation for Economic Cooperation and Development
OFTEL: Office of Telecommunications (UK)
ONP: open network provision
OSS: one-stop shopping

PBX: private automatic branch exchange (= PABX)
PC: personal computer
PCM: pulse-code modulation
PCS: personal communications system
PDI: personal disposable income
PDN: public data network
PHS: personal handyphone system
PPP: purchasing power parity
PMR: private mobile radio
PRS: premium-rate services
PSTN: Public Switched Telephone Network
PTO: public telephone operator
PTP: personal time preference
PTT: Postal, Telegraph and Telephone Agency
R&D: research and development
RBOC: Regional Bell Operating Company (US)
REA: Rural Electrification Act (US)
ROC: return on capital
RPI: retail prices index (UK)
Sat-PCS: satellite personal communication system (Mobile)
SDH: synchronous digital hierarchy
SEK: Swedish Krona
SIB: Securities and Investments Board (UK agency)
SITA: Société Internationale de Télécommunications Aeronautique
SLC: subscriber line charge (US)
SNG: satellite news gathering
SOC: social opportunity cost
SONET: synchronous optical network, also a set of standards developed for use with
          synchronous optical networks
SRMC: short-run marginal cost
STD: subscriber trunk dialling
STP: social time preference
SWIFT: Society for Worldwide Interbank Funds Transfer
SWOT: strengths, weaknesses, opportunities, threats
TCP/IP: transmission control protocol/internet protocol
TDM: time-division multiplexing
TFP: total factor productivity
TMA: Telephone Managers Association (UK)
TPMR: trunked private mobile radio
TQM: total quality management
TUA: Telephone Users Association (UK)
TV: television
UK: United Kingdom
UMTS: universal mobile telecommunications service
UPT: universal personal telecommunications
UTS: Universal telecommunications service (= universal service)
US: United States
USSR: Union of Soviet Socialist Republics

VANS: value added network services
VAS: value added services
VAT: value added tax
VPN: virtual private network
VDSL: very high rate digital subscriber loop
VSAT: very small aperture terminal
WARC: World Administrative Radio Conference
WATS: wide area telephone service (US)
WDM: wavelength division multiplexing
WIK: Wissenschaftliches Institut für Kommunikationsdienste (German research
    institute)
WLL: wireless local loop
WTO: World Trade Organisation

Chapter 1

Introduction

1.1 Telecommunications in a world setting

Telecommunications is an industry of world dimensions. The public voice networks, which by 1996 linked over 700 million people over telephones around the globe, make up the world's largest machine. Data are transmitted over other parts of the same physical network; their volume approaches that of voice traffic and is set to overtake it. The number of data terminals is unknown but must run into hundreds of millions. Cellular mobile telephones, which are linked to the PSTN and did not appear until the 1980s, numbered 146 million in 1996, heading for over 200 million by the end of the century. The number of pagers is similar. Satellite systems permit travellers in the most remote parts of the land, sea or air to position themselves accurately and keep in touch with the rest of the world.

Directly or indirectly the industry provides service to most of the world's population. It employs many millions and is an intellectual and creative challenge to the people who design, build and manage its networks, and to those who analyse its operations or frame policies for its governance.

Telecommunications is part of a wider industry which includes computing, information technology, films and broadcasting. It is concerned with the manufacture and supply of the means by which communication takes place within this wider group, the structure of which is illustrated in Table 1.1. Although telecommunications in this sense is separate from the supply of the actual communications, the markets run across industry borders. A company may provide basic telephony over lines leased from a network operator, and provide additional services which add value: cable television operators are offering telephony; new markets based on the Internet are being targeted by the telecommunications and computing industries.

The distinction between the content of services and the means of delivery is a useful one for regulatory purposes, where is it used in the drafting of laws and licences.

Telecommunications deserves special study, beyond its intrinsic interest, for four reasons:

(i)   It is becoming a central component in many commercial and government processes, including finance, transport and the management of public services.
(ii)   The industry is dominated by powerful firms. Regulation is needed to protect consumer interests.
(iii) Its economics exemplify issues, such as economies of scale, network externalities and cost allocation in a multiproduct industry, which arise in many other industries.
(iv)   It is widely seen as spearheading social change based on the use of new technology. Even on a minimal view it is a powerful facilitator comparable with transport and postal services.

Table 1.1 Structure of the wider industry

Layer	Product	Typical companies
Equipment manufacture	Telephones, exchanges, computers to other industries and to final users	Alcatel, IBM, Siemens
Networks	PSTN, private networks, resellers cable TV, broadcasting by radio	British Telecom France Telecom
Production	Business information, information for residential users, music, films	Reuters, Warner Bros
Distribution	Information suppliers, film distributors, radio and TV broadcasters	British Broadcasting Corporation

Telephone networks rank among the world's largest businesses and the wealth which they create commonly accounts for two or three per cent of total national output. The owners of the networks, usually called *operators* in what follows, or more fully as *public telecommunications operators* (PTOs), are among the largest enterprises in every country. Until the 1980s the general model outside North America was the *Postal, Telephone and Telegraph Adminstration* (PTT), usually a rather introspective state monopoly staffed by civil servants. Operations were conducted from policy positions rather than in relation to the market place. Telecommunications was then a world with its own language and rules and few lessons were drawn from external comparisons. Privately owned telephone corporations were the rule in most of the US and Canada, with postal services provided by a government body.

The industry is experiencing rapid technological change, which is a wealth-creation process of innovation of great interest in its own right. Many would say that the industry is driven by technology, independently of ownership or market structure. This begs the question of why change goes in one direction rather than another, but is true in the sense that technical advances are made because of the interest of engineers in improving existing systems and designing new ones. The importance of this element of personal satisfaction in carrying through the process of change is difficult to overstate. It is fair to ask whether a change from public to private ownership weakens the research dynamic or whether it redirects research to get better value. The same question arises with the effect of competition. It is certainly true, for example, that the research and development which was carried while British Telecom was in public ownership came to full fruition with digitalisation after privatisation and the introduction of competition, without being caused by either of them.

The chronic underinvestment found in many of the state-controlled monopolies of the developing world and the old command economies shows clearly enough that the potential for technological change is not enough to ensure that it happens, but this does not fully explain the changes taking place in the structure of the industry. Perhaps the most important factor driving change is the increasing complexity of the markets. The management of telecommunications networks in today's world requires a flexibility, focus and depth of knowledge which is not readily available to governments and their employees. For much the same reasons, no single operator can be expected to command all the skills needed to provide the rapidly widening range of services now available and probably, for the bigger services, not even one of them.

Trade and market liberalisation have widened the horizons of business planners and the level of business risk. Markets in developed countries are becoming more competitive. Operators engaged in foreign ventures have to carry out an analysis of market, political and economic country risk, sometimes in Third World countries where credit ratings, if they exist at all, are poor. Currency exchange-rate exposures may be significant and difficult to hedge at reasonable cost.

Equipment suppliers have made substantial technological advances in the products which they sell to the networks and directly to users, and they have experienced greater competition in their markets. New industries based on operations that add value to basic telecommunications services have grown up, using the network to provide information and facilities beyond the simple transmission of messages.

1.1.1 The political economy of telecommunications

Telecommunications policy is mostly in the area of *political economy*, to use an old term which acknowledges the place that the industry has in public policy. For example:

- telecommunications provides benefits to parties other than those paying for the calls;
- telecommunications networks provide benefits which are an essential part of the development of a modern economy; governments may have policies to force the pace of telecommunications development beyond that which would be produced by markets on their own in the belief that this will stimulate growth in the wider economy;
- large operators can dictate market terms if unregulated and they may be inefficient; even if they were fully efficient, competition would be preferred by many because it provides choice, which is a political as well as an economic freedom.
- in modern democracies only residential bill payers have votes, and an organisation with 20 million enfranchised customers must be subject to political pressures; this usually results in business users subsidising residential users.

The combination of size of turnover, extended customer base, market power, procurement need and strategic importance makes the management of telecommunications networks a matter of political as well as commercial importance. The balance between politics and economics varies from country to country, but the current worldwide trend is towards a more commercial orientation.

Telecommunications has been at the forefront of a worldwide movement towards the privatisation of public utilities and strategic industries. Privatisation has led to a spirited debate about whether change of ownership has any effect compared with the introduction of competition. There are, as some developing countries have found, times when privatisation on its own is not enough to overcome inertia, nepotism and other factors inhibiting change. Whatever it may have done to the performance of management, privatisation has brought major commercial and legal issues into the public arena, contestable in a way in which the state monopolies rarely were.

There has been a shake-up in the methods used to regulate the activities of utility monopolies. When the government owned them, they were largely self regulating and many of the people working for them had a strong sense of public service. They wanted to do a good job, but their antipathy to any form of market pressure was equally strong. The idea that efficient regulation could be achieved by nods and winks and a reliance on the public service instincts of employees and civil servants seems an odd one in retrospect. In the UK the telecommunications system was administered by the Post Office, without any qualified accountants, until the mid 1970s.

1.1.2 The researchers

Telecommunications policy analysis includes the rich fields of market structure, price, regulation and economic welfare. With so much of the industry in the thrall of markets where prices are seriously out of line with cost, any lack of precision in the science is fairly small in the context of the problems. Forecasting is a more subjective process, relying on sound methodology, market research and assumptions about the future behaviour of the regional and world economies. It involves propositions about what people will want or do.

Outside North America telecommunications economics has been a neglected subject. While telecommunications was in the UK public sector there was little domestic debate about its economics. Few commercial issues could be successfully contested in the political arena or the courts and research received little or no commercial sponsorship. Engineering economics was generally concerned with the supply side of the industry, although the economics of demand has been pursued vigorously by North American operating companies and their regulatory bodies, accompanied by a substantial academic interest. Telephone companies outside North America have published little economic work based on internal research.

Some of the newly established regulatory bodies have done better. The *Office of Telecommunications* (OFTEL) in the UK has a good publishing record, with many internal policy papers and studies by external consultants to its credit. In comparison with the *Federal Communications Commission* (FCC) in the US though, OFTEL has not published much basic statistical information because of the tradition and the legal framework of public service secrecy within which it works.

British Telecom's concerns about commercial confidentiality and ministerial concern for minimising regulatory demands have also inhibited publication.

A decade after privatisation was launched in the UK the interchange between government, commerce and the academic worlds was on a smaller scale in Britain than in the US, where the more contestable nature of the telecommunications market has lead to very lively public debate. Most of continental Europe was even further behind. Increasing competition is beginning to generate stronger commercial motivation in Europe and Asia, notably through disputes about the terms for exchanging traffic between networks, the closer alignment of prices with cost, more disaggregated price schedules and a greater use of volume discounts.

Public-choice theory applies economic methodology to the way in which resources are allocated through political institutions. The economics of public choice have been explored more rigorously in North America, and especially the US, than in Europe. It may be possible, with state-owned commercial organisations, to find an effective administrative solution to the problem of how prices can be set without making excessive profits in the absence of full competition, and how to make them behave efficiently in other ways. For many organisations, the problem has been resistant to solution.

1.2 Economic concepts

1.2.1 Macroeconomics and microeconomics

Studies concerned with the performance of the economy as a whole are described as *macroeconomic.* Those concerning the application of economics at the level of the individual business sector, such as the motor industry or telecommunications, are *microeconomic.* Government policy may attempt to influence events at either level.

At a microeconomic level, the large investments of capital and the dynamic effect of technological change, including the convergence of technologies in telecommunications and related industries, have attracted the interest of policy-makers. Some believe that the performance of the sector can have a positive effect on the performance of the whole economy, stimulating growth in developing countries and improving competitiveness among the industrialised nations.

1.2.2 Resource allocation and opportunity cost

Resource-allocation concepts are used extensively by policy makers and regulators. Resources have an *opportunity cost* which represents the value of their best alternative use. A manager may only have time to carry out project A or project B, neither of which can be carried out by anyone else. The opportunity cost of using the manager's output on project A consists of pay plus the loss of profit on project B. The opportunity cost of handling an additional telephone call in a slack period is likely to be small because most of the resources involved would have no alternative use.

Broadly speaking, resources are allocated (used) in an optimal fashion only if the total satisfaction derived from their use, as measured by what people are prepared to pay for them, cannot be increased by using them in some other way. For example, the use of labour and raw materials for the production of goods which cannot be sold at their production cost is not optimal if the same resources could have been used to make something else for which consumers are willing to pay the full cost.

The market price and opportunity cost of a resource will only be the same if the market price reflects the alternative use which is of greatest value. Accountants are generally concerned with actual movements of cash and for them the price originally paid (*historic cost*) is what matters. It is this that drives conventional financial reporting, with the strong justification that the figures will then be grounded in fact, actual receipts and payments. Opportunity costs are rarely known exactly and may be volatile or subjective.

Economists are interested in opportunity cost because of its implication for optimum resource allocation and the formulation of management problems and decisions. Imprecision does not invalidate the concept when used for analytical purposes, where it can be a powerful tool. If areas of significant uncertainty can be identified, it may be possible to get better data to strengthen the analysis.

1.2.3 Time and the consumer

The user invests time as well as money in the acquisition and use of a product. The number of minutes in a day, a year or even a lifetime are finite. Time is for all of us (so far), the ultimate scarce resource. Time shifting and time saving are major applications of current technology (witness the success of personal transport, air travel and, in the field of consumer goods, such things as the microwave oven and the video recorder). Time budgeting is an important element of consumer behaviour.

Direct dialling saves time because it is usually quicker than going through the operator. The importance of time as part of the purchase price is well demonstrated in telecommunications, where the introduction of *subscriber trunk dialling* (STD) for inland calls and *international direct dialling* (IDD) for international calls (not yet universally available) provides a direct stimulus to the calling rate. Some technologies have yet to reach the harvesting stage because user costs, in terms of the time which has to be invested to understand and use the new systems, are too high. The relatively slow initial take-up of *videotex* (e.g. Prestel and Minitel) and electronic mail, compared with the unexpectedly rapid growth of *audiotex* (telephone access to recorded information services) and facsimile transmission (*fax*) has illustrated this.

1.2.4 Time and the investor

Forecasting reflects the time dimension found in many economic problems in domestic and business life. A phrase which sometimes crops up in the prospectuses of companies raising share capital is '*in the foreseeable future*', which City analysts tend to regard as being up to 18 months. The presence of uncertainty concerning the future does not remove the need to plan or to invest, but it adds to the risk. Business and marketing plans are built on the best available judgements and assumptions, which are varied in a sensitivity analysis to identify the main risk factors. Of particular importance in these plans are assumptions about the future prices of goods and services to be bought and sold and the associated sales volumes.

Investment appraisal is usually based on the annual net cash flow from the project, discounted over the life of the project. Techniques of *discounted cash flow* (DCF) analysis can be found in accounting text books such as Brealey and Myers [1]. The key economic input is the discount rate, which must be at least as high as the cost to the firm of borrowing new capital. British and American regulators calculate the cost of capital for the companies which they regulate.

1.3 General characterists of telecommunications traffic

1.3.1 Calls

Calls, sometimes generically called *transmissions*, may carry *voice* (the spoken word) or *data*, in the form of coded on/off signals such as those between computers. Voice telephony has traditionally dominated network usage, although data is likely to overtake it. Voice and data transmissions may be *unenhanced* or *value added*. Value over and above that of the basic transmission may be added by providing some form of service for which the caller is willing to pay extra, such as call storage and forwarding or information about the weather, share prices or hobby interests.

Telephone calls may be measured by their number, duration, temporal distribution and route. Most calls are local. From an engineering point of view, local calls connect two phones served by the same telephone exchange, often called a *switch*.

Calls may be made from fixed or mobile exchange connections and terminated at either. Calls that are made from an ordinary telephone reach the local telephone exchange over an *exchange line*, often called an *exchange connection* (EC), or the *local loop*. Lines are often classified as *business* or *residential* according to their use. Calls taking place over longer distances pass from the local exchange to travel on an interexchange network which links local exchange areas. They have at least one additional stage of switching and are described as *trunk, long-distance, toll* or (in the UK) *national*. International calls are a fewer in number in most countries and their duration tends to be longer. They pass over the domestic network to *gateway exchanges* where they are connected to international cable or satellite links for delivery to the gateway in the country of destination, possibly via a *transit centre* in a third country.

Telephone exchanges experience traffic peaks at different times of the day, depending on the mix of business and residential traffic in their service areas. Calls made on business lines usually show morning and afternoon peaks from Monday to Friday and negligible use outside normal working hours. Calls made from the home often peak in the evening and at weekends. International calls tend to peak during the hours when offices at both ends of the line are open, if they are business calls, and when both ends of the line are outside normal working hours in the case of calls between residential users. The number of outgoing calls per line is the *calling rate*. Engineers express this as an hourly rate when planning voice networks. Business analysts use an annual figure because they are interested in revenue.

The duration of a call depends on such factors as the purpose of the call, the type of caller, time of day, price and distance. Table 1.2 shows some typical average durations.

Table 1.2 Call durations

	Year	Average duration (mins)
BT inland	1992/3	2.9
BT international	1992/3	4.7
US extralata		
intrastate	1994	3.8
interstate	1994	4.3
US international		
outward	1994	5.8
inward	1994	4.5

Note: Extralata calls have destinations outside the local access and transport area (LATA) of the operating company. There were 161 of these areas in the US in 1988. (Source: calculated from OFTEL and FCC statistics)

Individual calls vary greatly in length, with calls over longer distances tending to be of longer duration. The distribution of their durations is extremely skewed, with the maximum frequency peak at around a minute or so and a long tail stretching out to high values with no theoretical upper limit.

Signal quality affects the value of the call. *Quality-of-service* faults include noise on the line, fading, congestion, breaks in transmission and delays in fixing faults. Mobile installations are primarily used for voice transmission where signal quality is less critical than it is for data. Quality is improving and mobile data applications are of increasing importance. Network operators collect quality of service statistics but do not always publish them.

1.3.2 Prices

Call prices, like those for other telecommunications services, are often called *tariffs*, a designation widely used by monopolistic public utilities for a price schedule covering the full range of goods and services which they offer, and found in a few other trades. Strictly the tariff is the full schedule, not an individual price, but the looser usage is widespread. Business and residential exchange connections may have different tariffs, and tariffs for local calls often embrace a rather wider area than the local exchange, perhaps including that covered by adjacent exchanges. Calls are typically charged for in units of time – minutes, or intervals of a fixed number of seconds. The duration, temporal distribution and frequency of calls are all influenced by the tariff structure and its general level.

1.4 Sizing the industry

Economic analysis of the world telecommunications economy requires a statistical base in which there are dimensions and values.

1.4.1 The dimensions

There are many dimensions, including:

- Size and distribution
  public network infrastructure
  - voice
  - data
  - mobile
  - paging
  - cable
  - satellite
  private networks
  - same technical options as public
  volume of business
  - voice
  - data
  - value-added services
  - equipment supply;
- Market structure
  suppliers
  - services
  - equipment
  - ownership
  customers
  - business
  - private;
- Business growth
  trends
  - growth in volume
  - growth in value
  - prices
  drivers
  - world and regional incomes

population
lifestyles
technology;
- Performance measures
efficiency
service quality
trends
local and world rankings.

Publications by regional and national regulatory bodies, the annual reports of operators and suppliers and literature from academic, trade, conference and consultancy sources provide most of the data, although privatised operators tend to publish less information about their operations than do their state-owned predecessors. This text includes a few tables to illustrate the size and significance of the industry and shows where more information can be found.

The *International Telecommunication Union* (ITU) is an organisation which has been in existence since the 19th century, when it was set up for the co-ordination of arrangements to handle international traffic. The main purpose of the ITU is to agree international protocols for the exchange of traffic and payment for its conveyance, but it maintains a substantial statistical database which is published annually; summaries are available online.

The World Bank [2] publishes data summarised by world income groups and the ITU uses a modified version of these. The group structure is updated periodically. In 1997 it was as shown in Table 1.3.

Table 1.3 Income group structure for World Bank and ITU data

Income group	Average income ($ per annum per head in 1995)	Number of countries in the group:	
		World Bank	ITU
Low income	765 or less	63	63
Lower middle income	766-3,035	65	65
Upper middle income	3036-9,835	30	28
High income	Over 9,835	52	50
Total	**210**	**206**	

ITU modifications affect the treatment of American Samoa, Bermuda, the Cayman Islands, Channel Islands, the Isle of Man, Liechtenstein, Monaco and Taiwan.

1.4.2 World telephone penetration

Telephone penetration is related closely to living standards and is highest in four world regions, see Table 1.4. Between them, in 1996, the richer parts of North America, Western Europe and the Asia/Pacific Region had 66 per cent of the world's exchange lines but only 16 per cent of the population.

Table 1.4 Distribution of the world's telephones in 1996

Region/Country	Population (millions)	Main lines	
		000	per 100 inhabitants
North America:			
Canada	30.0	18 051	60.2
US	266.6	170 568	64.0
Western Europe:			
Austria	8.1	3 779	46.9
Belgium	10.2	4 726	46.5
Denmark	5.3	3 252	61.8
Finland	5.1	2 813	54.9
France	58.4	32 900	56.4
Germany	81.9	44 100	53.8
Greece	10.5	5 329	50.8
Ireland	3.5	1 390	39.5
Italy	57.4	25 259	44.0
Netherlands	15.5	8 431	54.3
Norway	4.4	2 440	55.6
Portugal	9.9	3 724	37.5
Spain	39.3	15 413	39.2
Sweden	8.8	6 032	68.2
Switzerland	7.1	4 547	64.0
UK	58.1	30 678	52.8
East Asia:			
Hong Kong	6.3	3 451	54.7
Japan	125.8	61 526	48.9
Republic of Korea	45.6	19 601	43.0
Singapore	3.0	1 563	51.4
Taiwan-China	21.5	10 011	46.6
Pacific:			
Australia	18.3	9 500	51.9
New Zealand	3.6	1 782	49.9
Total of above	904.0	490 865	54.3
Rest of the world	4874.1	252 797	5.2
World	5778.1	743 662	12.9

(Source: ITU [3], calculations)

ITU statistics for broadcasting show the comparable world penetration rates in 1992 as 36.3 radios and 15.7 television sets per 100 inhabitants. Both of these technologies are younger. Their higher penetrations are an indication of the relative value to the individual, at the margin, of broadcasting and telephony.

The penetration rate rises with income, as shown by Table 1.5.

Table 1.5 Telephones per 1000 inhabitants by income group, 1996

| Income group | Main lines | |
	000s	Per 100
Low income	79 685	2.4
Lower middle income	113 700	9.7
Upper middle income	58 824	13.4
High income	491 453	54.1
World	743 662	12.9

(Source: International Telecommunication Union)

The penetration rates in Tables 1.4 and 1.5 do not include cellular mobile telephones and pagers, which each numbered over 20 million by 1992. The number of cellular telephones grew at over 50 per cent per annum to reach 142 million by 1996 [3] and there was also a rapid growth in the number of pagers.

Table 1.6 shows that new lines were being added at a global rate of 24 million a year in the decade to 1993 and 46 million a year in the following three years. The latter figure represents annual investments totalling about £46 bn ($74 bn). In the context of world population the three-year total comes to eight lines per 1000 people and £8 per head. The acceleration in growth and its bias towards the low income countries are striking features of the figures.

In 1992, some 60 per cent of the world's exchange lines were in the hands of 17 operators in ten countries [4]. Five years later, after mergers in the USA, the same territories were held by 15 operators and comprised 58 per cent of the lines [6]. The concentration of lines among a small number of large operators is less than the concentration of business in some other sectors of the world economy, e.g. oil. In a 1996 advertisement, Coopers and Lybrand claimed that *45 per cent of the world's lines are owned by companies we audit* [6].

Table 1.6 Main lines added, 1983–96

	000s	1983-93 average annual %	1993-96 000s	Average annual %
By region:				
Africa	6 133	8.6		
US/Canada	42 714	3.1		
Other Americas	19 894	7.8		
Japan	15 580	3.1		
Other Asia	54 161	11.6		
Europe	94 184	5.0		
Australia/New Zealand	3 316	4.0		
Pacific	160	8.4		
By income group:				
Low income	25 966	15.3	45 518	32.6
Lower Middle income	46 999	8.0		
Upper middle income	36 927	8.6	91 058	7.0
High income	126 250	3.6		
World	236 142	5.0	136 576	5.0

(Source: ITU [3–5], calculations)

There are many forecasts available about different aspects of world telecommunications activity. In a situation where even gathering historical data can be difficult, these forecasts maybe of limited accuracy but they usually provide insights and are part of the common currency of ideas within the industry.

1.4.3 International traffic flows

The most heavily used telecommunications routes in the world are found in and between Europe, North America and Japan, see Table 1.7.

These account for about 13 per cent of global international traffic. The top 50 routes between the 220+ countries of the world account for just over 50 per cent of the total. On present trends, the great world trade and traffic flows will remain within and between these countries well into the 21st century.

Table 1.7 Thickest world telecommunications routes, 1995

Between	And	Annual minutes (total both ways in millions)
US	Canada	5 062
US	Mexico	2 749
US	UK	1 696
Hong Kong	China	1 664
US	Germany	948
US	Japan	893
Germany	Switzerland	791
Germany	Austria	761
Germany	UK	730
Germany	France	716

(Source: Telegeography [7]. Figures relate to total voice traffic between country pairs)

1.5 Critical trends in the world economy

1.5.1 The determinants of demand

The demand for telecommunications equipment and services is affected by trends in the national and international economy. The other main variables to which established services respond are price changes, demographic trends and broader elements of social change such as higher mobility, reduced household size and industrial restructuring. New products move along a fast rising curve until the market approaches maturity, when they respond more like established products.

National output is measured as *gross domestic product* (GDP). If net property income from abroad (interest, dividends etc.), usually only a marginal quantity, is added, the result is the *gross national product* (GNP). The distribution of world GDP is uneven, as shown in Table 1.8. There are three rich zones with nearly 80 per cent of the total world income shared among the 18 per cent of its population living in Western Europe, North America and Japan. These three zones are sometimes known as the *Triad*.

Table 1.8 World GDP in 1995

	GDP ($bn)	Population (million)	GDP per head ($)
By region:			
North America	7 819	293	26 714
Western Europe	8 854	383	23 111
East Asia:			
Hong Kong	140	6	22 629
Japan	5 134	125	41 009
Republic of Korea	456	45	10 165
Singapore	85	3	28 467
Taiwan–China	261	21	12 147
Australia/New Zealand	407	22	18 742
Rest of the world	5 538	4 775	1 160
By income group:			
Low income	1 488	3 180	468
Lower middle income	1 912	1 153	1 659
Upper middle income	2 031	438	4 635
High income	23 263	902	25 785
World	28 695	5 673	5 058

(Sources: ITU [3] (GDP), World Bank [2] (pop))

The world output growth trend is running at about three per cent per annum in real terms and rather less in real GNP per head of population. Growth rates vary between regions and are not converging, see Table 1.9. On present trends some parts of the developing world will catch up; others will not. For a number of countries there are difficult choices to make concerning ideology and wealth creation. The last few years have seen a striking shift towards more market-based economies among the ex-communist states.

Table 1.9 GNP per inhabitant: average annual growth rates

	1965-85 %	1985-95 %
By income group:		
Low income	2.9	3.8
Lower middle income	2.6	−1.3
Upper middle income	3.3	0.2
High income	2.4	1.9
Selected regions:		
Low income:		
India	1.7	3.2
China	4.8	8.3
Others	0.4	−1.4
Low and middle income:		
SubSaharan Africa	1.0	−1.1
East Asia and Pacific	n/a	7.2
Latin America and Caribbean	n/a	0.3
World	n/a	0.8

High income 1965-85 = 19 industrial market economies
Figures are in real terms, excluding inflation. Uses World Bank income groups
(Source: World Bank [2, 8])

Comparisons based on the use of a common currency can be distorted by unrepresentative exchange rates; real income indicators can be a useful substitute. Table 1.10 compares telephone penetration in a few countries with indicators for television, health provision and electricity. The television set, although a later invention than the telephone, has a higher penetration almost everywhere. Physical indicators are published less regularly than economic statistics and have other problems. There may be a large suppressed demand for consumer goods of certain types, or demand patterns may be distorted by prices which are unrelated to underlying costs.

The tables illustrate the association between income and telephone penetration, but they also show that other factors affect demand. This is illustrated by the relatively high growth rates among the poorer countries in Table 1.6, by the different penetrations of

Table 1.10 Comparison of real income indicators per 100 inhabitants, 1990

	Telephone main lines	TV sets	Doctors	Access to safe water
India	0.6	3.3	0.04	73
Hungary	9.6	41.0	0.29	98
Turkey	12.3	17.4	0.08	84
Germany	48.3	70.1	0.27	100
France	49.5	40.2	0.29	100
Finland	53.5	49.5	0.24	96
Canada	57.8	63.9	0.22	100

(Source: ITU [4], World Bank [9])

telephony, radio and the other real income indicators in Table 1.10 and by the more rapid growth of cellular radio services.

1.5.2 Location of the largest companies

In 1994, there were 107 companies in the world with a turnover of £12 bn ($18.6 bn) or more. The biggest telecommunications operators were AT&T, NTT, Deutsche Telekom and BT. If we include the Japanese trading aggregates (*Sogo shosha*), this top group grows to 115 companies, all of which are in the Triad countries. Table 1.11 shows how these companies are distributed geographically and by industry. Even the largest British manufacturing companies fell outside the group.The 20 largest information technology and telecommunications operating companies are all in Europe, North America or Japan.

1.6 Industrial structure

1.6.1 More industrial concentration or less

Modern business is highly concentrated, with a large proportion of total output coming from relatively few firms. Figures for the UK are shown in Table 1.12.

Table 1.11 Distribution of companies with an annual turnover of £12 bn or more in 1994

	World	UK	Other Europe	US	Japan
Finance	22	3	11	2	6
Energy	17	1	6	6	4
Vehicles	16	0	8	3	5
Distribution	8	0	1	5	2
Electricals	7	0	5	2	0
Electronics	7	0	0	1	6
Telecommunications	6	1	2	2	1
Chemicals	4	0	3	1	0
Computers	3	0	0	1	2
Food	3	0	2	1	0
Steel	3	0	1	1	1
Aerospace	2	0	0	2	0
Construction	2	0	0	0	2
Engineering	2	0	1	0	1
Railways	1	0	0	0	1
Other	4	0	2	2	0
Sogo shosha	8	0	0	0	8
Total	115	5	42	29	39

(Source: Times 1000 [10] plus Deutsche Telekom)

Table 1.12 Business size distribution in the UK, 1994

Annual turnover range (£000s)	Number of companies
50–99	356 278
100–249	358 754
250–499	154 315
500–999	90 358
1000–1999	49 865
2000–8000	40 394
8001–100 000	20 883
100 001–500 000	2 059
Over 500 000	552
Total	1 073 458

(Source: Business Statistics Office [11])

Concentration appears to be less pronounced in the US and some other European countries, but in all of them the telecommunications purchasing power which the largest firms command makes them prime marketing targets for telecommunications service providers.

The major telephone operating companies have global ambitions and threaten a worldwide domination of service, at least for large users. Meanwhile, developments in information technology encourage decentralisation and liberalisation permits the entry of specialist and niche operators. The trend may to be towards more or to less concentration within the industry. It is not clear which.

1.6.2 The growth of services, especially finance

Undeveloped economies have agriculture as the major part of GDP. As they develop, the balance of the economy shifts towards manufacturing, where they may build up strong trading position if local labour costs are low and the government follows an appropriate economic policy. In developed countries, service industries become dominant, typically providing twice as many jobs as manufacture. Tourism sometimes becomes a major contributor to the national income of an undeveloped country well ahead of other services. Developed economies where services have become dominant are sometimes (prematurely) called *post industrial.* The financial services sector has become important in most developed economies and is of special interest to telecommunications because of its substantial use of information technology and communications services.

1.7 The global firm

World business is becoming more internationalised and telecommunications shares in the trend. In the 1980s the largest telecommunications operators ranked high among the world's biggest companies, but they were predominantly local in their operations and had substantial monopoly powers. In this they resembled some of the other national monopolies — gas, electricity and transport — well adapted to managing a captive home market but inexperienced in the development of businesses abroad. A few had global ambitions but, with the exception of Cable & Wireless (C&W), their overseas ventures had made little money and more often lost it. Some of the companies in the equipment supply industries traded abroad more effectively.

Alcatel from France, Ericsson from Sweden, Northern Telecom from Canada and Siemens from Germany were among the most successful. None achieved the pre-eminence of the big manufacturers of electronic products in Japan and other parts of the Far East. The dimensions of the truly global firm are illustrated by a few examples in Table 1.13.

Table 1.13 Global firms

	Overseas turnover		Overseas investment	
	(£m)	(% of total)	(£m)	(% of total)
BAT 1989	11 724	67.0	344	32.5
BP 1988	18 678	71.7	2 409	41.3
Glaxo 1993	4 424	89.7	339	52.2
ICI 1989	10 254	77.9	643	79.1
Roche 1991	5 536	96.7	429	75.2

(Source: company annual reports)

Telephone operating companies, with their strong domestic market positions and lack of experience in dealing with competition, do not always find it easy to operate outside their home country. BT and Telefonica are two which have expanded their overseas asset base. C&W, on the other hand, has shifted towards the UK, although it still has the biggest proportionate global stake, see Table 1.14

Table 1.14 Percentage of PTO assets held abroad

	Early 1980s	1993	1995
Ameritech			6.0
AT&T			9.2
Bell Atlantic			4.6
BT	1.3	3.9	20.7
Cable & Wireless	84.9	51.5	34.8
GTE			5.4
KDD			1.7
NTT			0.3
Telefonica	0.2		9.7
USWEST			7.0

(Source: company annual reports, Winsbury [13])

1.7.1 Technological change and the market for services

Improvements in technology are changing the nature of the production process and the markets served. Relative prices are changing, creating an expansion of demand for products which are becoming cheaper but leaving other products at a disadvantage. The demand for telecommunications services is being affected as much as the process used to produce them. The first-order effects were obvious by the early nineties, with the expansion of information-based data services delivered over the ordinary telephone network to homes and offices. These were achieved by a combination of commercial innovation, deregulation and packet switching technology. The fax machine was an early beneficiary of falling cost, standardisation and ease of use. It soon accounted for a significant part of the demand for new lines, especially among business users. Electronic technology cheapened and increased the flexibility of products and services in many parts of the economy, so that their relative prices fell and demand for them increased. The demand for the telecommunications services used to produce or deliver them came through as second-order effects.

The world pharmaceutical industry is dominated by about half a dozen companies. Its research and development expenditures are among the most substantial of any industry and the time scales are long. Pharmaceutical products are of primary importance, providing better health, longer life and the relief of pain. The industry is not an especially large user of telecommunications, but by extending life spans it has a significant effect on the population age distribution and the pattern of telecommunications demand.

Biological sciences are reshaping the developing and developed world through improved agricultural yields and the introduction of products to improve the environment. New techniques of production going under the name of *Nanotechnology* and working at the atomic level enable extremely small components to be made, built up from very thin layers. Nanotechnology is seen to have applications in biotechnology (to which it is closely related), electronics, computing and medicine. It could change more general manufacturing processes in such a profound way that everyone could eventually have cheap access to most physical goods, leaving labour-intensive services to dominate family budgets. The service sector uses more telecommunications per unit of output than manufacturing does, so that this trend would be important in the growth of future telecommunications demand.

1.8 Telecommunications, technology and society

1.8.1 A dilemma

Telecommunications, although a large and rapidly growing industry, is a small but increasingly important part of a complex world. Its experience is typical of industries leading the new industrial revolution and encapsulates key uncertainties of the age. Powerful forces have yet to work themselves out. An enormously productive new information technology is not yet fully absorbed into lifestyles, products or factor prices. There are industry estimates of what could be achieved in terms of the delivery of new services and the enhancement of old ones and of the costs involved. No one can be sure which applications will prove popular, and markets will have to winnow them out.

Commercial and political interest in information *superhighways* exemplifies the dilemma. Telecommunications and IT companies see an investment opportunity in building powerful fibre-optic networks, but do not know how to stimulate enough useful information to fill the local residential loops which could be a major part of these networks. Politicians see a potential to improve national competitiveness and public wealth, recognising that entertainment and hobby interests are the areas most likely to generate cash from the public at large. For the majority, home banking has a long way to go before it repays its cost. Gambling and sex have proved strong growth areas but they sit ill with political initiatives.

1.8.2 The location of political power

The accessibility of information through radio, television, the press, entertainment media and electronic information services is now so extensive that the political power of lobbying groups and individuals has increased substantially, while conventional political structures have been losing support and credibility. Meanwhile, the power available to central governments has increased with the creation of nationwide, and in some cases international, databases containing information about individual citizens. These developments have positive and negative implications for personal freedom.

1.8.3 Wealth or leisure

The new technology has a vast output potential, but the marginal utility of electronic outputs is not necessarily large. A pocket calculator is rarely used to perform more than a tiny fraction of its potential because the calculations are not worth the time or trouble. Users invest time as well as money in obtaining service, and time is scarce.

There is a demand for labour-intensive personal services such as health and child care. Aging populations and improved medical technology have driven the rising demand in many western countries but there is, particularly in the UK and the US, no political consensus on how they should be paid for. As a consequence, workers displaced by improved technology cannot necessarily find jobs in these industries (even if they want them). Substitution of capital for labour is affecting managerial as well as manufacturing operations. Heavy investment in information technology has not always paid off, but the successes outweigh the failures by a large margin. The flattening of management hierarchies has been made possible by the creation of better information systems within companies and it has cost many managers their jobs.

A state of economic equilibrium would require that:

(*a*) Output expands to a level where, at the margin, each product provides equally good value to consumers, given its cost.

(*b*) The work/leisure balance is based on marginal leisure preferences. In broad terms, this means that the level of earnings would leave most people indifferent to whether they worked a little longer or a little less.

It was common, in the middle of the 20th century, to see mechanisation as a way of freeing people from the need to work so hard or long, while leaving them with incomes high enough to permit a satisfying use of their extra leisure. Many large businesses are expanding output by the use of more capital; most are not taking on more labour. New technology has increased the marginal productivity of capital and made the substitution of capital for labour a way of increasing profits. New technology is also enabling greater use to be made of existing assets. *Directory enquiry services* (DQ) provide an example of the substitution of capital for labour in telecommunications. At one time ranks of operators handled enquiries by looking up the numbers in sets of telephone directories. Later the domestic directories were replaced by microfiche and later still by computerised databases. Currently, the voice response comes from a computer, and large users can get the whole set on disk. The number of staff used in the service has thus been reduced with the increase in capital employed. *International directory enquiries* (IDQ) are still relatively labour intensive. They are more likely to rely on reference to printed directories which have been obtained from other countries.

New technology can raise productivity in any industry. The long-term trend may be to use it for the creation of more physical wealth without a reduction in the labour force, or it may be towards using it to replace labour. Although the greater marginal productivity of capital invites further job substitution, the way is still open for new products, unforeseen in current markets, to become a source of new job creation. Whatever the outcome, public and private sector personal services are likely to become more expensive but increasingly important as a source of jobs.

1.8.4 Training, retraining and labour-market structure

New technology has upgraded many skills and made others obsolete. Wordprocessing is more complicated than typing but requires less manipulative skill. In the old electromechanical telephone exchanges, an engineer's ears could detect the switch that was not working properly, but this is not possible with the later generations of silent electronic equipment. Manipulative maintenance skills lost their value, and the transition to the maintenance of electronic systems has required a shift to mental diagnostic skills and an effective retraining process, described for British Telecom in Reference 14.

The lack of equilibrium which leads companies to use capital rather than labour to expand output is not a new phenomenon. As in earlier centuries, the imbalance will take time to work itself out. Macro-economic measures are not likely to be effective in removing it, nor is retraining unless it is particularly well directed. A large pool of underused graduates and school leavers, unable to find jobs matching the skills for which they have been trained, has persisted since the mid 1980s. If there is to be a resolution, it is likely to involve more than flexible labour markets and adjustments to factor prices, necessary though these are. The evolution of new products must be a significant element.

1.9 Main themes

In describing the economics of telecommunications in a global context four main themes recur.

The first concerns the geodesic network as a model for change. The US Department of Justice commissioned a report on the scope for further competition in market structure which was established after the 1984 breakup of AT&T. The report was published as *The Geodesic*

Network [10] in 1987. Huber began by drawing attention to the role of the balance of cost between the two basic elements of a telecommunications network – the transmission of calls from node to node and the switching of the calls to their destinations at each node. With the traditional pyramidal architecture, calls from terminals (telephones, computers etc.) are collected at local exchanges and those for distant destinations are routed through higher-level area and regional exchanges. This reduces the number of circuits required and enables traffic to be concentrated on bigger routes, thus cutting cost. Switching is carried out at each exchange level in the pyramid (five in the case of the US). This is optimal when switching is an expensive manual or electromechanical operation and network intelligence is most economically kept within the network.

The development of computers, modern PBXs and other forms of intelligent terminal equipment with switching capabilities has led to the construction of local networks which increasingly link office block and campus site terminals directly, using ring or bus (linear) circuits, rather than routing traffic up the pyramid and back. At the same time, falling transmission costs and legal changes have encouraged the development of competition between long-distance carriers with their own networks. Long-distance, metropolitan and regional networks, often using microwave or satellite communications, have proliferated where market and regulatory conditions permit, and are particularly extensive in the US. It has become cost-effective to fit PBXs with equipment which can switch calls to the cheapest carrier. The new networks create many extra routes between terminals and add a geodesic web or mesh to the surface of the pyramid. They also diversify network ownership.

Huber saw powerful pressures for the integration of switching, local and long-distance functions, with the driving force coming from the marketing benefits of direct contact with the major customers - *account control* in marketing terms. The complexity of the choices facing businesses when setting up their telecommunications arrangements creates an opportunity for *systems integration* in the computer services and *one-stop shopping* (OSS), where a single operator offers a complete end-to-end national and international telecommunications package. Customers have a choice, instead of dealing with a monopolist. Huber thought that, in the long term, the choice would be made in a market dominated by large operators, each of which providing comprehensive network services. Similar developments in computing and equipment manufacture would lead to an ultimate comunications market structure dominated in each sector by a few large companies, competing with each other at all levels.

Although the links and nodes in a geodesic network with distributed intelligence need not be under the control of a single operator, it is vital that they connect with one another efficiently. The regulatory issues which arise from these relate to the maintenance of fair competition, service standards, the benefits of a unified network and interconnection terms.

The Huber analysis has not been fully accepted within the industry. Big national operators can use computers to provide network services which are difficult or impossible to obtain from terminal equipment. The economics of network operations may yet lead to the consolidation of monopoly. The industry continues to change, however, and the rapid growth of mobile networks since the Huber report was written has further increased the complexity of the network mesh and choices facing the customer. Whatever may follow in the long term, the geodesic network model seems likely to hold as a consequence of new technology and the opening of markets to competition. One of the themes of this book is to follow through the way in which these changes are reshaping the industry.

A second theme is the creative force that has been released by technological changes and it is here, perhaps, that there may be a resolution of the macroeconomic dilemma. In the last quarter of the 20th century there has been an explosion of new telecommunications products and services and in the use made of them, nationally and internationally. Mobile phones, the fax machine, desktop computers and the Internet are examples of things which hardly existed outside the largest business offices in the 1970s. Satellite and cable TV were little known in Britain. Travellers did not expect telephones on aeroplanes and war correspondents did not carry portable satellite dishes in their backpacks. The rapid spread of new electronic products and telecommunications services shows that the demand is there if the product is appealing and the price is low.

Manufacture has shifted towards developing countries but software and the management of systems remains under the control of the big economies of the industrialised world and this leads into a third theme — that of competitive advantage in an imperfectly competitive world. Companies do best by concentrating on activities where they have an edge over their competitors, but are tempted into wider-ranging ventures in the belief that they need to be global to succeed or that the converging technologies of computing, communications and entertainment provide a synergy strong enough to justify cross-industry mergers.

A fourth theme, providing a motor for much of the foregoing, is the endemic overpricing of many services by monopolistic operators, leaving opportunities for market entry as markets are liberalised. One question here is whether effective competition can survive in a telecommunications world where prices are more fully aligned to cost. The first three themes run through the book in four main contexts:

- the viability of competitive infrastructure in the local loop;
- the viability of competitive infrastructure in long distance services;
- interconnection problems and anticompetitive practices;
- the separation of network ownership from network use, to keep the benefits of scale in the network (if it was kept as a monopoly), while permitting competition in the provision of services over it.

A fourth theme is the effect of competition on public welfare. Experience so far suggests that there is no reason for concern and good reason to expect substantial benefits.

1.10 References

1 BREALEY, R, and MYERS, S.: 'Principles of corporate finance' (McGraw-Hill, London, 1984)
2 World Bank: 'World development report 1977' (Oxford University Press, Oxford, 1997)
3 ITU Basic Indicators (ITU Website, 17 March 1998)
4 'World telecommunication development report 1994'. ITU, Geneva, 1994
5 'Telecommunication indicators 1994/95'. ITU, Geneva, 1995
6 Coopers & Lybrand: 'Global telecomms business yearbook 1996' (Bell Sygma *et al.*, London, 1996)
7 STAPLE, G. C. (Ed.): 'Telegeography 1996/97' (Telegeography Inc, Washington D.C., 1996)
8 World Bank: 'World development report 1987' (Oxford University Press, Oxford, 1987)
9 World Bank: 'World development report 1994' (Oxford University Press, Oxford, 1994)
10 HUBER, P.: 'The geodesic network' (US Department of Justice, Washington D.C., 1987)
11 The Times 1000 1996 (Times Books, London, 1995)
12 Business Statistics Office: 'PA1003 Size analysis of United Kingdom businesses 1994' (HMSO, London, 1995)
13 WINSBURY, R.: 'How grand are the Grand Alliances?', *Intermedia*, **25**, (3), pp. 26–31
14 CLARK, J., McLOUGHLIN, I., ROSE H., and KING R.: 'The process of technological change' (Cambridge University Press, Cambridge, 1988)

Chapter 2
Telecommunications technology

2.1 Introduction

2.1.1 Networks

Most telephone calls pass over the *public switched telephone network* or PSTN. *Public* means that access to the network is open to anyone willing to pay the prescribed charges. *Switched* means that the network has call routing equipment which enables a caller to reach any required destination on the PSTN. Some national PSTNs are owned by the government, others privately. Calls pass freely between them so that each telephone on any national PSTN can reach any other telephone in the world.

Many businesses own and use *private networks* for internal telecommunications traffic. Private networks may be switched or unswitched. Access is allowed only within some form of a *closed user group* (CUG), usually the employees of the owner. Governments often have private networks for communications between officials. A private network need not be connected to any other network, public or private, but connection with another network requires a link through the PSTN. The conditions under which a link can be obtained depend on national regulatory law rather than any technical factor.

Networks use a variety of technologies of which three — the transmission of signals from point to point or between nodes, the signalling between nodes that is needed to set up a call and the routing or switching of signals at the nodes — are the most important. The power of these technologies has developed rapidly since the 1950s and there have been large reductions in cost, particularly for transmission systems. Networks may use *fixed links*, meaning that the termination points are at fixed locations whatever the means of transmission, or they may be mobile networks. In either case, they may be for voice traffic (and certain types of data traffic which can pass over voice circuits) or be specially constructed for data traffic alone.

Modern telecommunications services are becoming multidimensional. Although most calls are from one fixed point to another over fixed links in real time, mobile networks are growing in strength, adding an extra dimension to the service. New store-and-forward and voice-messaging services break the instantaneous time link. Another dimension runs from voice through data to interactive information services.

2.1.2 The conversion of sounds into transmittable signals

A sound passes through air at a constant speed in the form of a wave, or vibration. A sound wave has a *wavelength*, which is the distance between peaks, an *amplitude*, which is the height of the wave from trough to peak and a *frequency*, which is the number of peaks in a period of time. The precise pitch of the sound heard depends on the frequency. The volume depends on the amplitude. For measurement purposes each wave is called a *cycle*. The unit of frequency is the *hertz* (Hz), which is one cycle per second. Higher units are the kilohertz (kHz, 1000 cycles per second), *megahertz* (MHz, one million cycles per second) and *gigahertz* (GHz, one billion cycles per second).

Sound waves of different wavelength and amplitude may be sent out simultaneously as they are, for example, by a musical instrument such as a violin, or by the human voice. The human ear can detect frequencies from around 30 to at the most 20000 Hz. The key elements of speech fall in a smaller range and intelligibility can be achieved by transmitting only sounds in the range from about 300 to 3400 Hz, and this is what telephone voice circuits are usually set up to carry. The range of frequencies carried, 3100 Hz in this case, is called the *bandwidth*.

Sounds have to be converted into electric currents or optical signals for transmission from a telephone and reconverted into sounds at the other end of the line to be heard. The movement of the diaphragm of a microphone and the waves in the current have a pattern compounded of all the different sound frequencies detected. The electrical signal is thus said to be an analogue signal.

2.1.3 Methods of transmission

Signals are transmitted using apparatus at each end of the circuit. With *analogue transmission* the electric currents are transmitted to a coil at the other end of the circuit, where they are used to vibrate a diaphragm and regenerate the original sound. The transmission can be without frequency modification over short distances and this is the method normally used in analogue local loops. It is known as *baseband transmission* and was used over larger areas in the early days of telephony. Speech signals can be added to a higher frequency *carrier wave* before transmission, a process known as *modulation.* At the other end of the circuit the higher frequency can be removed to recover the original speech signal. A cable may be capable of carrying a much higher bandwidth than is required for a single voice circuit. Cable capacity can be split up into separate circuits, or *voice paths* by *frequency-division multiplexing,* (FDM), in which a number of different carrier frequencies are used simultaneously. If these carriers are spaced out at intervals of 4 kHz, each one can transmit the 3.1 kHz required for a voice circuit without interfering with its neighbours. For example, a circuit may have carriers at 30, 34 and 38 kHz. With the speech bandwidth added, these cover bands of 30 300–33 400, 34 300–37 400 and 38 300–41 400.

Analogue transmission is used extensively in the local loop but has been largely replaced by *digital transmission* over longer distances; circuits can be set up to handle analogue or digital transmission. Digital technology replaces the analogue waveform with a series of pulses representing binary numbers, illustrated in Figure 2.1.

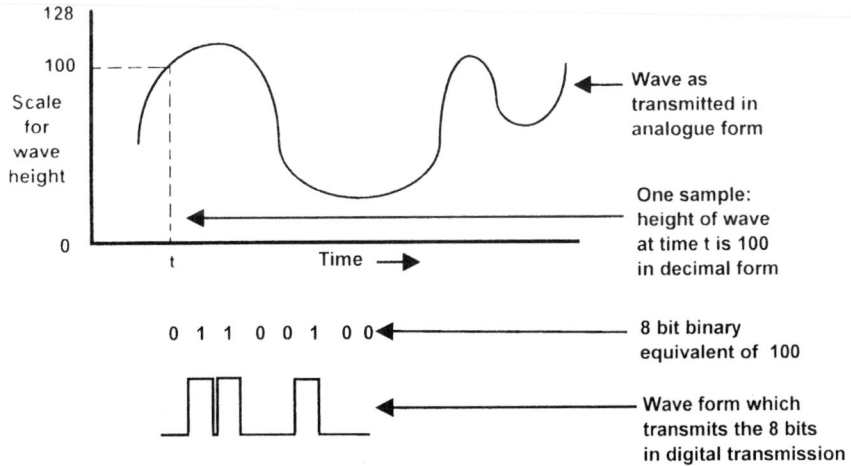

Figure 2.1 Analogue and digital transmission

The waveform of the current generated for transmission and shown in the top half of the Figure is sampled at a regular rate. Samples are close together. It can be shown that a sampling rate of double the highest frequency will give a virtually exact reproduction of the original.

Each sample reading is converted to a binary number using a *codec* (coder/decoder). With binary arithmetic any number can be represented as a series of 0s and 1s. Each 0 or 1 is a *bit* and the set of bits representing the whole number is a *byte*; an eight bit number is usually used for coding. Binary numbers are transmitted at a constant rate, separated by a short interval to show where each byte ends, and the process is called *pulse code modulation* (PCM). A codec is used to reconstruct the original waveform at the other end of the line. The speed at which the bits are transmitted is the *bit rate* and is measured in bits per second or multiples such as 64 kbit/s (64 000 bits per second), 2 Mbit/s (two million bits per second) etc. Each bit can be represented, in transmission, by the presence or absence of an electric or optical pulse in a particular time interval defined by the bit rate. The capacity of a cable is usually measured in terms of the bit rate which it can handle.

The basic bit rate required for digital voice traffic is, from the foregoing, $4000 \times 2 \times 8 = 64\ 000$ bits per second and is the unit of capacity used in digital network planning. As with analogue transmission, bulk capacity can be divided into separate voice or data channels by multiplexing. The time between the transmission of consecutive samples in a particular digital channel can be filled with samples from other channels, a technique called *time-division multiplexing* (TDM).

Digital transmission has cost and operational advantages [1]. Costs have been falling for the electronic equipment required with evolving technology, proving an increasing cost advantage against analogue systems, and time-division multiplexing is cheaper than frequency-division multiplexing. Multiplexing costs are of increasing importance because the carrying power of cables is rising, permitting more channels to be fitted into each. All signals are subject to corruption from background noise during transmission and the corruption rises with transmission speed. Because binary digital signals consist of only two states, on and off, corrupted signals can be reconstructed, unlike those on an analogue network. Signal regenerators along the circuit need only to detect the presence or absence of a pulse to make a perfect reconstruction and, as a result, the quality of transmission is almost independent of distance.

The rising importance of data and video traffic, which are usually less tolerant of transmission errors, provides another source of advantage. Digital signals can use sophisticated error-detection and correction techniques which are outside the scope of analogue techniques.

In North America and Japan, 24 64 kbit/s telephone channels are interleaved to produce a primary multiplex group transmitted at 1.544 Mbit/s, marketed in the US as a *T1*. The European primary multiplex group contains 30 channels transmitted at 2.048 Mbit/s. These may be transmitted over ordinary twisted-pair cables originally designed for a single analogue telephone channel. Higher-order multiplexers may transmit up to 1920 channels on coaxial cables or microwave radio links and 7680 channels at 565 Mbit/s over optical fibres.

With *synchronous transmission*, the data are sent in blocks and there is a clock signal to align the blocks. The *synchronous optical network* (SONET), developed by the US Bell Operating Companies, uses synchronous transmission to provide transmission speeds in multiples of 51.84 Mbit/s up to the multigigabit range. The *synchronous digital hierarchy* (SDH) is a set of synchronous transmission standards laid down by the *ITU–T* at 155 Mbit/s and above and consistent with SONET. The ITU–T, which was formerly the *CCITT* (International Telegraph and Telephone Consultative Committee), is a specialised agency of the United Nations. Operators are beginning to use SHD to set up or reconfigure circuits remotely in software, so that customers can buy whatever capacity they need at short notice. This facility is known as *bandwidth on demand* and is of growing importance in high-volume data applications.

A reduction in the bandwidth or sampling rate used for speech will reduce the bit rate required for transmission at the expense of some loss in quality. The capacity of a digital circuit may be further increased by compression techniques. These remove or code the redundant elements of the message so that it takes up less space. Examples of redundancies which can be removed are the pauses during ordinary speech, the white background of a typed message on a piece of paper and any element which has not changed between two successive frames of a film or video. *Digital compression and multiplexing equipment* (DCME) is used by some operators to fit around five voice channels into a 64 kbit/s circuit. The compression is performed by using software.

Technology continues to evolve. *Wavelength division multiplexing* (WDM) has the theoretical ability to increase the carrying power of optical fibre by a factor of 10 000 or more. The single frequency used to transmit channels with TDM could be augmented by thousands of others, effectively combining the principles of FDM and TDM in a single optical fibre.

2.2 Transmission media

Three transmission media are now in common use. These are copper cable, optical-fibre cable and radio (terrestrial or via satellite). An end-to-end connection between two telephones is called a circuit and may pass through several links; an individual cable or radio link may carry many circuits. Terrestrial and satellite radio links can be used in fixed-link networks where the telephones are physically attached to the network in homes and offices. They are essential in mobile applications, where the telephones are not physically connected to the network.

The cost of transmission is a function of the degree to which the full capacity of a cable is utilised by the degree of multiplexing employed. This is an important factor in the development of both public and private networks and in some forms of network competition.

2.2.1 Copper cables

Telephones are usually connected to the local exchange by a pair of copper wires twisted together, called a *pair* or *twisted pair*. Twisted pairs are capable of carrying individual telephone conversations and most can have their capacity increased considerably with *digital subscriber loop* (DSL or XDSL) systems.

Extension of digital transmission beyond the local exchange to customers' premises enables customers to send and receive either speech or data at 64 kbit/s. This permits the development of an integrated services digital network (ISDN), which enables customers to obtain both services over the same pair of wires to the same local exchange without the need for special equipment to convert digital signals to analogue form. ISDN is considered in more detail in section 2.4.4. Other varieties of DSL are:

(i)   *ADSL* (asymmetrical digital subscriber loop): delivers 2 Mbit/s in one direction and 64 kbit/s in the other. Works over distances up to 3.5 km on a single copper pair,

(ii) *HDSL* (high-speed digital subscriber loop): delivers 2 Mb/s in each direction, works over distances up to 3 km and needs three copper pairs, possibly two with further development.

(iii) *VDSL* (very-high-rate digital subscriber loop): delivers very fast bit rate over a single copper pair but over a shorter distance than that for ADSL.

The main use seen for these systems is the delivery of broadband services including video over the existing local network of copper pairs, a market not yet developed in the late 1990s.

Coaxial cables have a greater capacity and are used extensively for long-distance transmission. They are made with an inner copper wire separated from an outer copper tube by an insulating medium. TV cables are a familiar domestic example.

2.2.2 Optical fibre cables

Optical fibre is replacing coaxial cables on long-distance and international routes and is being used in parts of the local network. Optical-fibre cables have a much higher capacity than copper as signals are transmitted as light from a laser rather than electric currents. Their economic importance lies in the huge carrying capacity of a cable of small size. They cheapen the cost per circuit, enable more capacity to be packed into congested spaces, reduce weight and have lower maintenance costs. And there are countries where optical fibre compares well with copper because it is worthless to a thief!

The economic viability of fuller deployment of optical fibre in the local loop depends on extra traffic volume being generated to match the extra capacity, and on the extent to which maintenance costs will be reduced.

2.2.3 Radiocommunications in fixed networks

Radio has long been used in long-distance networks, particularly over stretches of water and over terrain where it is difficult to lay cables. Microwaves have a line-of-sight range, so repeater stations are often required on hill tops and tall towers are needed in cities. Microwave links and coaxial-cable systems have similar channel capacities.

Radio can be used instead of a copper or optical fibre link from the local exchange to the home or office. Such links are sometimes called *radio tails* or *wireless local loops* (WLLs). They require towers placed in an open position and authorised transmission frequencies. The towers are equipped with transmitters and receivers, and much the same kind of other equipment as would be found at the end of a cable on the same route. Radio links in the local loop require a station in the home, to which a handset can be attached. Radio lacks the capacity of optical fibre but compares well with copper. It is subject to interference from rain and snow, but the importance of this depends the strength and frequency of the transmitted signal. Cordless telephones use radio to enable handsets to be moved around a customer's premises.

Trunk radio transmission is a long established technology; the development of local radio systems is more recent. It has been driven by the desire of PTOs to extend service into remote or underdeveloped areas quickly and cheaply, and by competitors seeking an economic alternative to the copper local loop of established operators. Much of the expense is in the towers used to gather calls from the customers' stations, and various network configurations are used to minimise these.

2.2.4 Satellite communications

Satellites can be used to relay signals over short or long distances. They are used in the PSTN and in private networks as an alternative to cables. The *very small aperture terminal* (VSAT) is a small satellite dish, typically around 60 cm across and used in private networks to receive Ku–band (11.7 to 12.7 GHz) data transmissions from geostationary satellites. Mobile network operators use satellites to provide global marine and land communications and global positioning systems. Broadcasters use them for *direct broadcasting by satellite* (DBS), which requires a 30 cm dish for reception.

Communications satellites carry equipment which can exchange signals with terrestrial *earth stations*. They have *transponders* to send and receive messages. The satellite sector has four components:

(i)   The *earth segment*, terrestrial earth stations which transmit or receive signals, ranging from major transmission installations to the small dishes used to receive DBS television services.
(ii)   *Uplinks*, the transmission channels to the satellites.
(iii) The *space segment*, satellites.
(iv)   *Downlinks*, from satellite to the earth segment.

Satellites can be put into an orbit some 35 800km (22 400 miles) above the equator which leaves them always in the same apparent point in the sky and enables each one to cover about one third of the earth's surface. This is the *geostationary* (*geosynchronous*) *orbit* (GSO). The GSO follows a narrow path much favoured by satellite operators. Satellites need to be spaced out along the orbit so that they will not interfere with one another. The number that can be placed in the GSO without mutual interference depends on the technology of the time, creating a potential economic value derived from the scarcity of positions on it, which are known as *slots*. Some parts of the orbit are more crowded than others. A satellite in GSO cannot reach the most northerly and southerly parts of the globe, which are hidden by the curvature of the earth. There is a significant transmission delay because of the distance that the signal has to travel, which detracts from their usefulness for voice telephony.

Launch costs are substantial and satellites carry a significant, and insurable, risk of failure. They may drift out of position and can be steered or relocated during their working lives. After ten years or so, their orbit decays and they fall back to earth. Satellites can be launched into lower orbits more cheaply but they will not then be geostationary. *Low earth orbit* (LEO) satellites typically operate at an altitude of 500–1 500 km (300–1 000 miles) and are only visible for a short time from any one point on earth. *Medium earth orbit* (MEO) and *Intermediate circular orbit* (ICO) satellites operate at a higher altitude, some 7000–12 000km (4400 to 7500 miles) up. Both have lower launch costs but require far more satellites to be in use. A LEO network covering the world requires a minimum of 40 to 70 units to maintain radio contact and needs switching facilities around the globe because of the lower altitude. Satellites may also be put into a *highly elliptical orbit* (HEO), an elliptical orbit which at its furthest could be around 32 000km (20 000 miles) up. Because of its higher altitude continuous cover for northern (or southern) areas can be obtained with as few as three satellites. LEO, MEO and HEO satellites can cover the northern and southern areas out of reach of the GSO. The HEO was used by the former Soviet Union to reach the north of its territory.

Communications networks based on LEO systems need multiple satellites transmitting to each other as well as to the earth segment if continuity of signal cover is to be maintained. In planning a new system, operators have to balance five factors:

(i)   Technology risk. GSO technology has a proven track record.
(ii)   Coverage. LEO, MEO and HEO systems can reach parts of the world which are out of GSO range.
(iii)   Transmission quality. GSO systems present more substantial speech delay and echo problems.

(iv)   Satellite cost. GSO systems need fewer satellites, though each satellite may cost more.
(v)   Transmission cost. LEO and MEO systems need less powerful transmission equipment, offering potential cost savings.

Conversion of satellites to digital technology is increasing the capacity of the international systems and reducing the cost per channel.

2.3 Switching

2.3.1 Switches and their functions

Early telephone exchanges worked manually. The lines served by the exchange terminated at sockets on a board controlled by an operator and the operator made a connection between two lines by plugging in a cord which linked the two sockets. Automation has eliminated most of the functions of operators in modern public exchanges and the cords have long since gone. On some private networks, operators still handle incoming calls from the PSTN for internal extensions.

The original automatic equipment was electromechanical. It was worked with substantial moving parts, the movements of which were determined only by the numbers dialled by the caller. The mode of operation was called step-by-step and the system was named after its inventor, Almon B. Strowger, a Kansas City undertaker. Strowger equipment was robust and long lived; some plant remained in service for 60 years. Secondhand equipment found a market in third-world countries where the absence of electronic components made it easier to service with the skills locally available.

A later generation of electromechanical equipment called *crossbar* replaced the operators and the cords by a matrix of contacts, or small switches. To connect lines 6 and 8 the switch at the intersection of lines 6 and 8 had to be closed. Texts such as Martin [1] explain the principles more fully. Crossbar exchanges are more powerful and cheaper to maintain. They were widely installed from the 1960s onwards while electronic digital exchanges were still being developed.

After some initial experiments with valves, the earliest successful electronic exchanges were brought into use with electronically-controlled reed relays. Digital electronic exchanges were brought into use in the late 1970s. A very readable account of the history of switching is in Meurling and Jeans [2], which describes the evolution of the Ericsson family of switches. The modernisation of exchanges with electronic switches could be cost justified in most cases, enabling crossbar systems to be replaced before the end of their working lives. The electronic systems themselves were usually given relatively short accounting lives to allow for possible obsolescence from future systems. Digitalisation and computer control greatly increased the power of switches and enabled whole networks to be controlled. This paved the way for the introduction of many new services and improvements in service quality.

All the systems so far described use circuit switching, which ties up the complete circuit end to end for the duration of the call. In parallel with these developments, data traffic began to develop large and complex network requirements, where *packet switching* technology is used. Data signals are broken up into packets (i.e. pieces of short length) which can be addressed individually and serially numbered. This has been widely used in data networks, typically working at 64 kbit/s or less. The packets can be transmitted independently along different routes, using whichever circuit is free, and reassembled in the right order at the other end. Special packet switching exchanges are required, but packet switching reduces cost by enabling circuits to be loaded more fully. Switches for packet networks are designed to cope with particular protocols (modes of working), so that separate switches are needed for each system.

Specialised switches are used in mobile and private networks. Those used in private networks and offices are called *private (automatic) branch exchanges* (PBXs or PABXs). The smallest, which may serve only two or three internal lines, are *key systems*. All such switches are sized by the number of exchange lines which go into them and the (larger) number of extension lines that they serve. Blocking can occur if there are insufficient exchange lines for the incoming traffic. Direct dialling of internal extension numbers from the PSTN is possible with modern PBXs.

Switching technology continues to evolve. With digital technology, voice and data are conveyed in essentially the same way, making it difficult, if not impossible, to tell which is which or to measure the relative volumes. Voice and data are still priced separately where they can be identified. The integration of voice and data traffic and the addition of video transmission to form new *multimedia* services in new high-speed digital networks will require switches that can handle a wide range of transmission speeds. Multimedia services combine voice, data and video elements. Optical switching systems which are under development promise further increases in power and reductions in cost.

2.3.2 Signalling

Signals have to pass between the originating and destination telephones and their exchanges before a call can be established. These are codes covering such matters as the destination of the call and whether the distant line is engaged. With electromechanical exchanges these signals are sent over the same channel as the call itself, but with digital systems there is a separate signalling channel, permitting more powerful systems which can provide a greater range of information about the nature of the call, allowing the connection of more complex services. CCITT No.7 is a *common-channel signalling* system recommended by the *ITU–T* for use between digital exchanges.

2.4 Fixed link network structures

2.4.1 Network elements

Fixed-link telecommunications infrastructure has five basic elements:

(i)   Terminal apparatus, which may be a telephone or a device such as a fax machine or a computer.
(ii)   The access network, which consists of cable or radio links between terminal apparatus and local switching centres.
(iii)   The switching centre, or telephone exchange. Switching centres operate at various levels in the network. Most provide local service but there are others in the network between local exchanges (Section 1.3.1).
(iv)   The transmission network, with its associated plant. Long-distance traffic travels over the *trunk network*. Traffic between local exchanges and between them and the trunk network goes over a *junction network*. Radio links may be used.
(v)   The international network of cable and satellite links, connected to the domestic network by gateway exchanges.

Private networks may use cable or radio links as transmission media. These are usually leased from PTOs.

2.4.2 Private and cellular mobile networks

Private mobile networks have been used for many years; some are local, others regional. The required radio frequencies are issued under licence and transmission strength is limited so that the frequencies can be reused elsewhere. A central transmitter sends out radio messages to the vehicles or handsets using one of the frequencies assigned to the network. Messages can be returned, but a simultaneous two-way transmission may not be possible. Speech quality is often poor, although good enough for the purposes of the network. A later development, *trunked private mobile radio* (TPMR) enables frequencies to be used more intensively in local areas and has significantly eased congestion.

Public mobile services require an infrastructure of local transmitters/receivers linked by radio or leased lines to each other and to the PSTN. The basic system is *cellular mobile telephony*, in which the territory covered is divided into cells, each of which is served by a base station. Radio frequencies are allocated and dedicated to the system. The use of low transmission power within the cells enables frequencies to be reused many times in different cells. The cells vary in size from 20 km across in rural areas to a few hundred metres in city centres. The *backhaul* network is made up of fixed radio or leased line links to the local mobile switching centres and on to the PSTN.

The user buys a handset which contains a unique identifier in its circuitry, enabling all calls through it to be appropriately billed. There has to be an effective method of passing a call from cell to cell (*hand-off*), so that a moving caller can pass over a cell boundary. Failure to hand the call over, or *drop-off*, results if all frequencies are already being used in the cell that has been entered. Early cellular networks were analogue, but these are now being replaced by digital systems, and GSM uses a digital technology developed in Europe by the *Groupe Speciale Mobile*. In current usage, GSM stands for global system for mobile communications and the system incorporates a special database which permits a customer of one GSM system to make calls on another, a facility called *roaming*. Roaming requires practical problems, such as how calls are handed over and who pays what to whom, to be solved. GSM is competing with American and Japanese digital systems in the world market.

Cellular mobile systems require a means of allocating a channel to a caller. There are several. *Frequency-division multiple access* (FDMA), which was employed in some of the early systems, uses a different frequency for each channel and works through a dialogue with the base station. The GSM system uses *time-division multiple access* (TDMA) in which time slots are allocated on a single channel serving the base station area, which can provide more channels than FDMA. Both may give way to *code-division multiple access* (CDMA), which was developed by American manufacturers and uses the spectrum more efficiently than either of the other two options.

Microcellular systems use smaller cells but require more base stations, and the handsets need not be so powerful, so they can be lighter and the batteries may last longer. Such systems are suitable for urban areas. A *digital european cordless telecommunications* (DECT) system is being developed for use in the urban local loop. Many base stations would be needed because of the short range of the system.

2.4.3 Data networks

Morse is usually transmitted by hand at a speed slow enough to be decoded by those listening to it. Faster data transmissions, such as those from a computer, can be transmitted as sounds over a voice network using an *acoustic coupler,* but they are more usually converted into modulated frequency form with a *modem* (modulator–demodulator) if they are to be transmitted over analogue circuits. The tranmissions then have to be demodulated using a modem at the other end of the line. A fax transmission is an example of data traffic that can pass over analogue voice circuits in this way.

A digital local-access line is needed if the use of a modem and the potential signal corruption from an analogue access line are to be avoided. Most PTOs offer *public data network* (PDN) services. In the past, PDNs have had a poor reputation for quality, although this seems to be changing.

When digital data traffic is exchanged, *protocols* governing the exact form of the transmission are needed. A protocol is an agreed set of formats and procedures for exchanges of information between terminal equipment. One used widely for the exchange of data on and between computer networks is *transmission control protocol/internet protocol* (TCP/IP), which is fully consistent with telecommunications networks and is used extensively for online information services. Protocols include error-detection and correction procedures for transmissions – one missing digit in a set of financial data can involve millions of pounds.

Separate digital networks have been established to handle data traffic. In a data network the real-time continuity and sequence of a signal which is essential for acceptable voice messages need not be maintained. Voice traffic is continuous on a two-way channel of fixed capacity, whereas data traffic is in bursts. It often only needs one-way transmission but has a highly variable requirement for channel capacity. Data traffic can show much greater volatility than speech, with short peaks of extremely high usage, such traffic being called *bursty*. Where the traffic level justifies it, voice and data networks are designed differently to handle these differences in the most efficient way.

Various data technologies and protocols have been developed, some by international agencies and others by operators and equipment suppliers. The CCITT produced the X-series of recommended data transmission protocols in 1980. One of these, *X.25*, was for the attachment of data terminals to PDNs. Packet networks are often known as X.25 networks. *Frame relay* was introduced for use on digital circuits in 1991. It uses packets of variable length and is faster, offering transmission rates of up to 2 Mbit/s and improved network efficiency. For more information see for example, Flood [3] and Motorola [4]. Higher transmission speeds can be provided by *asynchronous transfer mode* (ATM) which is based on *cell relay*, where the packets, or cells, are all of the same length and are of short duration. ATM is designed to work with digital transmission at a fixed speed of approximately 150 Mbit/s, into which the cells are interleaved. Slower transmissions are handled by changing the intervals between cells. *Asynchronous* means that the clocks at each end of the transmission need not agree on the time, although they must run at the same speed.

2.4.4 ISDN

The digitalisation of trunk networks has been based on the construction of 64 kbit/s paths, the value standardised for high quality digital voice traffic and also usable by data traffic. An *integrated services digital network* (ISDN) has the 64 kbit/s paths extended all the way to the customer. Conversion of an ordinary analogue line to ISDN can be carried out in the exchange by replacing the circuitry controlling the line. Almost all existing copper pairs are good enough to provide the final link to the customer, although a few need upgrading. The basic ISDN connection has two 64 kbit/s paths called *B channels* and a 16 kbit/s *D channel* which is used for signalling. The arrangement is called 2B+D and each B channel can be used for a voice or data call. Low-speed data can also be sent down the D channel, which has more capacity than it needs for signalling. The two B channels can be run together for higher quality music and low-quality video transmission. In Europe larger users are offered a bundle of 30 B channels and a 64 kbit/s D channel, known as a primary-rate connection. In the US the offer is 23B+D. ISDN lines remain part of the PSTN and telephone calls (although not data) can pass freely between them and analogue lines.

ISDN is not a service but a facilitator of improved performance, mainly for data. No modem is needed and transmission rates are several times faster than over an analogue line; fax transmission can be faster with suitably-designed machines. Some of the benefits are being overtaken by the improvement of analogue services, for example, modems working at 33 kbit/s and 56 kbit/s are replacing the 9.6 kbit/s modems which were common when ISDN was devised.

2.4.5 Cable TV networks

Early British cable networks (broadcast relay) were *narrowband* systems. They were designed to provide service in areas outside the reach of broadcast signals and, using twisted copper pairs, could only carry about four TV channels. Except for those in the UK, cable networks are usually not connected to each other. American cable networks, which are the world's most extensive, and all modern cable TV networks are *broadband*, usually using coaxial copper cable to get enough bandwidth for a good signal for up to 40 channels [5, 6]. They have three basic elements:

(i)   A *head end*, from which signals are transmitted.
(ii)   Cables in the street.
(iii)   Individual customer equipment: cable from the street to the TV set and a *set-top box* to decode the signal.

The commonest layout is *tree and branch*, with the root at the head end and the individual TV sets on the twigs. Signals travel one way only and are not switched.

Cable network layout can also be made interactive, allowing transmission back to the head end. Such systems are more expensive and involve choices about the bandwidth of the return channel. When cable franchises were distributed in the UK in the 1980s, better terms were offered to franchisees willing to install interactive *switched-star* systems using optical fibre and switching points within nodes on the local network. These offer the potential for higher performance at a higher cost. Not many were built. Basic telephony can be provided by electrical integration into the same cable, or by laying separate twisted pairs alongside the coaxial TV cable. The latter is the normal practice in Britain, where *cable-TV telephony* was well developed by the late 1990s.

2.4.6 Broadband networks

A digital network which can provide data transmission speeds of 2 Mbit/s upwards which are required in some business and video applications is called a *broadband network*. It is also known as *broadband ISDN* (B-ISDN) or an *Integrated broadband communications network* (IBCN). Broadband networks are still under development, but could use SDH as a framework, with switches designed for ATM. DSL techniques and cable networks could link the network to customers. Access lines capable of working at speeds of up to 600 Mbit/s and based on optical fibre technologies are under development.

The cost of deploying such a system as part of the PSTN and the range of services for which it is likely to be used are as yet unclear, although some see broadband networks as essential in a future *information society*. Many business users already see them as filling a public need and have private broadband networks.

2.4.7 Local, wide and metropolitan area networks

A *local area network* (LAN) is a private data communications network that links together computers on a local site such as an office block or university campus. Data traffic can be sent from a LAN to a public network using suitable protocols. A *wide area network* (WAN) links computers over a wider area and may use leased lines to connect sites. *Metropolitan area networks* (MANs) are used to collect business traffic in large cities.

TCP/IP can accomodate broadband services (although not very well) and has a greater penetration than ATM for this purpose in LANs. TCP/IP is mainly used in the Internet, and some see it as a serious rival to ATM in the public and private networks that provide Internet services.

2.4.8 Intelligent networks

The first use of computers in the control of public telecommunications networks was at individual telephone exchanges. Computers were soon being applied to network management at various levels up to national network control centres and they were used for many other functions such as billing. The control of networks by computers permits a good deal of flexibility, e.g. it permits more versatile number translation, using software databases instead of mechanical equipment. Number translation facilities in the network can map the number used by the customer onto a possibly different number used for routing calls. *Intelligent networks* (IN) carry the use of computers further to provide:

- existing network services such as personal numbering more cheaply;
- network alternatives for services provided by terminal equipment;
- faster development of new services;
- new features.

INs can be provided by putting intelligence into the main network or by building a special network overlay. For example, a network can be made self healing by putting fault-detecting software into computers at the nodes. Calls due to pass through a part of the network which is detected to have failed can then be rerouted from the node instead of being lost.

2.4.9 Networks for information services

Information services are distinguished from basic voice and data telecommunications because they provide information of a standard kind, for which the caller may pay a premium. Voice-based information services can use the PSTN for access and are then called *audiotex*. The service provider builds up a database of recorded messages or, in some cases maintains a staff of operators to take calls. Chat lines, which provide a *lonely-hearts* facility are examples of the latter. The PSTN is required to split the call fee between the operator and the information provider, a facility provided by an intelligent network or a special overlay network such as the *derived-services network* (DSN) used by BT to deliver its initial *premium-rate services* (PRS). International extensions require administrative as much as technical problems to be solved. They involve bilateral agreements between operators in different countries defining how call revenues are to be collected and transferred to the right recipients.

Videotex services deliver data on a television or computer screen. They can be accessed over the PSTN, but they are transmitted as data and require a modem on analogue lines. Unless provided free, or by the PTO through which access is obtained, they require the IN facility of transferring part of the call charge to the line being called. Similar services can be transmitted over television networks by using spare capacity in the TV channel. Some, such as CEEFAX and ORACLE in the UK are well established and useful, although they are slow in operation and have small databases. The most important part of all networks is the information, not the network used to carry it.

2.5 Numbering

Telephone numbering is basically hierarchical. Its purpose is to give a unique identification to each telephone so that calls can be routed across the network. Guidelines for national numbering schemes are provided by the *International Telecommunication Union* (ITU). and the basic structure is

service prefix + country code + area code + exchange code + local number

Numbers can be shortened. The international service prefix and country code are only needed for international calls, and in most countries the trunk service prefix and area code do not need to be dialled for local calls. In the UK, for example, local calls are made with a six or seven digit sequence comprising of the exchange code and local number. Numbers need not uniquely define routing. *Number translation* facilities can be built into electromechanical and digital switching systems to enable calls to be routed independently of the numbering scheme [3, 7]. Such a system has been used in large British metropolitan areas since the 1930s.

Numbering plans work within constraints on the number of digits in each part of the structure, and they differ in the efficiency with which they use numbers. The balance of the structure varies between countries. For example, in North America all customers' numbers have the same number of digits but many other countries have numbers of varying length. For practical reasons only a few per cent of the available numbers are actually assigned, the remainder not corresponding to combinations used for the designations of countries, areas, exchanges and lines.

When the exhaustion of the telephone number stock in cities such as London and New York requires the introduction of a new area code, there are several options:

(i) Geographical: e.g. inner and outer London, north and south of the Thames, east and west of 5th Avenue, north and south of 42nd Street etc.

(ii) Temporal: existing customers keep their codes, new customers get the new code.

Unique blocks of numbers may be assigned to new services and operators; assignation can be by the use of a high-level digit in the number. For example, from April 1995 the UK moved to eleven or twelve digits by adding a new (second) digit to the service prefix for inland calls. The extra digit enables a wider range of service indications to be made. As at September 1997 the full service prefix scheme proposed by OFTEL [8] was:

00   international
01   geographic numbering range with five, six or seven-digit local numbers
02   geographical numbering range with eight-digit local numbers
03   reserved for future geographic numbering
04   reserved for future use
05   reserved for corporate numbering
06   spare for future use
07   personal numbering, mobile and paging
08   special services range for freephone and other calls priced no higher than standard national tariffs
09   premium-rate numbering range and for multimedia services

The proposed numbering, if implemented, will give a clearer indication of the appropriate call tariff. Corporate numbering has been called *virtual numbering* [9] because it would enable a company to have a range of nongeographic numbers in place of geographic ones. Many numbers would have to be changed.

Numbering systems are more transient than may at first sight be supposed. It is not difficult to find British lines which have had four different numbers in around thirty years, the changes reflecting provision for system growth and technological change. OFTEL [10] includes a wide-ranging discussion on numbering issues.

2.6 Billing

The revenue of a telecommunications operator is made up of a huge number of low-value transactions with customers who typically outnumber those of any other single business. Billing systems are expensive to devise and introduce, but they are a shop window for the operator. Billing is the commonest source of customer complaint to the regulator [11], and poor performance can lead to a loss of credibility with customer and regulator alike.

It is only readily possible to collect information about the origin, duration, time of day and destination of a call. Electromechanical exchanges generate regular pulses which are used as charging intervals. With *periodic pulse metering* the pulse rate is different at peak and off-peak times, and the pulses may be counted to arrive at the call price. The average number of charged units in a call may differ from the number in the average call duration, depending on how part units are treated in the charging structure, because most calls will begin and end part way through an interval between pulses. Digital exchanges can achieve a greater precision and per-second charging is common. The introduction of exact electronic measurement systems will then change the charged length of a call, as happened in the UK.

Digital exchanges can collect a wider range of information at less cost. The liberalisation of telecommunications markets has created a need for more sophisticated billing systems and brought many small new operators into the field. Specialised companies now provide customised billing systems to mobile, cable and other operators at home and abroad. At least 15 were operating in the UK in 1995.

2.7 Plant capacity and grade of service

Most of the equipment used to handle telecommunications traffic has two related economic characteristics:

(i)   Plant capacity is designed for the load expected at the busiest period of the day, the *busy hour*. The traffic load varies and may, on occasions, exceed the capacity provided. The proportion of calls lost because of this happening is known as the *grade of service*, a derivation for which is given below. It can be reduced by installing more plant, but only at an extra cost. The grade of service used in equipment planning is therefore of economic significance to the network operator as a cost and to the consumer as a purchase.

(ii) Plant requirements usually rise less quickly than traffic. Most big installations have lower costs per unit of output for a given grade of service than do smaller installations. There are various reasons for this but one of them goes back to the statistical theory of queuing.

The basic theory has its ancestry in the *Poisson distribution*, a formula used to describe the frequency distribution of randomly distributed events and published by Poisson in 1837. It is the limiting case of the more general binomial distribution which describes the distribution of random events that are not necessarily large in number. Poisson's distribution gives the probability of 0, 1, 2, 3 etc. random events within a time interval of fixed length as successive terms in the series

$e^{-m}, e^{-m}m, e^{-m}m^2/2!, e^{-m}m^3/3!$ etc.

where m is the expectation, i.e. the average number of events occuring in an interval. See, for example, Yule and Kendall [12] for a fuller exposition of the theory.

Such an infinite series requires modification for use in trunk occupancies in telecommunications, where the number of channels, switches or other pieces of equipment is finite in number and may be very small. No modification is needed for call arrivals. The most commonly used adaptation was derived by the Danish traffic mathematician A. K. Erlang, who gave his name to the unit used for traffic measurement. One erlang is the load presented by 60 minutes of line occupancy (not necessarily all on the same line) within one hour. Line occupancy starts from moment the telephone is picked up. Erlang expressed the probability $P(x)$ for x calls in progress as

$$P(x) = (m^x/x!)/(1 + m + m^2/2! + m^3/3! + \ldots + m^N/N!)$$

where m is the traffic in erlangs offered to the group of N lines.

Any traffic arriving while all N switches are already engaged will be lost. The grade of service is then probability, $P(N)$, of all N switches being in use. Practical grade of service calculations may have to take into account more complicated aspects of system design. A fuller exposition can be found in References 3, 7 and 13, and in technical engineering literature.

For the immediate purpose it is sufficient to look at the example in Figure 2.2. This shows how, for a given grade of service, the average occupancy of the switches or channels rises as the traffic load increases, so that the average cost falls. About 20 circuits would have to be provided to handle traffic totalling ten erlangs and the average circuit occupancy is just over 50 per cent. Only about 30 circuits are needed to handle 20 erlangs and average occupancy rises to about 65 per cent.

At low traffic levels the attainable average occupancy falls sharply as the grade of service is improved, but the difference is less marked at higher traffic levels.

Grade of service is one of a more general bundle of characteristics which make up the quality of service. Some of these, such as fault rates, are related to engineering. Others, such as the length of waiting lists, are features of management. For all of them, there is a trade-off between cost and quality. In most markets for goods and services there is a choice. Some customers go upmarket, for the better apple, some go down. The phenomenon of decreasing returns tends to make marginal quality improvements progressively more expensive. At some point the marginal cost of obtaining an improvement exceeds the marginal improvement in value to the user. Because individual circumstances vary, this balance point occurs at different quality levels for different people.

Quality of service economics and economies of scale are discussed at more length later in the book. A fuller exposition of technical and marketing issues affecting quality is in Reference 14. Further theory and other essentials of the technology of telecommunications networks can be found in References 3 and 15.

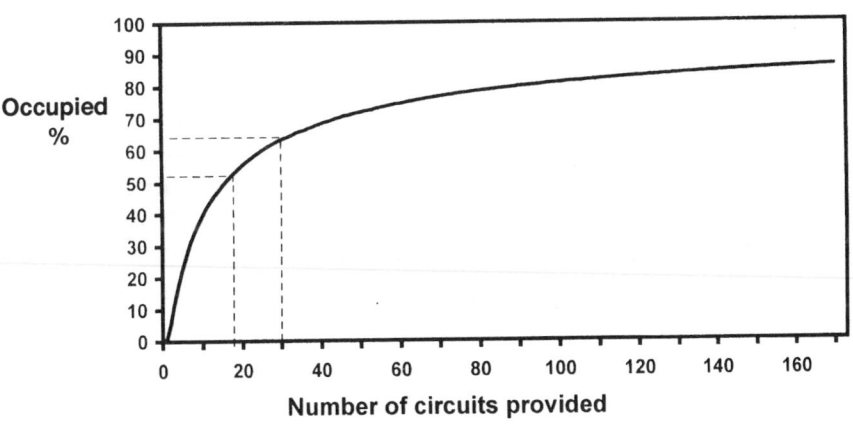

Figure 2.2 Average circuit occupancy with 0.005 grade of service

2.8 Telecommunications dynamics

2.8.1 Capital and labour costs

Telecommunications is often thought of as a capital-intensive, high-tech industry but this is not completely true. Current account labour costs usually form a higher proportion of the inputs than capital consumption. Capital assets are in some cases, very labour-intensive products. For example the duct network under our streets, which makes up one of the largest elements of the capital base, was mainly created by pick and shovel work and bricklaying. Cable TV companies can be seen digging trenches for their own duct networks in many towns.

Operating companies may spend large amounts on research and development in absolute terms, but in relation to turnover the figures are more modest. In the year to March 1994 BT spent £227 m on R&D, some two per cent of its operating cost and well short of the percentages spent by leading chemical and pharmaceutical companies. In the same year C&W, BT's main UK competitor, had no central R&D facility. Manufacturers of telecommunications plant and equipment usually spend more in relative terms, but their sales total less than a quarter of the market for services.

2.8.2 Migration of added value from hardware to software

Early telecommunications equipment was electromechanical, and its manufacture and installation were labour-intensive. All the added value was in the hardware. Modern systems are making increasing use of computer control. The code used to control the computers has become critical and much of the added value has migrated from the hardware to the software. The software represents labour in another form, but the skill requirement has changed from hand to brain and the workplace moved from the factory bench to the computer workstation. The shift is part of a bigger change in industrial and social structure.

2.8.3 Convergent technologies, divergent markets

In the 1970s and 80s there was a good deal of talk of convergence in the telecommunications and computing industries. The convergence was based on the use of a common digital technology, the increasing use of computers to control telecommunications equipment and the introduction of data networks. In practice the convergence did not extend to the markets into which the outputs of the two industries were sold, and attempts to create mergers between the two were largely unsuccessful. The synergy needed in the market place to reduce joint marketing costs did not exist. In addition, there was unease on public interest grounds about allowing the biggest firms in computing and telecommunications to merge with each other.

In the 1990s talk has turned to the convergence of telecommunications and computing with broadcasting. This again is based largely on the use of common technology and potentially common networks. It remains to be seen how successful these are, and whether TV and telecommunications operators can develop sufficient expertise about each others' business and markets to create effective and permanent market entry.

2.8.4 Technologies for the future

Future economic growth depends fundamentally on technological developments. Electronics is likely to be the base for further advances, although there are probably other areas where breakthroughs have yet to come and where there is a market waiting. Some of these are in telecommunications, most will affect its markets. Growth generates new telecommunications demand, the amount depending on which industries and applications are growing fastest.

Computers are increasing in power in absolute terms and in terms of the computing power offered per kilogram of weight or per cubic centimetre of physical size as well as per £ and per $. Some of the increase derives from more powerful chips, following the empirical *Moore's law* named after its proposer, G. E. Moore, that the amount of information storable on a given amount of silicon chip has been roughly doubling every three years since 1970 and will continue to do so into the 21st century. In the early 1970s the circuits in most devices on computer chips measured five microns (0.005 mm) in width. Twenty five years later some were only half a micron wide, with the expectation that they would get smaller, possibly reaching a limit of 0.05 micron set by the size of atoms (see e.g. Reference 16).

A further increase in power comes from parallel processing, running different parts of a calculation simultaneously through different subcomputers. More powerful systems include *neural computing*, based on logic processes in the human brain. Such systems will improve the machine/man interface, speed up programming, data processing and transmission, and improve systems for decision making, information, navigation and similar applications. The use of telecommunications in these processes will become cheaper, increasing demand. *Computer-aided design* (CAD) and *computer-aided manufacture* (CAM) will cut production costs and make more use of telecommunications by remote working.

Battery power is a limiting factor in the usefulness of most portable and mobile electric devices, including telephones, computers and modems, power tools and vehicles. Mobile telephony is growing quickly and is a new dimension in telecommunications, but incoming calls are frustrated when users switch off their handsets to extend battery life. Some portable equipment is designed to conserve battery power by switching off features not in current use. Cheaper and lighter power packs with a longer life between charges would stimulate further growth.

Opto-electronics enables electronic circuits to work faster and optical fibres to carry more signals. Weight savings will permit increases in the power of portable electronic devices.

2.9 References

1   MARTIN J.: 'Telecommunications and the computer' (Prentice-Hall, Englewood Cliffs, 1976)
2   MEURLING, J., and JEANS R.: 'A switch in time' (Telephony Publishing Corporation, Chicago, 1985)
3   FLOOD, J. E. (Ed.): 'Telecommunications networks', 2nd Edition (The Institution of Electrical Engineers, London, 1997, 2nd edn.)
4   Motorola: 'The basics book of frame relay' (Addison-Wesley Publishing Company Inc., Reading, MA, 1993)
5   Home Office and Department of Industry: 'The development of cable systems and services Cmnd. 8866' (HMSO, London, 1983)
6   CORNFORD, J., and GILLESPIE, A.: 'Cable systems and the geography of UK telecommunications, PICT policy research paper No.21' (PICT, Uxbridge, 1992)
7   MORGAN, T. J.: 'Telecommunications economics' (Technicopy 1976, Stonehouse, 1976, 2nd edn.)
8   OFTEL: 'National numbering conventions – consultation on revisions, Part B: public switched telephony network' (OFTEL, London, 1997)
9   HANNINGTON, S.: 'Virtual private numbers to give boost to VPNs', *Commun. News,* August 1997, p.16
10   OFTEL: 'The national numbering scheme, a consultative document' (Office of Telecommunications, London, 1996)
11   OFTEL: 'Annual report' (Office of Telecommunications, London, 1993)
12   YULE, G. U., and KENDALL, M. G.: 'An introduction to the theory of statistics', (Griffin, London, 1950, 14th edn.)
13   COX, D. R., and MILLER, H. D.: 'The theory of stochastic processes' (Methuen, London, 1965)
14   OODAN, A. P., WARD, K. E., and MULLEE, A. W.: 'Quality of service in telecommunications' (The Institution of Electrical Engineers, London, 1997)
15   WARD, K., (Ed.): 'Telecommunications engineering, a structured information programme' (The Institution of British Telecommunications Engineers, London, 1991–8)
16   'Can Moore's law continue indefinitely?', *Computerworld,* January 1996

Chapter 3
The world telecommunications economy

3.1 Telecommunications in the national income accounts

Traded goods pass through the economy in a cycle which starts with production and finishes with consumption, generating income on their way round. The national income, or gross domestic product, can be measured in three ways, each of which should give the same answer, within the limits of measurement error, since they are different ways of measuring the same flow. GDP is measured at factor cost, i.e. excluding indirect taxes:

(i) *Output measure of GDP:* This measure is the total of the *value added* by all producers of goods and services. Value added is the value of sales less the value of goods and services bought in to produce them. It is also called *net output.* The pay of the producer's labour force of the company is part of the value added, not a deduction. Output is analysed by type of product.

(ii) *Expenditure measure of GDP:* This totals up *final expenditure.* Final expenditure at market prices is expenditure which is not part of an input to other sales. It includes indirect taxes; GDP is calculated by netting off indirect taxes and imports, as below:
consumer expenditure (personal expenditure)
+ *public current expenditure*
+ *gross domestic fixed capital formation* (GDFCF), which is investment
+ *changes to stocks* held by producers but not yet sold
+ *exports*
= *total final expenditure* (TFE) at market prices

– *indirect taxes* (called the *factor cost adjustment* (FCA))

– *imports*

= *gross domestic product at factor cost*

(iii)   *Income measure of GDP:* This is the total of *factor incomes.* The two main *factors of production* are capital and labour. The incomes of these factors are the gross trading profits of companies and the pay of individuals, before tax.

The net output of telephone operating companies is part of the output measure of GDP, and the net output of companies producing telecommunications equipment is also included in this figure. Most equipment sales are the purchases of other companies and are netted off by them as input costs to avoid double counting. The profits of operating companies and equipment suppliers and the incomes of their employees go into the incomes measure of GDP. *Personal disposable income* (PDI) is personal income after the deduction of tax and social security contributions. It is an important variable in the analysis of telecommunications demand from the personal sector.

Empirical studies of telecommunications demand distinguish between sales to private individuals and sales to businesses, because the personal and business markets have different characteristics in marketing and forecasting. The expenditure measure of GDP includes telecommunications expenditure by individuals for their own use in consumer spending, which is an alternative to PDI in personal sector demand studies. Consumer spending, the current account spending of the public sector, purchases of capital goods by all sectors, and export sales are included in the total of *final expenditure* in the expenditure measure of GDP. It is *final* because the purchases are for consumption and not for manufacture, or for resale.

Some telecommunications expenditure for business purposes is included in the public, investment and export sectors of final expenditure. Most of it is not included directly but represents *intermediate expenditure,* being used as inputs in the production of other goods and services. The sales come through to final expenditure as part of the bill for the products. Thus, consumer expenditure on apples includes payment for the telecommunications input costs passed on by farmers and distributors. An analysis of intermediate expenditure, showing who buys what from whom, appears in *input/output tables,* which are discussed further in Chapter 4.

3.2 World telecommunications services markets

3.2.1 World services revenue and investment

Telecommunications services totalled $414 bn in 1992, equivalent to 1.9 % of world output. The proportion was highest in the high-income countries, although with a good deal of variation. Operators invested $128 bn. This is part of world GDFCF and equal to 0.59 % of GDP. The proportion was much the same for countries at all income levels. Both sets of figures are analysed in Table 3.1.

Table 3.1 Telecommunications services and investment, 1992

Country/ Region	GDP per head ($)	Revenue % of GDP	GDFCF % of GDP
World	3 980	1.90	0.59
Low income	309	1.11	0.67
Lower Middle income	1 472	1.34	0.49
Upper middle income	3 522	1.55	0.74
High income	21 007	2.03	0.57
Europe	13 311	2.06	0.71
Australia	16 886	3.04	0.75
Canada	21 635	1.92	0.60
US	22 245	2.14	0.37
Japan	27 118	1.63	0.59

(Source: calculated from ITU statistics)

The method of calculation uses a mixture of output and expenditure GDP components. It overstates the contribution to GDP by being based on gross rather than net inputs, but this is to some extent offset by the exclusion of equipment manufacture. The combined contribution of services and equipment for all purposes is likely to be between two and three per cent in the high-income countries.

World telecommunications revenue rose by 12.1 % in 1993, and the number of lines rose by 5.8 %. The world market for telecommunications services is expanding at over five per cent per annum by volume, made up of at least five per cent from new fixed lines plus five per cent or more from higher calling rates and mobile services. Service prices are falling in real terms, but annual revenue growth appears to be at least ten per cent. The market for equipment is growing at a similar rate (Table 3.5), perhaps a little more slowly.

3.2.2 Cable television networks

Although present cable TV networks are mostly one-way services and not connected with each other, the infrastructure is changing, and they are becoming more important in the telecommunications sector. In 1995, 64 million US homes had cable TV. China had the second largest total with 40 million, followed by India with 17.5 million and Germany with 14.6 million [1]. Cable TV operators in the UK are permitted to offer telecommunications services over their networks, and OFTEL initially showed some impatience with the slow development of residential local-loop provision by the cable operators. The operators later speeded up their telephony provision and by January 1998 they were providing 403 000 business and 3 039 000 residential lines, about 13 per cent of the residential market, and growing quickly [2]. There has been extensive market entry by US and Canadian operators into British cable TV, and those companies usually also provide telephony.

Market entry of telephone companies into broadcasting has been small owing to regulatory restrictions. BT initially had stakes in several of the cable TV operators but later disposed of them. In 1997, Cable & Wireless merged its UK cable interests and merged them with its domestic telephone subsidiary to form a new company, Cable & Wireless Communications (CWC), which it partly owns.

Other cable-based telephony operations exist or are planned in Belgium, the Netherlands and Spain [3]. Operations have been started in the US, through a cable subsidiary of one of the network operators.

3.2.3 Data services

With the increasing computerisation of information and data transfer systems, national and international data services are growing more rapidly than those for voice. Most voice traffic travels over the PSTN. Data traffic can travel over the analogue parts of the PSTN via modems or, more rapidly, over the ISDN (see Section 2.4.4). Private circuits are are used extensively for data transmission because there are higher quality requirements and fewer restrictions on use. Measurement of relative traffic volume is conceptually difficult because the circuits used for data vary in their capacity and transmission speed. Volumes can be measured by disparate methods such as the peak-capacity requirement, number of characters conveyed and minutes of use. Little information is available about the balance between the voice and data volumes. The data proportion is low in the PSTN but over 50 per cent on some routes if private circuit traffic is included.

Cable modems capable of handling one-way data speeds up to 10 Mbit/s are now available for cable TV networks, giving them the potential to be serious competitors in the broadband data market. In May 1997, there were over 20 000 US and Canadian cable modem customers [4].

3.2.4 Value-added services

Telephone calls form the basis of other services which can command a higher price. Value is added, for example, by storing a call for forwarding at a later time. Such services are called *value-added services* (VAS). The amount by which the value is enhanced varies and may be a negligible amount. The distinction between value-added and other calls derives its main importance from a common restriction of monopoly rights, if granted, to basic (unenhanced) services. Value-added services have been liberalised almost everywhere. Value-added services are of four main types:

- information services;
- call processing;
- messaging;
- network management.

Messaging includes *electronic mail* (E-mail), *electronic document interchange* (EDI) and *electronic funds transfer* (EFT). The simplest value added services are not interactive. Applications such as basic EDI and e-mail use techniques developed in the computer industry to move data between different individuals and businesses. Data input costs are high in relation to the value added and they have not been the basis of any substantial businesses.

Greater value can be added by using telephone connections interactively. For example, whereas a central database can be used to provide a chain of motor dealers with information about the catalogue specifications of vehicles and their spare parts, a more powerful application is to use it to search all dealer stocks for the location of a particular kind of vehicle wanted by a customer in the showroom. Travel agents use this type of service extensively for booking holidays. Credit, debit and charge card validation, to control fraud, also operate interactively and are common in shops. The computerisation of charge-card handling procedures has cut costs and made them easier to use. Televoting, where viewers telephone yes/no responses to questions put out from TV studios, generates sharp traffic spikes and needs a combination of call-processing and network-management techniques to control congestion.

There are numerous companies supplying value added services at both levels, and many of the more successful have approached from the application back to the network rather than the other way round. Competitive advantage does not usually lie with network operators. Successes and failures in the value added field illustrate the subtlety and variety of the market. Although most value added services serve domestic markets, there is a significant international trade. High-value information services travel across national frontiers and often cut the transmission cost by using private circuits. The Internet offers international access for the same price as domestic telephony by exploiting tariff anomalies. Websites may become one of the the biggest international value-added service providers.

Information services are provided by numerous large and small *information providers* (IP), and are the largest group in revenue terms. BT's Prestel system, like Minitel developed in France, was difficult to establish on a commercial basis and did not become as large or profitable as systems developed by information providers outside the telecommunications industry. Both have gained strength through distribution on the Internet. The French government saw Minitel as an important venture in information technology. When France Telecom launched the system in 1980, simple terminals were issued to users free and the White pages of telephone directories were loaded onto it. Initial plans to withdraw paper directories were later dropped. By 1984 there were over 100 million hours of use annually from 20 million users on 6.5 m terminals. Directory enquiries accounted for 41 per cent of a total of 1.9 billion calls and revenues of 6.6 billion French Francs (Minitel). A full account of the genesis of Minitel is in Marchand [5].

Information services with a higher added value have been provided for special interest groups. Reuters is an example of a company serving the financial markets with a combination of high quality real-time information, dealing positions and software.

3.2.5 Home shopping

Mail order is a long-established form of shopping. The customer receives a catalogue by post or sees goods advertised in a magazine or newspaper. The goods are then ordered, paid for and delivered by post. Home shopping is an interactive information service in which telecommunications can replace the post for all but the delivery of physical goods. The extent of replacement depends on the extent to which it is seen as a cost-effective substitute.

Examples of possible replacements are given in Table 3.2:

Table 3.2 Replacement of post by telecommunications in home shopping

Postal	Telecommunications
catalogue or advert	Teletext, Internet
ordering by mail	ordering by telephone
reply paid envelope	freephone
payment by mail (cheque or card)	payment by credit card over the telephone
delivery of software by mail	downloading of software over a data network
delivery of videos	downloading of videos over a telephone network

A customer can choose between personal and remote shopping. Remote shopping saves time and travel costs, and may offer enhanced product ranges and lower prices, but incurs communication and delivery costs, carries risks about product acceptability (goods cannot be physically examined before purchase) and money transfer may be more prone to fraud.

The introduction of freephone services has accelerated the growth of telephone ordering and encryption systems are making money-transfer systems more secure. The delivery charges are likely to be much the same whether the earlier stages of remote shopping are carried out by post or telecommunications. Typical examples are the £2 postage raised by booksellers and the £5–£15 courier fees charged by computer software sellers for goods ordered by telephone. A US store charges $6.95 plus five per cent of the order value for the delivery of orders on an interactive grocery service [6]. A service with a £5 delivery charge from 27 stores was opened by the UK food retailer Sainsburys in 1998. Retailers have scope for making savings to offset the delivery costs – wharehousing costs less than the display of goods in a shop and higher sales volumes enable better bulk purchase terms to be negotiated.

At one time the high quality of mail-order catalogues was beyond anything offered by electronic media and a major marketing advantage. This may still be true for ranges of goods such as cheap clothing, but there are now two powerful electronic competitors:

- TV shows demonstrating and selling products;
- high quality video images on broadband networks.

Experience has shown that for some goods image quality may not be important, for example if the actual product can be seen in a local store. Remote shopping works best with goods of a known quality and fairly high, but substantially discounted, unit value. Data services operating over ordinary telephone lines and TV-based data services such as Ceefax and Oracle in the UK were early into the marketing field, offering travel and other popular services. The booking of TV-based services could be carried out over the telephone in the absence of interactive facilities. The electronic media create a remote shopping market for new types of goods and services as well as competing with old ones. The market share of conventional mail order peaked before their arrival. Table 3.3 shows market shares in 1994.

Table 3.3 Retail market shares in the USA, 1994

	$ bn
Online shopping	0.2
TV shopping	3
Mail order	70
Total retail	2 000

(Source; Buckley [7])

These figures appear to underestimate the amount of ordering by telephone. As many as 25 per cent of long-distance calls in the US were freephone calls in 1993 [8] and they generated $250 bn in terms of consumer demand [9]. The Internet has become a measurable competitor, with some 25 000 companies carrying out business and sales estimated to have reached $500 m in 1995 [10]. The sales work out at about $17 per user, a remarkably high figure for such a new medium. Industry projections of $100 bn sales by the end of the century (see for example, Reference 11) would still cover less than two per cent of the retail market. There is some overlap with freephone sales because Internet advertisers often give freephone numbers. Business purchases are becoming more common on the Internet, creating another area of overlap in the figures.

The ceiling for electronic market penetration is unknown. It will probably be boosted by *virtual reality* systems when these become cheap enough. Some industry observers have put a limit at around 20 per cent of retail sales. Personal and remote shopping are not perfect substitutes for each other and both seem assured of a long-term future.

Remote ordering facilities are not necessarily profitable after they have become established. Advertising and sales revenue projections in Reference 10 show rapid growth in both but low sales per dollar of advertising expenditure.

3.2.6 Call-processing services

Call diversion is a facility which has long been available within companies through customer premises equipment. Telephone answering bureaus are common in the US, where they handle incoming calls on business and private lines while the owner of the line is away. In another variant of call handling, a telecommunications operator or a computer-based service provider may allocate a personal number to the customer, who keeps the service provider informed of the number to which *call diversion* is required. All incoming calls are then automatically diverted. No intelligent network feature is needed and it works with fixed or mobile services.

Many modern services work by charging part or all of a call to some other party or by transferring part of the call revenue to the line being called. Collection of the actual cash may be complicated and can be a major practical limitation especially for cross-border services. With *Freephone (0800)*, the calls are charged to the line *receiving* the call. Examples of more complex charging requirements are:

(i) *Premium-rate services:* an enhanced charge, typically for information services, is billed to the caller and the revenue is divided between the network operator and the service provider.

(ii) *Universal personal telecommunications* (UPT): this could theoretically allow customers to roam worldwide and use many types of service on any telephone anywhere, and have them all charged to one account (an automatic and generalised version of transferred charges). The service has yet to be developed. Global charging and billing arrangements would offer some of the most difficult problems to be solved and a global numbering framework would need to be devised [12].

(iii) *Calling cards:* the card issuer arranges for calls to be made from other telephones at homes and abroad, sometimes to third countries, and charged to a special card account. For example an American in Italy might call Japan and have the call charged to his US account. Calling cards are discussed further in Chapter 17.

Call-management centres (call centres) are now widely used by organisations to handle incoming calls. The centres commonly filter calls through an *interactive voice response* (IVR) system for the initial routing to a department, recorded information service, voicemail or other internal line. A large installation may be able to handle hundreds of calls simultaneously. *Automated call distribution* (ADC) switching systems put the incoming calls into orderly queues and may transfer them to other call centres. Company employees answering calls may have access to online information about the caller's business through *computer telephony integration* (CTI) and may, through *calling line identification* (CLI), be advised of the caller's identity before speaking so that they can call up relevant files. Large productivity increases are claimed for such systems, which save the caller, as well as the company, time and other resources.

Call centres may generate substantial volumes of business in their own right, examples being telephone banking and their use in *telemarketing* to sell products such as double glazing. They are becoming a significant employment sector. In the UK, for example, there were at least 150 000 call centre employees in 1998 and there might have been as many as 1.2 million in Western Europe [13]. Since telecommunications costs are largely independent of distance, the centres can be located in places where employment, property prices and tax incentives and the general business environment are most suitable. Favoured locations in the British Isles are Glasgow, Dublin and North-West England. There is competition from other European countries and the international dimension can be expected to strengthen as international call prices fall. The larger call centres each employ thousands of people on a single site.

3.2.7 Network management

Managed network services (MNS) are provided to private network users and replace corporate management staff by facilities of the service provider. The management of private networks is a substantial business which is increasingly achieved through software.

Virtual private network (VPN) and other configurations provide secure private access with guaranteed service quality by computer control rather than by providing a physically separate circuit.

3.3 Equipment and service suppliers

The equipment supply industry is large and diverse. Most of it has always been privately owned in western countries and much of the rest is being privatised. The main markets served are:

- switching and transmission equipment and cables for the PSTN, sold to PTOs;
- telephones and other terminal equipment for attachment to the PSTN, sold wholesale to PTOs for distribution or retailed direct to business and residential users;
- switching and transmission equipment and cables for local area and other privately-owned networks;
- aerospace equipment, sold to PTOs, ISOs and private satellite operators;
- mobile services infrastructure, sold to mobile operators;
- telephones and other terminal equipment for mobile networks, marketed to users through mobile operators and retailers,
- private network design and installation;
- retail and wholesale distributors, selling to installers, businesses and residential users.

The companies supplying these markets tend to specialise in the technology employed, rather than the service provision market where it will be used. Thus the General Motors subsidiary Hughes Electronics will build satellites for ISOs, design and build private satellite networks and has other markets in the aerospace, defence and motor industries. The manufacturers of switching equipment for the PSTN usually have a range of products for use in private voice and data networks. Telecommunications cable manufacturers such as BICC are also major suppliers of power cables to the electricity industry, competing with companies such as the French manufacturer Alcatel Alsthom for business in markets at home and abroad. BICC has interests in civil engineering through its ownership of Balfour Beatty and is the largest manufacturer of optical fibre cable outside the US.

Table 3.4 gives an overview of the world equipment supply trade in 1991. World exports amounted to $63 bn. The total market, including domestic sales, was estimated at $75 bn in 1988, of which $29 bn, or 40 per cent, was sold in the US and $9.8 bn in the then Soviet Union (NTIA, 1988). By the mid-nineties the total had grown to around $100 bn. The low and middle income regions are net importers of equipment but some of their member countries, notably China, Hong Kong, Malaysia and Singapore have substantial export businesses. Japan and the US between them had 43 per cent of the world export trade. Japan had by far the largest trade imbalance.

Table 3.4 Shares of world equipment trade in 1991

Country/region	Exports %	Imports %	Net exports $m	Net exports %
China	1.11	3.09	−1 241	−1.98
Other low income	0.21	2.21	−1 258	−2.00
Total low income	1.32	5.30	−2 499	−3.98
Malaysia	2.38	3.00	−390	−0.62
Other lower middle income	1.68	5.34	−2 301	−3.67
Total lower middle income	4.06	8.34	−2 691	−4.29
Korea (Republic)	3.30	1.99	826	1.32
Russia	0.34	1.95	−1 011	−1.61
Other upper middle income	1.80	7.75	−3 732	−5.95
Total upper middle income	5.44	11.68	−3 917	−6.24
France	4.43	4.11	202	0.32
Germany	7.75	8.44	−435	−0.69
Hong Kong	5.89	5.29	377	0.60
Japan	28.39	3.72	15 479	24.66
Singapore	4.13	4.45	−196	−0.31
Sweden	3.89	1.63	1 424	2.27
UK	5.03	5.70	−420	−0.67
US	14.59	19.89	−3 328	−5.30
Other High Income	15.09	21.46	−3 996	−6.37
Total High Income	89.19	74.68	9 107	14.51
World	100	100	0	0

(Source: calculated from ITU figures)

Powerful scale economies exist in some parts of the equipment supply industry, particularly switching, aerospace and computing. The research and development costs needed to bring a new type of telephone exchange to the market may exceed £500 m. Development costs for the for the Ariane 5 rocket came to nearly ecu 6 bn (£4.5 bn) [14]. In the mid 1990s, the computer chip manufacturer Intel was spending up to $2.4 bn (£1.5 bn) on a factory that just does wafer processing. Facilities for chip assembly and test costs are extra [15]. In 1993 there were 20 manufacturers with global equipment sales in excess of $1.5 bn and four with sales over £10 bn. These companies compete across the world for contracts to provide equipment for the industries which they serve. High development costs have led to many mergers. It is a common industry view that there is not room for more than five or six manufacturers of exchange equipment in the whole of the world market. Figures for some of the largest equipment manufacturers (excluding aerospace and computing), in Table 3.5 show an annual growth rate of ten per cent.

Scale economies in manufacturing are less pronounced than in research and much less than in development. It is common practice for the larger manufacturers to meet national demands for domestic supply by establishing local manufacturing subsidiaries. Most countries import their equipment or manufacture it locally under licence. Table 3.5 partly masks this.

Table 3.5 Major telecommunications equipment suppliers

Company	Sales ($bn)			Average annual growth (%)
	1986	1988	1993	
Alcatel		10.5	17.4	10.6
AT&T	10.2		11.8	2.1
Ericsson		3.3	7.8	18.8
Fujitsu		2.5	4.4	11.9
NEC		5.7	8.7	8.9
Northern Telecom		5.1	7.8	8.9
Siemens		6.6	11.9	12.5
Total			69.8	10.1

Notes: Telecommunications equipment only. Average percentage weighted by 1993 sales. Figures may be affected by difference in sources

(Sources: NTIA (1988), US Department of Commerce (1990), ITU indicators, 1994/95)

The ITU [16] has estimated annual investment requirements of $58 bn for the closing years of the century, based on a cost of $1500 per new line. World demand is summarised in Table 3.6. The figures only include exchange lines and associated equipment; mobile, cable TV, data and other services would make the total larger.

3.4 Public telecommunications network operators

Telecommunications operators are among the biggest commercial enterprises both in the world and in their home countries. Table 3.7 ranks the largest by turnover. They trade extensively with each other in the exchange of international telephone calls but, in spite of their size, their business is basically domestic and usually exercised with considerable market power. The ranking is therefore largely a

Table 3.6 World exchange line investment requirements 1993–2000

	Investment total $m	Main lines annual growth %
Africa	18 414	9.0
US and Canada	66 190	3.2
Other Americas	38 927	0.5
Total Americas	105 117	3.7
Japan	26 940	3.4
Other Asia	134 387	3.7
Total Asia	161 327	7.1
Europe	134 050	4.8
Pacific	5 911	4.2
Former USSR	41 335	6.7
World	466 154	
Low income	86 006	14.4
Lower middle income	80 481	9.0
Upper middle income	90 547	7.8
High income	209 120	3.7
World	466 154	5.0

(Source: International Telecommunication Union [16])

comparison of dominant operators working at home. It is influenced by currency exchange rates and local tariffs. NTT, for example, has the highest revenue per line among the local operators in spite of providing no international services. German tariffs are relatively high. MCI and Sprint are mainly long-distance operators.

Operators owning their own networks, as all of these do, are called *facilities-based* operators, to distinguish them from service providers who operate over leased lines. Between them, the 18 operators in the table had nearly 60% of the world's exchange lines and a similar proportion of global revenues.

3.5 Satellite operations

Geostationary satellites have operational uses inside and outside the telecommunications field; some are used wholly or partly for broadcasting. Satellite links are used as an alternative to terrestrial

Table 3.7 Largest telecommunications network operators

Operator	Country	Revenue ($m)	Lines 000s	Revenue per line $	Trunk services	Intl. services
NTT	Japan	84 080	59 580	1 411	yes	no
AT&T	US	47 277	0		yes	yes
Deutsche Telekom	Germany	46 151	40 706	1 134	yes	yes
France Telecom	France	29 613	32 000	925	yes	yes
Bell Atlantic	US	26 837	36 920	727	no	no
BT	UK	22 786	27 300	835	yes	yes
SBC	US	21 712	30 000	724	no	no
Telecom Italia	Italy	18 463	24 854	743	yes	yes
BellSouth	US	17 886	21 123	847	no	no
GTE	US	17 374	18 527	938	no	no
MCI	US	15 265	0		yes	yes
Ameritech	US	13 428	19 057	705	no	no
Sprint	US	12 765	6 700	1 905	yes	yes
US West	US	11 746	14 847	791	no	no
Telefonica	Spain	11 008	15 095	729	yes	yes
DGT	China	10 457	40 706	257	yes	yes
Telstra	Australia	10 431	9 200	1 134	yes	yes
Telebras	Brazil	9 388	12 100	776	no	no
Total		426 667	408 715			

(Sources: Global Business Yearbook 1996, Telegeography 1996/7, company news reports and news releases, calculations. Figures are the latest available in 1996. Bell Atlantic and Nynex merged 15.8.97, Bell South and PacTel merged 1.4.97; both mergers included above, using 1995 data; US West excludes domestic cable interests acquired in 1997.

international telecommunications cables. A single satellite can provide service over more than one country but may be used for more limited purposes, such as to provide domestic communications over difficult terrain or in remote areas. They provide links between the scattered communities of the Asia/Pacific region.

By 1988, about 30 operational GEO satellites were used for domestic services in the US. These were mostly owned by Alascom, AT&T, Comsat, Contel ASC, GE Americon, GTE Spacenet, Hughes Communications, IBM and MCI, and provided a variety of leased line, private network and other services [17]. A world total of 170 commercial satellites provided service in 1997, 40 per cent of which were constructed by Hughes [18]. The fixed, mobile and broadcasting services provided by satellites were worth $11.4 bn in 1992 and expected to grow to $38.3 bn by 2002.

The satellite launch industry, like the construction industry is very concentrated. Arianespace was making around half of the commercial launches in the mid 1990s [19].

LEO and other nonstationary systems are being used in global networks for mobile communications, and can provide a single unified system which permits the same handset to be used on the network from anywhere in the world, a facility known as *global roaming*. Satellite news-gathering (SNG) has become a well established method of sending in news from battlefields and remote areas of temporary importance. Small portable dishes are used to transmit speech and pictures by satellite back to the newsroom. Solar power can provide the necessary energy.

3.6 Ownership of satellites

3.6.1 International satellite organisations

Satellite systems that provide international telecommunications services have a variety of ownership structures, generically known as *international satellite organisations* (ISOs). The main ones are listed in Table 3.8.

NTIA [17] contains a summary of the early history of the ISOs. The domestic satellite communications industry of the US has been open to competition since the implementation of the *open skies* policy in 1972. The satellite networks are privately owned and are free to compete with other domestic telecommunications operators.

A number of other countries use satellites for domestic communications, including Australia, Brazil, Canada, China, France, India, Indonesia, Italy, Japan, Mexico, Germany and the former Soviet Union (NTIA, 1988). Some, such as MEASAT (Malaysia), PALAPA (Indonesia, also serving adjacent countries) and THAICOM (Thailand) are part of the basic infrastructure and owned or controlled by the largest operator. Others are private ventures. The degree to which private operators are free to compete is usually more limited than it is in the US, although increasing through deregulation, as in Australia and Europe.

The services provided by regional and domestic systems could, in principle, be supplied by others covering the area, and more cheaply to the extent that there are economies of scale. INTELSAT, although basically set up for international operations, does provide transponders for domestic services in some countries. The

Table 3.8 Main international satellite operators, 1997

Global consortia of operators and their governments:
  INTELSAT (International Satellite Organisation)
  INMARSAT (International Maritime Satellite Organisation)

Regional systems controlled by operators:
  ARABSAT (Arabian countries)
  ASIASAT (Asia)
  EUTELSAT (Europe)
  INTERSPUTNIK (former Soviet Bloc)
  PALAPA (Southeast Asia)

Private systems:
  ORION (transatlantic)
  PANAMSAT (transatlantic)

---

considerations which keep a local alternative in the sky may, as with airlines, have a political element, especially in Asia [20].

The owners of an ISO, like other users, have three distinct types of interest:

(i)   Direct access to circuits at wholesale prices without a distributor's mark-up.
(ii) Membership of the consortium provides a degree of control over operations and some representation on its board.
(iii) The consortium is an investment within their business sector.

ISOs have considerable market power if they are controlled by investors who also control most of the alternative infrastructure. Customers wanting direct access to circuits might not want to become investors in the operation even if they were allowed to do so. They would not have to do this when leasing land lines or motor vehicles, for example, and might not see it as relevant in the context of their overall investment strategy.

Some ISOs are under pressure to restructure their operations to meet or create more competitive market conditions. Restructuring is necessary to avoid a conflict of interests where a reseller obtains capacity from the owners of the satellite and then competes with them. Options ranging from the creation of a privately-owned susidiary to full privatisation were being considered by INTELSAT in 1997 and its future was being discussed in US Congressional hearings [21]. In May 1998 the operator spun off New Skies Satellites to provide video and other services in competitive markets. INMARSAT has established a subsidiary to handle its global mobile radio plans.

3.6.2 INTELSAT

INTELSAT is the largest ISO, with 19 satellites in geostationary orbit in 1998, after spinning off New Skies satellites, and it provides coverage to every part of the world. It was the first to be established and set an organisational model for later ISOs such as EUTELSAT and INMARSAT. INTELSAT works within the framework of an agreement between governments (known as *Parties*) who nominate operators (*Signatories*) who sign an operating agreement, own shares and have the right to use INTELSAT capacity. Mongolia became the 142nd signatory on 5 September 1997. In the case of INTELSAT, shares are also held by *investing entities* (37 of these with an investment share of 6.5 per cent at 31 March 1997) which are smaller international operators allowed by their home governments to take shares in INTELSAT and have direct access to its circuits. All the investors have shares roughly proportional to usage, which they pay for by investing capital and by paying for capacity used. Investors can and do resell capacity to third parties, sometimes at a considerable mark-up.

At one time, private users could only get access to INTELSAT facilities through one of the signatories. Direct access was made possible from 1994 in European Union countries by an EU satellite liberalisation directive; Germany and the UK had relaxed the rules earlier. All EU governments now allow specified users, known as *appointed customers,* to have direct access to INTELSAT. There were, at 31 March 1997, 361 appointed customers.

New investment is funded out of retained earnings and by contributions from the signatories and investing entities. (the net cash flow is usually in favour of the investors, giving them a positive cash flow from their investment). INTELSAT finances for the years 1992 to 1996 are summarised in Table 3.9. They reflect the rapid growth of international traffic over the period.

In 1996, voice and data services contributed 74 per cent of the operating revenue, the remainder coming from video services.

3.6.3 INMARSAT

The INMARSAT (international maritime satellite) organisation was established by a convention signed in 1979, primarily to provide maritime services by geostationary satellite. It has replaced earlier terrestrial radio systems for most large sea-going vessels and many small ones. By 1997 there were 79 member countries and land-mobile services were also being provided. INMARSAT then had five satellites of its own and was leasing capacity from the European Space Agency

Table 3.9 INTELSAT finances

	1992	1993	1994	1995	1996
US $ in millions					
Telecommunications revenue	616.3	658.2	706.3	805.4	911.4
Operating expenses	311.8	328.3	401.5	438.1	477.4
Operating income	304.5	328.3	401.5	438.1	477.4
Total assets at 31 December	2686.9	2813.1	3224.9	3525.4	3469.2
Per cent of average total assets					
Return	18.9	18.5	16.0	17.3	20.2
Depreciation			7.7	8.6	9.5

(Source: INTELSAT annual report 1996)

and INTELSAT. The system can be used by land and sea travellers virtually anywhere in the world without being dependent on the local infrastructure.

In 1990, INMARSAT was handling 21 million telex and 29 million telephone minutes annually and had revenues of $182 m. Three years later, revenues had risen to nearly $400 m and assets were worth over $1 bn.

3.6.4 Regional satellite systems

Geostationary satellite systems may be owned by regional operators (such as EUTELSAT) or by private commercial consortia (such as PANAMSAT and ORION). Some provide alternatives to INTELSAT satellite circuits, others are intended to provide private users with a competitively-priced alternative to the PSTN. The private systems can only provide service where local regulation permits them to do so.

3.7 Mobile services

3.7.1 Origins

Radio was in use for mobile communications in the early years of the century [22]. The British Post Office opened a ship-to-shore radio station in 1908 and took over Marconi coast stations in 1909. Crippen was famously arrested for murder after a radio message in 1910. A hand microtelephone which combined a transmitter and receiver in one handset was introduced in 1929, but growth in mobile use was slow. UK private mobile radio services were first licensed in 1947. The Post Office did not open a public radiophone service for vehicle users until 1959; a radio service to aircraft came in 1961. The original equipment was heavy and, although mobile, was usually built into vehicles. Private and military networks used equipment that was more portable where they had to.

Once installed, PMR is cheap to run and it is widely used by taxi drivers and some other small businesses and public services such as airport security systems and police forces. PMR systems are for groups of private users and are not connected to the PSTN.

3.7.2 Land-mobile systems

Modern land mobile communications began with the development of the cellular mobile technology described in Chapter 2. This was lighter, cheaper and provided a more efficient use of the spectrum and was connected to the PSTN. It reached the market at a time when people were moving about more, in both their business and their personal lives. The first European systems were opened in Norway and Sweden in 1981, and the Scandinavian countries had achieved high penetration rates by 1990, the highest being 5.2 per 100 inhabitants in Sweden [23]. By 1997 more than 30 per cent of Norwegians and only slightly fewer Swedes had a mobile phone and people without them were said to feel at a social disadvantage [24]. The high level of penetration in Sweden has been ascribed partly to the large number of Swedes with second homes. Penetration growth has been slower in most other countries.

3.7.3 Satellite mobile systems

Mobile telephony has attracted a number of consortia of commercial interests intending to provide global mobile systems based on LEO and MEO satellites, several of which are open or under construction. They are known generically as *satellite personal communication systems* (Sat-PCS) and most are meant as premium services for the top few per cent of the international mobile services customers, for whom the facility is worth the price.

IRIDIUM was first in the field and opened for service in 1998. It is a consortium led by the mobile terminal equipment manufacturer Motorola. It has a system of six spare plus 66 operational LEOs (originally 67, the number of electrons in the element Iridium, hence the project name).

ICO is backed by INMARSAT and based on the provision of global mobile telephony using 12 satellites which, while not geostationary, are in relatively high orbit and hence fewer in number than for a LEO system. Launch costs have been estimated at $2.6 bn [25].

Other global systems that have been reported include:

- GLOBALSTAR ($1.8 bn, 48 + 8–16 spare LEOs) for personal mobile services;
- TELEDESIC ($10 bn, 288 satellites plus spares) for PSTN services to remote areas and high-speed data services where required.

The exact cost and specifications of these and other systems have varied during the planning process. An industry estimate of $35 bn annual revenues by 2005 would, at current growth rates, be equal to just over two per cent of global fixed-link revenues. Mobile service revenues at that date are very speculative. If mobile penetration reaches 25 per cent of fixed-link penetration, and if mobile revenues per line are double those in fixed networks, $35 bn would be four to five per cent of mobile revenues. At this level of income, more satellites would presumably be needed to provide the service. If half the income was needed to cover depreciation, and if the average asset life was five years, gross assets would be in the region of $35/2 \times 5 = \$87.5bn$, considerably more than the reported cost of currently planned systems.

Tariffs charged for usage are likely to be above those of a fixed link where it exists, because of higher terminal and infrastructure costs, but even here a market opportunity may be provided by tariffs which are out of line with costs. For some types of call, such as international mobile to mobile between remote areas, local mobile and fixed link, alternatives may not exist. Even where they do exist, convenience factors, including the value of the time saved and hassle avoided by the user, may justify a price premium.

The premium market may, even so, not be big enough to support the cost of the systems, and other markets are being sought. There are applications as alternatives to the PSTN in remote or scattered parts of the developing and developed world. The pattern of demand over the longer term is unknown and may be quite different from expectations in spite of careful market research.

3.8 Private networks

3.8.1 Economics of the sector

Most inland data traffic and a good deal of traffic for international destinations is carried over private networks. Voice traffic is also carried. Network types include LANs and the specially configured systems used for longer distances. The over pricing that commonly exists within PSTN tariffs leaves room for private networks to handle in-house traffic at a lower cost than the public tariff where regulation permits and leased lines are priced close to cost. Self provision of circuits may sometimes be a viable alternative to the use of leased lines but is often not allowed.

Many large companies and public bodies have telecommunications managers, sometimes linked with the computer services function, to provide communications requirements. Typically, they build private networks from leased lines and purchase data and/or voice switches to provide services more cheaply than those obtainable by sending all traffic over the PSTN. They may also provide better transmission quality for data.

Operators recover some of the revenue lost to the PSTN through the revenue from leased lines, but part of the value added (which includes profit) is transferred to the owner of the private network. Operators try to win revenue back by offering network-management services to replace the telecommunications managers, and VPNs to replace the leased lines, in deals that will be cheaper than the in-house alternatives (Section 3.2.7).

3.8.2 Satellites and VSATs in private networks

Private networks use geostationary satellites for data and for voice transmission where regulation permits. Access is through VSATs installed on the premises of the user. Typically, a business will buy or lease the VSATs from an equipment manufacturer and lease the satellite circuits from an operator. Some operators will set up the whole network for the user. Satellite infrastructure is becoming cheaper for operators to buy, although the cost is still usually above that of providing service through terrestrial links. The costs passed on to users in tariffs are often attractive because of the excessive prices charged by operators for long-distance services over the PSTN alternative. Also, VSAT access in a private network may bypass regulatory restrictions or be subject to less rigid control than that applied to terrestrial links.

The planning of private networks based on satellite links through VSATs takes the pattern of traffic flows and type of communication into account. Communications may transmit in one direction only (*one-way*) or permit a *two-way* exchange. A common configuration well suited to the economics of satellite technology is *one-to-many* and two way. As an example, a company head office may have radial satellite links to regional offices in other countries and exchange data with them. The regional offices would use the PSTN to communicate with each other (*off-net*), or more usually keep the calls *on-net* by routing them through VSATs to the head office hub. The US has extensive private satellite business networks. Most installations are one-way only but about a quarter offer two-way facilities. VSAT links are widely used in data transmission.

3.8.3 Private-network consortia

Some private networks are collectively owned. One of the largest is SITA which is owned by the world's airlines and by 1996 was operating in over 225 countries and had a turnover of $873 m. Its main operations are with flight management, passenger reservations and freight. SWIFT is an international money-transfer network owned and used by banks. All large airlines and banks have networks of their own. These are competitors to the collective networks for some services, while being customers for others. SITA has won network-management contracts for airlines and it has a subsidiary which offers services to other sectors.

3.8.4 The Internet

The Internet is made up of linked packet networks accessed over the PSTN, usually for the cost of a local call. It differs from other telecommunications networks in being a largely unregulated federation of interests with worldwide connections and open to anyone willing to pay the fees. The name is slightly confusing because *internet* is a generic term for a network of computers. There are many internets. *The Internet* started life as a data network supported by the US government. The National Science Foundation continued to provide funding until 1994, but the Internet has since been taken over by commercial interests. It connects a large number of independent internets together and has network capacity and organisational features of its own. An Internet user dials up through a computer to access information, which can be searched, read, downloaded or printed out. There are a substantial number of academic users and their interest led to many of the current commercial applications. Most of the other users are business employees but private interest is growing. By the mid 1990s there were about 30 million computers linked to the network and the number was increasing rapidly. Many companies, including telecommunications operators, manufacturers and service providers, have reference pages which give information about themselves. Like many other government departments, the *Federal Communications Commission* (FCC) in the US and OFTEL in the UK use the Internet to make announcements and provide the text of press releases and policy documents. Law firms, consultants and other businesses provide information pages, and the complete text of substantial documents such the US Telecommunications Act 1996 have been made available. Electronic mail and news groups facilities for those with special interests are among other options. A great deal of the information on the Internet is advertising matter.

Figure 3.1 outlines the way in which information and charges circulate on the Internet. The actual settlement of bills may be by any paper or electronic means jointly acceptable to both parties. The customer pays the telephone company for each call made and pays a flat rate or usage-related subscription to an *Internet service provider* (ISP) which is the point of contact with the network. In 1997, there were several hundred ISPs in the UK and over 3000 in the US, although rationalisation is expected to reduce the numbers. Customers may provide information by setting up special pages (*sites*) identified by a name or acronym. They may supply the information free, or make a charge. ISPs may provide information services and may charge extra for access to them. They connect their customers to other

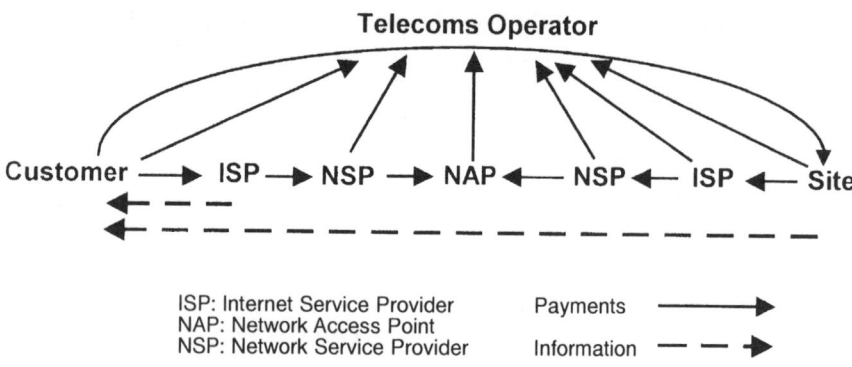

Figure 3.1 Internet transaction flows

parts of the Internet through *network service providers* (NSPs) and pay the NSPs for this. The lines they use are of high capacity, so that the unit cost of transmission is usually less for them than it is for the customers which they connect. The maximum transmission speed that they can handle depends on the capacity of the link. Switching facilities are provided at *network access points* (NAPs). The owners of the switches get paid by the NSPs for the traffic that they handle. There are also international links.

The pattern of ownership represented by these arrangements is extremely diverse. The ISPs are agencies, usually privately owned, and some of them are PTOs. The backbone circuits may be leased lines or, if the NSP is a facilities-based operator as some are, circuits of its own. The switches have a variety of owners. In the US these include the government, operators and equipment manufacturers. Operators get paid for use of any facilities which are leased from them by the other parties, as well as being paid for the phone calls to the ISPs, so they understandably see the Internet as good for business. The international extension of the Internet has added to its organisational complexity. It is a good illustration of the nature of Huber's geodesic network (Section 1.8).

Internet sites have unique names but, unlike telephone numbers, their organisation and assignment is not centrally organised by the network. A private agency manages the process on a *de facto* basis, without any statutory support. Site names have a commercial value, e.g. *Superco* to a company of that name (if there were one) and many have been registered by opportunists who see the chance of selling them on to the companies and organisations using them as trading names.

Internet usage has been stimulated by the development of efficient search and data manipulation software and the adoption of standard methods of presenting data. The development of services for the Internet has become a substantial industry, with a 1997 global turnover in the region of $20 bn (industry estimates). Electronic mail was popular from the outset, and other uses have been evolving fast with new growth areas such as:

- *intranets* for the free and secure circulation of data and messages within closed user groups, typically companies, as an alternative to a corporate private network;
- *extranets* linking corporate partners;
- *e-commerce*, which is trade between companies and their customers within a closed user group;

- shopping facilities for personal customers, for whom effective measures for fraud control and safe electronic money transfer have to be created,
- voice telephony.

The rapid growth of the Internet owes much to two features:

(i)   The various services have proved useful, or if not useful, at least interesting. E-commerce is becoming important for the sale of catalogue goods such as computers and office supplies to organisations. E-mail is cheap, easy to use and works well.

(ii) Access is often perceived as cheap. Many users are corporate or academic employees who do not pay their own bills. A local call secures access and in much of the US local calls are free, so there may be no phone bill to pay, even where an overseas site is accessed.

Its growth has been encouraged by the telephone companies, which saw a new source of revenue that could offset the maturing of their basic networks, and by *information society* enthusiasts of many persuasions who saw a potential for enhancing the quality of life. The Telecommunications Act 1996 contains a finding of the US Congress that:

> 'The Internet and other interactive computer services offer a forum for a true diversity of political discourse, unique opportunities for cultural development, and myriad avenues for intellectual activity' (Sec.230).

The pricing of access by the ISPs is based on the tariffing of data networks and is not without problems. It will be discussed more fully in Chapter 9, as will the economics of Internet voice telephony.

3.9 New types of operator

3.9.1 Facilities-based carriers and retailers

The liberalisation of markets has attracted entrants of several new types. Some have an infrastructure that could be enhanced for use as a telecommunications network and can become facilities-based operators after further investment. They derive economies from the joint provision of telecommunications and other services and these may be enough to enable them to compete successfully against larger and better known operators. A few have spare or disused capacity on their own networks, which could be used to provide service, although this situation is uncommon. Others have no capital equipment of their own but see an arbitrage opportunity. The main groups are:

- Local
  cable TV operators with local networks
  companies with local ducts and tunnels, sometimes disused
  electricity, gas and water companies with local distribution
  networks
  urban canal and sewer owners
  Competitive access suppliers (CASs) and competitive *local exchange carriers* (CLECs) now bypassing *incumbent local exchange carriers* (ILECs) in the US. Some provide direct links to trunk networks for large users, others operate switched data networks;
- Long-distance
  electricity companies with long-distance pylon networks
  railway operators with long-distance rights of way and sometimes telecommunications networks of their own
  canal companies with long-distance waterways.

Some examples of attempted entry are as follows:

An obsolete duct network which had been used to power hydraulic lifts in central London was sold for £1 m to a new entrant (Mercury) for local access. Underground railways in London and Paris have suitable capacity.

British Waterways offered canal capacity for long-distance cables, which could be laid in or beside the water. British Gas also considered plans. Neither of these came to fruition, but two gas utilities in Portugal have allied with an electricity utility to bid for an operating licence.

Electricity companies have been active in several countries. ENERGIS uses telecommunications cables wound round the earth cables of pylon networks for long-distance transmission and is the only electricity company in Britain to have opened for service. Companies in Britain and elsewhere have considered providing service over the local power distribution network. Electrical integration of local power supply and telephony is technically possible. Systems already exist for distributing music in this way [26] and there are plans for Internet services. Tokyo Electric Power considered entry in Japan. Other ventures exist or are planned in Austria, Finland, Germany, Ireland, Italy, Sweden and Switzerland.

Mercury laid cables along railway land when it entered the UK market. Railways in Austria, Denmark, Germany, Ireland, Italy, The Netherlands, Spain, Sweden and Switzerland have been seeking partners in telecommunications ventures.

3.9.2 Resellers and others not facilities based

Resellers lease capacity from operators, break it down and sell it retail. Resale of domestic services is an active, if marginal, business in the US where transcontinental distances are large. In European countries, where the internal distances are shorter, trunk tariff reductions have made domestic resale a largely unprofitable activity. International resale is important and will remain so as long as international calls are priced well above cost; it is analysed in Chapter 17.

Aggregators first appeared in the US. They collect traffic together from small and medium-sized businesses to qualify for larger bulk discounts from the network operators. Like other retail systems, the operation owes much of its profitability to the over-pricing of calls to small users. It may add value through better marketing, but it would probably play a smaller part if call prices were better related to costs.

Call-back and other operations based on *calling cards* are increasing. These arbitrage differences between tariffs at each end of the same call and mainly involve international services. They are discussed in Chapter 17.

3.10 References

1   BERENDT, A.: 'Cable TV: do investors have the courage and conviction for the long term?', *Intermedia*, 1995, (6), pp.14–16
2   Cable Communications Association: 'Thanks a million! Phone users flock to cable in 1997'. CCA press release, London, 9 March 1997
3   Commun. Int., October 1997
4   IDATE News, 1997 3rd quarter, (17), p.7
5   MARCHAND, M.: 'The Minitel saga', Larousse, Paris, 1988
6   JACKSON, T.: 'Peapod's vision', *Financial Times*, 4 December 1995, London
7   BUCKLEY, N.: 'Slow stroll to the virtual mall', *Financial Times*, 24 April 1995, London
8   'On the line to a wider service', *Financial Times*, 29 July 1993, London
9   McCLELLAND, S.: 'An interactive world', *Telecommunications*, September 1997, p.S1
10   BERMEJO, M. A., and MARTINEZ DEL CERRO, F. J.: 'Internet: a new space for business'. Paper presented at ITS 11th Biennial Conference, Seville, 1996
10   Statistics Canada, catalogue 56–205
11   STIPP, D.: 'The birth of digital commerce', *Fortune*, 9 December 1996
12   HOLMES, T.: 'Universal personal telecommunications'. Paper presented at the European Telephone Numbering conference, London, 11 May 1992
13   'Reporting Britain'. *Financial Times* survey, 23 April 1998
14   OWEN, D.: 'Moment of truth today for Europe's space hopes', *Financial Times*, 30 October 1997, London
15   MOORE, G. E.: 'Can Moore's Law continue indefinitely?' *Computerworld*, 7 January 1996
16   'World telecommunication development report 1994'. ITU, Geneva, 1994
17   'NTIA Telecom 2000' (National Telecommunications and Information Administration Special Publication 88–21, Washington D.C., 1988)

18 COLE, G.: 'Satellite Loads Soar', *Financial Times*, 28 October 1997, London
19 SAPIO, B., and BONANZINGA, P.: 'Strategic alliances for the provision of new telecommunications services: a methodological approach to a European satellite system'. Paper presented at ITS *11th Biennial Conference*, June 1996, Seville
20 'Ground control over Asia's space', *Comm. Int.*, June 1997, pp.73–76
21 CLARK, B.: 'Intelsat seeks time for its restructuring plan', *Financial Times*, 31 July 1997
22 'Post Office telecommunications statistics' (General Post Office, London, 1962)
23 'Mobile communications guide to European subscribers to mobile systems', *Fintech Mobile Communications*, 13 September 1990, *Financial Times*, London
24 CANE, A.: 'Hooked on phones', *Financial Times*, 13/14 September 1997
25 'Inmarsat's Lundberg to spearhead new drive to mobile telephony', *Financial Times*, 14 March 1995, London
26 CANE, A.: 'Power lines may deliver telecoms', *Financial Times*, 8 October 1997, London

Chapter 4

The structure of demand

4.1 Theoretical framework

The formal economic analysis of demand is based on a few propositions about consumer behaviour. Consumers are assumed to have limited resources so that there is a *budget constraint* on their expenditure. They derive *utility* from goods and services purchased. Utility is what the goods bought are worth to the consumer and is measured in cash terms for analytical purposes. Consumers are assumed to be *rational* and have the maximisation of utility as their expenditure objective. Each single additional purchase by the consumer is a *marginal purchase* and provides *marginal utility*. The marginal utility of an extra quantity of a particular good or service usually falls, and never rises, as the volume of purchases increases. Expenditure decisions are based on the marginal utility of the marginal purchase compared with its price. No more purchases will be made if marginal utility of an extra purchase is less than the price. A consumer's total utility is the sum of the utility of all purchases.

The existence of a budget constraint means that at the margin consumers will have to choose between one purchase and another. Rational consumers will choose the combination of purchases which maximise utility within the resources available. Each consumer is supposed to have a *demand function* of the form

$$x_i = f(p_1, p_2, .., p_n; M) \qquad (4.1)$$
$$i = 1, 2, .., n$$

in which expenditure on the each good, x_i depends on the prices, p_i of all n available goods and on M, the total budget. Some of the p_i relate to telecommunications goods and services. The function symbol f

means that the set of market prices, p_i, and the budget limit, M, between them determine the volume of each of the i available products that the consumer will purchase. The exact form of the relationship may be complicated. The budget constraint is determined by income, which is often used in place of the budget, in empirical work.

Real price changes are the price changes for a commodity relative to changes in the general rate of inflation. Thus, if the price of long-distance calls remains unchanged when prices in general rise by five per cent, long-distance calls become about five per cent cheaper in real terms. For more precise work this is measured as a price ratio rather than a price difference. In the example, the exact real price change is

$$100*((P_{c1}/P_{c0})/(P_{g1}/P_{g0})-1) = 100*(1/1.05-1) = -4.8\ \% \qquad (4.2)$$

where P_{ci} is the price of calls in year i

P_{gi} is an index of the general price level in year i

Demand functions are usually taken as being independent of the overall price level, i.e. the volume purchased is assumed to be unchanged if all prices and the budget limit or income are changed by the same percentage. This property is described as the freedom from *monetary illusion*. Freedom from monetary illusion is a useful analytic simplification, although consumers are not completely free of it in practice, especially under conditions of rapid inflation.

The relationship between price and quantity purchased can be graphed as a *demand curve* in which price is measured on the vertical (dependent variable) axis, as shown in Figure 4.1.

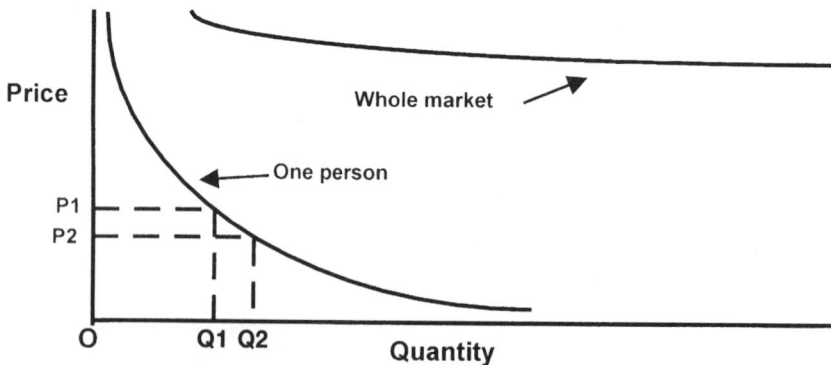

Figure 4.1 Personal and market demand curves

Demand curves are of three types:

(i)   Each individual has a demand curve or *demand schedule*. The curves will vary with personal preferences and incomes. Individual consumer curves are not usually observable.

(ii) Individual demand curves can be added together to form a *market demand curve* in which the vertical axis is unchanged but demand as measured on the horizontal is now greater for any given price. This type of curve, representing the totality of demand facing the supply side, is the one usually used in demand analysis. The market may be *segmented* into groups of consumers with similar functions for greater precision.

(iii) An individual firm in a market where there are many competitors faces an own-price demand curve, measuring changes in demand for its products arising from price changes made when all other firms make no price change.

4.1.1 Price and income elasticities

The proportionate response of demand to real price changes is called the *price elasticity*. It is measured as a ratio of the proportionate volume and price changes associated with a small change in price. In Figure 4.1 this is the ratio

$$(P_1/Q_1)((Q_2-Q_1)/(P_2-P_1)) \tag{4.3}$$

which in the limit becomes

$$(P/Q)(dq/dp) \tag{4.4}$$

Thus if a one per cent price increase results in a 0.1 per cent drop in sales volume the price elasticity is $-0.1/1.0 = -0.1$. If the price elasticity is less than unity in absolute value, the demand is said to be inelastic. Note that the addition of identical individual demand schedules to form the market demand curve will not change the price elasticity, which depends only on proportionate changes. In practice, the composite will be a rough average of all the individual price elasticities in the market.

Price elasticity may vary from point to point along the curve. In the special case where it is the same at all points the curve is called a *constant elasticity demand curve*. A curve of this type would be given by the relation $Q=P^{-a}$, which would plot as a straight line with slope $-a$ on paper with a double logarithmic scale — see Figure 4.2.

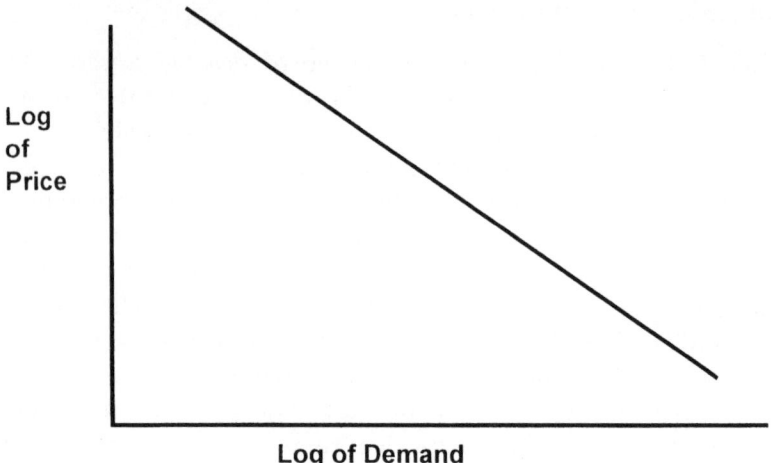

Figure 4.2 Constant elasticity demand curve

The slope of the market demand curve is the *market-price elasticity* collectively faced by the supply side. The response of the demand faced by an individual to changes in a firm's own price is the *own-price elasticity*. Unless the firm has a monopoly, this will be steeper than that presented to the market as a whole. In an extreme case, all sales could be lost to competitors as a result of a small price increase even if total sales in the market are not much affected by a change in the prevailing market price. The response of demand to a change in the price of another product is called the *cross-elasticity* of demand with respect to price.

Some goods in the demand function are a near substitute for others and these will usually have the largest crosselasticities. A telecommunications example is the demand for calls at different times of the day. Traffic may move from a peak period to a cheaper off-peak period if the off-peak tariff is reduced. Cross-elasticities have the opposite sign to elasticities. The demand for a good will rise if there is a rise in the price of the substitute good but fall in response to an increase in its own price. The steepness of the demand curve facing a firm in a competitive market is another example of a cross-elasticity.

The long-distance services offered by two competing operators such as BT and Mercury or AT&T and MCI may be close substitutes for one another, so a small change in the price of one can have a large effect on the demand for the other.

The response to a change in real income is called the *income elasticity*.

Typically, a 1 per cent increase in real income or GDP may raise demand by up to 1 per cent. A 10 per cent reduction in real tariffs may raise the volume of long-distance calls by around 5 per cent, sometimes more.

If income rises and the budget constraint is eased, there will be an increase in demand for some or all of the goods and services in the demand function. The increase may vary from product to product, some could be replaced by an alternative of better quality. Those for which demand falls with an increase in income are called *inferior goods*. The others are called *normal goods* in demand theory. Demand functions evolve with changing real incomes and the availability of new products.

A change in the price of a single product will have two effects:

(i)   A *substitution effect* as the consumer adjusts an expenditure pattern to the changed price, so as to maximise utility with the new set of prices.

(ii) An *income effect*, which is the net change in utility after this has been done.

4.2 Demand for services

4.2.1 Market sectors

In the early stages of economic development most lines will be taken by business users, but in the richer countries most are for residential use. This is illustrated by the history of the UK, where the business users' share of Post Office and British Telecom exchange connections fell from 59 per cent in 1931 to 38 per cent in 1961 and 24 per cent in 1995 [1, 2]. In the US, 34 per cent of the connections were for business users in 1994 [3].

Calling rates depend on tariffs, income levels and a variety of regional and social factors. The calling rate over the fixed network in a typical European country is around 1000 calls per year. In North America the number is higher, nearer 3000. Calling rates are highest on business lines. Calling rates in developing countries may be high or low depending on the mix of business and residential users and the prices.

Business and residential markets differ in one fundamental characteristic. Residential customers maximise utility but businesses maximise profit. In maximising profit, businesses pay attention to cost minimisation and the possibility of using the telephone as a substitute for other inputs. Most business telephone calls are paid for corporately, weakening the link between the decision to call and the cost of calling.

Most residential customers do not make enough calls to use the full potential of the exchange connection, and the use could be shared with other customers on party lines. This facility was once regularly offered, especially on rural areas, but had become less common by the 1980s. Most businesses are small and some have similar usage characteristics to the larger residential users; in the UK, the majority only have one telephone. Larger businesses can share lines among groups of people within a company. These differences lead to the following differences in demand patterns, most pronounced for large businesses in the service sector:

(i)   Calling rate: business customers make more intensive use of individual exchange lines and are more likely to make international calls.

(ii)   Price sensitivity: residential customers tend to be more affected by price changes.

(iii) Time of day: calls from business lines peak in business hours, those from residential lines peak in the evenings and weekends.

(iv)   Service requirements: a wide range of services has been developed for business users.

(v)   Quality requirements: many business users expect a higher quality of service and will pay for it.

(vi)   Private networks: large business users often lease or own private circuits and networks to handle their traffic.

(vii) Terminal equipment. Business customers use more call-management equipment, use advanced services such as videoconferencing and have sophisticated business machines for image transmission and processing.

Not all of these differences arise from the specific characters of the two markets, some reflect imperfections in the general market structure. As noted earlier, private network use is encouraged by the over-pricing of long-distance calls in relation to the price of leased lines. The spread of computer networking equipment and fax machines into the home puts residential demand into the corner of a more general picture including data network usage and terminal equipment. Neither customer group is homogeneous. The requirements of an international finance house, a local government authority and a farmer differ greatly, as do those of a well-to-do suburban household, a teenager and a pensioner close to the poverty line.

Economic geography can provide useful insights to the pattern of demand. In most public networks, one of the problems encountered in studying regional telecommunications traffic is the lack of telephone data below national level. Canada, Germany and the US are the main exceptions. Airport passenger data, which pick up business and personal links between countries, can be useful proxys when studying international telecommunications traffic. The figures reveal, for example, a strong German/Turkish connection which is present in both sets of statistics and arises from the large number of Turkish workers in Germany. Other links, such as those between the UK and Commonwealth countries, are also present in both sets of data and may be more familiar. Demographic data which include the ethnic affiliations of EU residents, are available through the EU and is another source of information about potential international telecommunications traffic flows. They tie in reasonably well with the airport data.

4.2.2 The business market

Some sectors of the economy use a higher proportion of telecommunications inputs than others. *Input/output tables* are published in most industrial countries. These, as explained in Chapter 3, show the flows of intermediate expenditure around the economy. They include a statement showing the total value of telecommunications bought by each business sector and the total costs and turnover of each sector. From this it is possible to calculate the *tele-intensiveness* of the sectors, defined here as telecommunications inputs as a percentage of turnover. Tables 4.1 and 4.2 give information drawn from the input/output tables for the UK in 1990. They show how the service sector, and especially financial services, dominates the pattern of usage.

The degree of service sector dominance varies from country to country. Table 4.3, for Germany, displays the less developed nature of the German financial services markets compared with the UK at the time.

US industry has been increasing its telecommunications share of inputs, by three per cent of the base percentage per annum [5]. This is at constant prices. If allowance is made for falling telephone charges

Table 4.1 The most tele-intensive UK sectors in 1990

Sector	Tels input £m	Total sales £m	Tels % of sales
Insurance	1 416	20 827	6.8
Auxiliary financial services	275	5 086	5.4
Estate agents	158	3 941	4.0
Banking and finance	1 073	40 758	2.6
Legal services	173	7 245	2.4
Office machinery	33	1 667	2.0
Personal services	83	5 159	1.6
Computing services	153	9 960	1.5
Accountancy services	85	5 698	1.5
Other business services	288	19 761	1.5
Metal ores	3	219	1.4
Transport services	209	15 465	1.4
Stone/clay/gravel	4	317	1.3
Vehicle repair	234	18 877	1.2
Railways	43	3 535	1.2
Sea transport	50	4 229	1.2
All Sectors	8 288	969 340	0.9

(Source: Central Statistical Office, input/output tables for 1990)

most of the growth disappears. Telecommunications is an increasing input to educational services in the US, but the growth has been less rapid. Usage is given as 1.4% in 1963 and 2.3% in 1991, an annual increase of some 1.8% of the 1.4% base at constant prices. After allowing for falling telephone charges the share shows a fall.

4.2.3 Economic geography and business demand

The usefulness of input/output figures can be increased if they are combined with sector employment to produce figures for telecommunications usage per employee in each sector.

A regional dimension can then be added by the use of sector employment statistics for individual local authority areas, such as those available in the UK.

Table 4.2 The largest UK telecommunications services markets

Sector	Tels input £m	Share of business market (%)
Insurance	1416	17.1
Banking and finance	1073	12.9
Wholesale distribution	494	6.0
Hotel and catering	295	3.6
Other business services	288	3.5
Auxiliary financial services	275	3.3
Vehicle repair	234	2.8
Construction	222	2.7
Transport services	209	2.5
Recreational and welfare services	197	2.4
Retail distribution	185	2.2
Legal services	173	2.1
Estate Agents	158	1.9
Computing services	153	1.8
Road transport	149	1.8
Other professional services	141	1.7
Printing and publishing	108	1.3
Agriculture	105	1.3
Accountancy services	85	1.0
Total Business Market	8288	

(Source: Central Statistical Office, Input/Output Tables for 1990)

On a larger scale, the European Commission has published figures for employment and GDP for the European regions. Two of the more interesting statistics that can be calculated from these are GDP per head by region and GDP per square kilometre by region. The second of these is of most interest to telecommunications analysts, since a geographical concentration of GDP indicates a similarly concentrated demand for telecommunications business services. The concentration makes this customer group cheaper to serve. Their custom is above average in profitability because the cheapness is not fully reflected in tariffs. Concentrated areas of demand are prime targets for new

Table 4.3 The most tele-intensive German sectors in 1990

Sector	Tels services input (DM m)	Sales (DM m)	Tels % of sales
Education	1 511	71 324	2.12
Retailing	2 480	160 112	1.55
Banking	1 818	120 527	1.51
Other transport	1 924	127 602	1.51
Hotels	926	79 722	1.16
Insurance	626	54 714	1.14
Wholesaling	2 353	207 693	1.13
Precision and optical	326	37 560	0.87
Local government	3 196	372 622	0.86
All Sectors	34 216	5 252 790	0.65

(Source: German Statistical Office [4])

entrants to the market; much business in the City of London has been lost by BT to competitors. Table 4.4 shows the EU regions in which GDP is geographically most concentrated.

With the exception of Ceuta y Melilla, which is a Spanish enclave in North Africa, all of these lie in a banana-shaped area stretching from the Rhineland through Belgium, Holland and Paris to northern England, which includes the rival financial centres of Frankfurt, London and Paris. The international telecommunications traffic flows between the five EU countries in the group are, apart from the link between Austria and Germany, the largest in Europe.

The demand for business telecommunications is highly concentrated in another dimension, that of the individual business. The degree of business concentration varies between countries and is remarkably high in the UK. As seen in Chapter 1, there were fewer than 600 UK companies with a turnover of more than £500 m in 1994. Although few in number, their average size is such that they account for a large slice of business demand for telecommunications services and the profit that can be made from them. In circumstances where long-distance and international call prices are well above cost the companies are prime marketing targets among rival service suppliers.

Table 4.4 GDP per square kilometre in EU regions

Region	GDP (Ecu m)	Area (sq km)	GDP/sq km (Ecu m)
Brussels	25 771	161	159.7
Greater London	139 716	1 579	88.5
Hamburg	57 309	755	75.9
Berlin	57 359	889	64.5
Bremen	18 440	404	45.6
West Midlands County	35 073	899	39.0
Ceuta y Melilla	1 155	31	37.3
Greater Manchester	33 033	1 287	25.7
Ile-de-France	291 187	12 012	24.2
Merseyside	14 901	652	22.9
Dusseldorf	114 104	5 288	21.6
South-Holland	55 487	3 446	16.1
Darmstadt	109 869	7 445	14.8
West Yorkshire	27 397	2 039	13.4
Utrecht	18 670	1 434	13.0
Brabant	42 610	3 358	12.7
Antwerp	32 661	2 867	11.4
Cologne	82 716	7 365	11.2
North-Holland	44 533	4 042	11.0

(Source: calculated from Eurostat [6])

Few companies publish information about the size of their telecommunications bills, but they do publish employment data. This can be combined with knowledge of the business sector of an individual company and average telecommunications usage per employee in the sector to produce working estimates of the value of its telecommunications bill. Sometimes this information can be collected directly. Guldmann [7] gives results for an input/output study based on a large sample of calls within a US *local access and transportation area* (LATA) in 1990. Mean calls, call minutes and call durations per employee are shown in Table 4.5, together with the figures for calls from residential telephones.

*Table 4.5 Telecommunications-employment ratios in the US
         LATA call sample, February 1990, outgoing calls*

Sector	Call volume %	Messages per employee	Minutes per employee	Average duration (mins)
Agriculture	0.32	7.2	19.3	2.7
Surface mining	0.05	20.9	57.8	2.8
Construction	2.81	14.6	40.5	2.8
Manufacturing	5.87	7.0	19.1	2.7
Transport, communications, public utilities	2.85	15.7	43.3	2.8
Wholesale trade	2.91	13.5	31.3	2.3
Retail trade	8.74	9.8	23.2	2.4
Finance, Insurance, property services	8.69	31.9	89.0	2.8
Other services	18.24	17.0	49.0	2.9
Public Administration	2.81	5.9	17.5	3.0
Residential (per household)	46.71	13.6	23.3	1.7
Total	97.19			2.6

(Source: Guldmann [7])

Residential calls were, on average, shorter. The workers making most use of the telephone were in financial services. Surface mining is capital intensive, accounting for the high calling rate among the few workers employed. Intrasector traffic was highest within transport, communication and public utilities (32.7%), financial and property services (38.2%), public administration (38.2%) and residential customers (52.8%).

4.2.4 The residential market

Under conditions of relatively stable economic growth the growth of network access among residential customers tends to follow the S curve found with the take-up of other new products, similar in form to a logistic curve, although the logistic itself may not fit well. There is a period of slow growth after the initial introduction, an acceleration to a period of steep growth and then a slowing down as penetration moves to saturation. Some products (like candles and gaslamps) follow the curve to a peak and then go into decline as they are replaced by something better doing the same job. So far this has not happened with domestic telephony, although the future might see radical changes in the delivery system. Figure 4.3 shows how the system developed over a period of 113 years in the US. The period of development was long – the telephone was a 19th century invention – and some products have achieved fuller penetration more quickly. The domestic television set, invented in the 1930s, is an example.

The growth in real GDP over the whole of this period followed a long-term trend of a little more than two per cent per annum, the most substantial deviations being in the recession of the 1930s and the 1939–45 World War. The take-up of telephone service, like that of most other innovations, begins with the well-off and spreads down through the income scale.

The relationship between income, price and telephone connection has been exhaustively studied in North America, where the numerous local companies have struggled to understand their markets and the regulatory systems have encouraged the publication of large amounts of relevant data. Taylor [11] provides a useful summary and analysis. This Chapter describes models and demand characteristics. The fitting of models is discussed in Chapter 16. The introduction of competition has tended to shift the balance of prices and raised the question of whether some of the poorest customers will leave the system. This is discussed further in Chapter 14.

Residential system size can be studied as a stock or as the net effect of two flows – joiners and leavers. Since people join and leave for different reasons the stock-and-flow analysis is best for most purposes

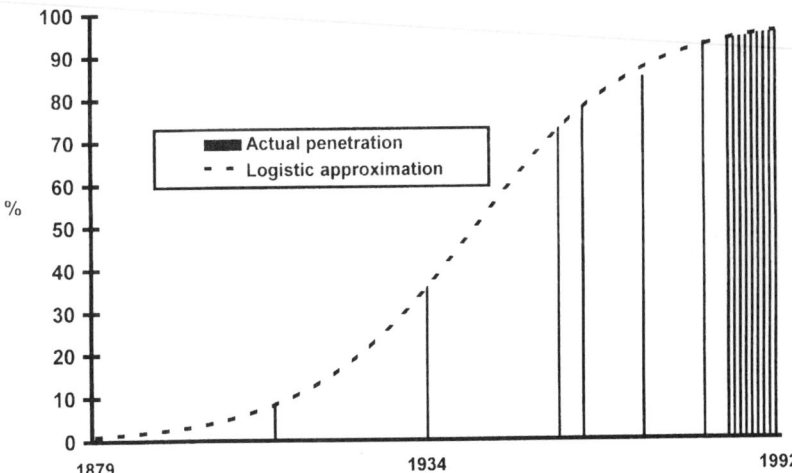

Figure 4.3 Penetration curve: households with access to a telephone US 1879–1992
Source: calculations from figures in Cooper [8], Mitchell [9] and Temin [10]

and it lends itself to *choice analysis*, in which demand is considered as a set of individual binary choices influenced by income, age, price and other factors. The theory of residential demand identifies (following Taylor [11]) four features:

(i)   The starting point is demand for an exchange connection (access). The main, but not exclusive, component is *option demand*, that is the option to make and receive calls, even if none are actually made. Taylor argues that option demand, some of which will be for emergency services, may be more important in rural areas and other areas served by small exchanges, where the immediate neighbours live further away and incomes tend to be lower.

(ii)   Calls are usually analysed through the calling rate once access has been obtained.

(iii) Each exchange connection provides a *network externality* (sometimes called the access externality) by being part of the set of lines that can be called by other users on the network. Each time a new line is added the network becomes potentially more useful to those already on it, since another line can now be called. Calls create a *call externality* by providing utility to the person being called. The network and call externalities make preferences interdependent between customers.

(iv)   Calls differ by type, time of day, distance, duration and quality, with differing income and price elasticities.

The demand for access can be examined at three levels:

(i)   Globally, through a cross-section analysis of the penetration rate in countries with different average incomes as in Reference 12. This is effective for telephony as a whole but difficult for residential users since statistics of residential EC stock are not usually available.

(ii)   As a time series in an individual country or state, using the housing, household, residential-connection statistics and service prices, with or without an income variable.

(iii) Incrementally, using short-term economic and demographic data to model flows of joiners and leavers, as in choice analysis.

Price affects access demand in a fairly complex way:

•   the decision on whether to join or leave the system depends partly on the rental, although the elasticity is relatively low on its own;

- the price of calls, and hence the size of the total bill, will often have an influence;
- the balance between rental and usage charges may be important;
- a connection charge which is paid to join the system can have a big effect, as least in the short term;
- price elasticities tend to vary with age and income, being highest for the young and the poor.

Most studies have found the short term *price elasticity of access*, defined as the proportionate response of a change in system size to a change in real price, to be low, in the range −0.01 to −0.1 in developed countries. This is not surprising, given the many factors which are involved. In the longer term a higher elasticity is to be expected, especially in the developing world.

Cracknell (*pers. comm.*) suggests that in developing countries the price elasticity might be more like higher elasticities shown by those newly joining the system, rather than that of system size, which nets off leavers and joiners. Support for this view comes from cross-section studies of telephone penetration, which rises in broad terms in income per head, from the poorest countries to the richest. An increase in the price of access reduces the spending power available for other purchases. Reactions to it might therefore be expected to be similar to those to an income change. ITU statistics show that the rentals charged for access tend to rise with the average income of the country, which is consistent with there being a significant price elasticity for access in the less developed countries, although it is hard to observe because of frequent underprovision and consequent long waiting lists.

Option demand for cellular mobile services has received less attention in the literature, but is known to be important for some customers. One example is the market for equipment which a motorist can use to ring for help without having to get out of a vehicle that has broken down.

The incomes of individuals can be used in choice analysis to obtain a greater understanding of how the network grows at the penetration at each level of the income scale. Income effects can also be studied at an aggregate level in relation to real consumer's expenditure or personal disposable income, for which data can readily be obtained. Since all the levels in an income distribution are likely to change by much the same percentage over short periods, and the price and income elasticities within income groups are likely to be stable, this is an efficient method for short and medium-term forecasting of aggregate demand.

Traditional patterns of residential-connection growth are being disrupted by new factors, such as the advent of cable systems offering telephony, networked home computing and various forms of wireless telephony. The household penetration rate seems likely to creep up further, although the rapid growth of the middle years will not be repeated. Experience in eastern Europe shows that rapid periods of growth can occur where past provision has lagged far behind demand.

4.2.5 Residential bill-size distributions

Whereas the revenue from rentals is dominated, in developed countries, by residential users, call revenues are more evenly divided. Business users may provide the majority of revenue because of their higher calling rates and more extensive use of long-distance and international services. Residential call bill size distributions are strongly skewed, with the peak frequency relatively low and a long tail of higher users. There are three measures of central tendency with a skewed distribution. These are:

- the *mean*, which is the arithmetic average;
- the *mode*, which is the maximum frequency or peak;
- the *median*, which is the halfway value, i.e. the one for which there are as many higher as there are lower.

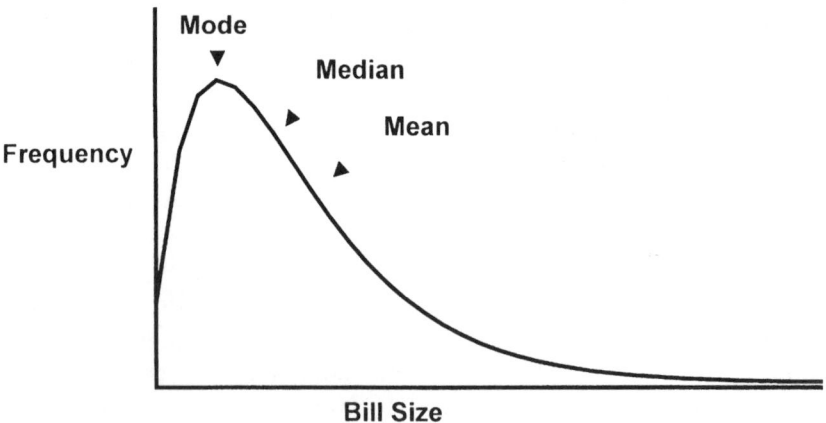

Figure 4.4 Typical call bill distribution

Only in a completely symmetrical distribution would all three be in the same place. In a skewed distribution like the one in Figure 4.4, the mode is the lowest followed by the median and then the average. This type of distribution is commonly found in call bills. OFTEL [13] used estimates of medians, based on the average usage of all residential customers for connection charges and of all customers in the middle 50 per cent of the call distributions, to calculate median bills. Table 4.6 shows figures for 1989 and 1993. For the use of median bills in price regulation, see Section 12.5.4.

Table 4.6 Median quarterly bills for residential customers of BT

	1989 £	1989 %	1993 £	1993 %
Connection charges	1.74	5.0	1.35	3.0
Exchange-line rental	13.95	40.4	19.54	43.0
Local calls	10.33	29.9	14.56	32.1
Long-distance call	7.18	20.8	8.23	18.1
International call	1.36	3.9	1.74	3.8
Directory enquiries			0.35	
Total	34.56	100	45.42	100

(Source: OFTEL [13])

Highly summarised data in Reference 14 gives the bill distribution in Table 4.7.

Table 4.7 Bill size and household income in the UK, 1995

Gross Household Income Level	Number of Quarterly Bills (m) <£35	£35–70	over £70	Total
Over £7500 p.a.	3	9	4	16
£7500 or less p.a.	2	3	0	5

(Source: OFTEL [14], figures rounded)

4.2.6 Empirical results for call price, income and system size elasticities

Call price elasticities vary with call price. In North America, where long-distance calls are banded and priced by distance, it has been found that price elasticity tends to be higher for the more expensive calls. A similar effect has been found with international calls. Local calls usually have a low price elasticity, –0.1 or less, reflecting their cheapness. Even where local calls are free, there appears to be an upper limit to demand which is in the region of 3000 calls per year, indicating a price elasticity of zero. A call still costs time, even if it does not cost money.

Although the main reaction to a price change is reflected immediately in demand, there are factors which can take longer to come through. Social habits of telephone usage evolve slowly and involve more than just the telephone itself. The substitution of telephony for travel involves personal and public transportation costs, time adjustments and choices about home and employment locations. The long-term elasticity has been found to be higher than the short-term elasticity in a number of studies. Breslaw and Pizante [15] found that 75 per cent of a price effect came through in nine quarters for Bell Canada business access service and that 95 per cent came through in two and a half years for residential long-distance calls. Cracknell [16] found a lag of up to six years in UK local call data.

Some empirical values for call price elasticities are shown in Table 4.8. They can be taken as total market elasticities since, at the time the studies were made, the operators concerned faced no significant competition. Applebe *et al.* [17] obtained *unidirectional* and *bi-directional* elasticity estimates, defining unidirectional elasticity as being when there was a price change at the originating end of the route only, and bidirectional as being the effect of an equal percentage change at both ends. The other studies relate to uni-directional elasticities. Call price elasticities tend to be higher for calls over longer distances and for more expensive calls. A fall in international-price elasticities in recent years may have been caused by large reductions in their price. Cracknell [18] noted that estimates of the price elasticity of international calls from the US, summarised in Taylor [19] and Munoz and Amaral [20], showed a fairly consistent decline from 1962 to 1992. They were still falling and were now firmly in the inelastic range, i.e. <-1. Cracknell's paper advances the theory that international call-price elasticities are highest in the middle part of the call-bill size range, citing evidence from BT. Hackl and Westlund [21] found that the price elasticity for international calls from Sweden to Denmark, Finland, Germany, Norway and the UK had risen (i.e the elasticity had increased) from 1976 to 1991, whereas on the US route it had fallen. By 1991 it was closer to zero on the US route than on any of the five European routes.

Applebe *et al.* obtained estimates of *reciprocal calling coefficients,* defined by them as the response of traffic from B to A to a change in the price-deflated traffic level from A to B. These were positive and in the range 0.24 to 0.72, which is consistent with the existence of a network externality and apparently one of the few studies to quantify any aspect of it. Other elasticity studies are discussed in Chapter 15.

Table 4.8 Call price elasticities

Study	Date	Call type	Country	Elasticity
Local Calls				
Cracknell	1994	local short term	UK	−0.17
		long term		−0.47
Davis *et al.*	1973	local	US	−0.21
Turner (unpublished)		local calls	UK	−0.06
Long-distance business:				
Larsen & McCleary	1972	long run	US	−0.98
Waverman	1974	long run	Ontario & Quebec	−1.03
Long-distance residential:				
Dobell *et al.*	1972	short run	Ontario & Quebec	−0.3
		long run	Ontario & Quebec	−1.9
Khadem	1973	short run	Trans-Canada	−1.28
		long run	Trans-Canada	−2.58
Larsen & McCleary	1972	long run	US	−1.01
Rash	1972	long run	Ontario & Quebec	−0.94
Waverman	1974	residential revenue per phone long run	Canada	−1.16
Long-distance business and residential:Appelbe *et al.*	1988	full-rate		
		unidirectional	Canada	−0.21 to −0.48
		bidirectional	Canada	−0.36 to −0.73
		discount rate		
		uni-directional	Canada	−0.39 to −0.49
		bi-directional	Canada	−0.59 to −0.75
Gatto *et al.*	1988	long run	US	−0.72
GPO CSD Report 41	1965	full rate trunk	UK	−0.52
GPO SBRD Report 2	1968	cheap rate trunk	UK	−0.77
GPO SBRD Report 12	1971	all trunk	UK	−0.17
GPO SBRD Report 27	1973	all trunk	UK	−0.18
GPO SBRD Report 73	1976	cheap rate trunk	UK	−0.096
GPO SBRD Report 89	1977	full rate trunk	UK	−0.114
Waverman	1974	trunk per tel.	Sweden	
		short run		−0.29
		long run		−0.58

continued

Table 4.8 continued

Study	Date	Call type	Country	Elasticity
Waverman	1974	trunk per phone	UK	
		short run		−0.41
		long run		−0.72
International:				
Appelbe *et al*	1988	full-rate		
		uni-directional	Canada-US	−0.43 to −0.49
		discount Rate		
		uni-directional	Canada-US	−0.45 to −0.53
Berstein *et al.*	1977	international	Canada	−1.391
BT CPR	1979	UK-France short run	UK	−0.211
BT CPRD	1979	UK-France long run	UK	−0.378
BT CPRD	1979	UK-New Zealand	UK	−0.816
BT CPRD	1979	UK-US	UK	−0.936
Craver & Neckowitz	1979	UK-US	US	−0.515
Craver & Neckowitz	1979	UK-US minutes	US	−0.325
Deschamps	1974	Trunk	Belgium	−0.24
Drew	1973	Calls and letters	UK	−0.86
Khadem	1977	International calls	Canada	
		short run		−1
		long run		−1.5
Kwok, Lee & Pearce	1975	short run		−1.7
		long run		−2.71
Lago	1970	International	International	−1.25
Nace	1974	International	Japan	−2.28
GPO SBRD Report 28	1973	International	UK	−0.117
Yatrakis	1972	46 nations in 1967		−1.03

References in this table are from summaries by Littlechild [22], Taylor [11] and Beesley [24] and papers by Applebe *et al.* [17] Cracknell [16], and Gatto *et al.* [23]

Estimates of income elasticity have tended to be somewhat more stable than those for price elasticity. Real GDP has proved a reliable basic measure, although it may be improved in some cases by a related variable such as consumer's expenditure for calls from residential lines, and by a variable such as service sector output for business calls. The robustness of the relationship is illustrated by Figure 4.5, which plots the percentage growth of UK call volume (including changes in duration) against real GDP. A least-squares regression fit to the 1980–92 data gives a trend growth of 4.3 per cent per annum plus 0.88 × the per cent GDP growth, i.e. an income elasticity of 0.88. Figures in later years relate only to British Telecom and so understate the call growth. The trend growth includes a contribution from the growth in the number of exchange lines which is also income related but in a less precise way and over a longer period. Typical income elasticities in empirical studies are in the range 0.5 to 1.0 for domestic traffic although rather higher elasticities have been found in the international traffic on some routes, see Table 4.8. Cracknell [16] gives an income elasticity of 0.63 against GDP for local calls.

System size can be included in the analysis by using calls per line as the independent variable, or more flexibly by including system size as one of the independent variables. Cracknell [16] found a system size coefficient of 0.69 (i.e. less than unity) for local calls in the UK when using residential system size as an explanatory variable.

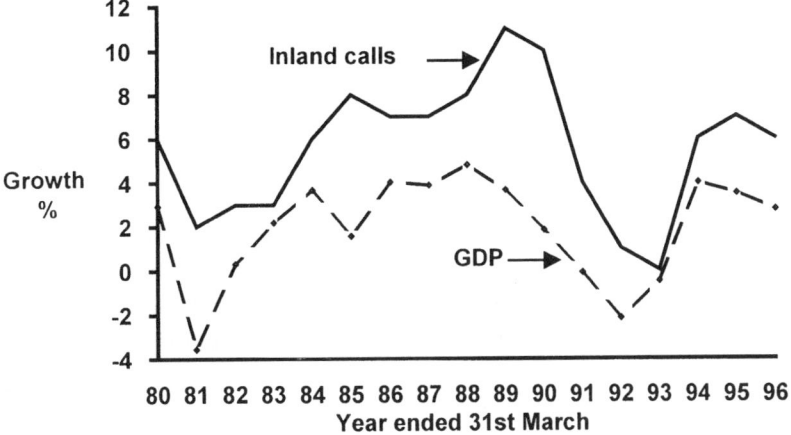

Figure 4.5 GDP and call volume growth in the UK
Source: Post Office and British Telecom Annual Reports, Central Statistical Office

4.3 Mobile services

4.3.1 Cellular mobile services

Typical customers of mobile cellular networks were initially business users, with rather limited penetration into domestic markets. This is not surprising with monthly rental and usage charges ranging from £25 per month upwards (UK average £32.96 in 1997 [25]), several times the size of a typical residential phone bill. But as with the petrol bill for the car, network expansion and stronger competition are bringing the cost down. Residential penetration has been increasing in response to falling prices and the development of new price schedules more suited to the pattern of private calling. Private use may become dominant at some stage, as it already has for ordinary telephony.

Customers of mobile networks in the UK make fewer calls than those on the fixed network; annual call minutes averaged about 1000 in 1994 [26]. The average mobile user received 428 calls [27]. Almost 30 per cent of customers use their mobile phones at least once a day but 18 per cent only use them for emergencies [25].

A rapidly evolving digital technology began to replace analogue systems within fifteen years, and currently there are widespread industry expectations of a third system, *universal mobile telephony service* (UMTS) which could provide a wider range of services, including high-speed data but make the first digital system obsolete. Equipment and usage costs have fallen with expansion of the market, and scale economies have increased. However, technology changes and in some cases the method of marketing have kept the churn rates high. Some obsolescence costs can be reduced by multimode equipment, such as a handset that contains circuitry to work on more than one system.

Penetration in the business sector has been accompanied by changing patterns of work. Mobile phones have increased the productivity and effectiveness of workers in construction and transport, where there is no fixed place of employment, as well as helping salesmen and company executives to make better use of their time. Mobile communications help *hot desking*, in which executives share desks and spend most of their time out of the office. Executives may also see mobile phones as part of a pay and benefit package, conferring status and reward. Owners of small businesses will be more conscious of the cost of mobile telephony but often see it as an essential facility.

Personal users pay their own bills. They are more directly aware of the cost and can make more fully informed purchase and usage decisions. Developing markets may eventually reach a situation where mobile telephony is cheap enough to be seen as an essential feature of domestic life, but in the initial stages demand is more for niche uses, of which these are some examples:

- to provide temporary telephony in a second home,
- to provide security in vehicles, especially for women travelling on their own in cars,
- for well-off families with work and social activities which frequently take them out of the house, sometimes as a second mobile phone where one or other of the members of the family has a company phone,
- for birdwatchers and some other hobbyists with urgent mobile communication needs,
- self-employed people who can combine business and leisure use.

4.3.2 *Mobile data and paging*

Paging systems deliver short data calls to paging devices which are small, light and fit in the pocket. The calls are transmitted from a hub which may cover a small area such as an office complex or which may cover a whole country. Modern pagers can accept and display short messages, which can be scrolled through on a liquid crystal display. The original market for pagers was confined to business users, but they are now bought by private individuals as a simple and relatively cheap form of mobile telephony.

The business use of paging systems needs no further description. The spread of private use has come at a time of wider interest in mobile data systems. In the UK, for example, there are two paging networks for birdwatchers. They transmit information about the exact location of the rare birds that some observers pursue, from databases which are updated several times during the course of a day as new observations and updates come in. A birdwatcher who is willing to drive 200 miles or more on the chance of finding something new may rate the cost of the pager as one of the smaller expenses of the hobby, and the information obtained as of great value in avoiding wasted journeys. This is a small example of the way that the spread of information technology is transforming the daily lives of individuals. It illustrates the value of information and of time.

4.4 Impact of information technology

The maturing of basic telephony markets in developed countries has slowed the growth of traditional sources of telecommunications service growth. Information technology is providing new potential sources, which operating companies have been trying to exploit. Coming from the other direction, telecommunications facilitates changes in society at large. In business markets, data traffic over the PSTN and private circuits has been growing faster than voice and with a strong international component. Computing and telecommunications technologies have come together to globalise financial markets. Chains of transactions link London, New York, Tokyo and their related markets in what is effectively a 24-hour trading floor. Margins have become finer and volumes larger.

Better communications are facilitating the transfer of work from the office to the home, or the car. Leisure activities involving telephony come from entertainment online information services and the Internet. Operating companies are anxious to move into all of these and see themselves well placed to provide networks for cable television.

4.5 References

1   'Post Office telecommuncations statistics 1962' (General Post Office, London, 1962)
2   'Report & accounts for the year ended 31st March 1995'. British Telecom, London, 1995
3   'Statistics of communications carriers'. FCC, Washington DC, 1995
4   'Input–output–tabelle 1990', Statistiches Bundesampt, Wiesbaden
5   CRONIN, F. J., GOLD, M. A., BURNHAM MACE, B., and SIGALOS, J. L.: 'Telecommunications and cost savings in educational services', *Inf. Econ. Policy*, 1994, **6**, (1),pp.53–75
6   'Basic statistics of the European Union' (Office for Official Publications of the European Union, Luxembourg, 1995, 32nd edn.)
7   GULDMANN, J.-M: 'Input–output analysis of regional telecommunications flows', *Inf. Econ. Policy*, 1993, **5**, (4), pp.311–329
8   COOPER, M.: 'Universal service: a historical perspective and polices for the twenty-first century'. Benton Foundation and Consumer Federation of America, Benton Foundation website, 28 April 1998, last updated 9 December 1996
9   MITCHEL, B. M.: 'Utilization of the U.S. telephone network (Rand, Santa Monica, 1994)
10 TEMIN, P. with GALAMBOS, L.: 'The fall of the Bell system' (Cambridge University Press, Cambridge, 1987)
11 TAYLOR, L. D.: 'Telecommunications demand: A survey and critique' (Ballinger, Cambridge MA, 1980)
12 Jipp A.: 'Wealth of Nations and Telephone Density', *ITU Telecommun. J.*, July 1963
13 'The telephone bill of the 'typical' residential customer, Statistical Note' (OFTEL, London, March 1994)
14 'Pricing of telecommunications Services from 1977, Consultative Document' (OFTEL, London, 1995)
15 BRESLAW, J. and PIZANTE, G.: 'Lag structure in telecommunications demand analysis', *Inf. Econ. Policy*, 1989/90, **4**, (4), pp.324–345
16 CRACKNELL, D. R.: 'Growing the telecommunications market – lessons from the past', Paper presented at ITS *Regional Meeting*. September 1994, Crete
17 APPLEBE, T. W., SNIHUR, N. A., DINEEN, C., FARNES, D. and GIORDANO, R.: 'Point-to-point modelling, an application to Canada-Canada and Canada-United States long distance calling', *Inf. Econ. Policy*, 1988, **3**, (4), pp.311–331
18 CRACKNELL, D. R.: 'The demand for international telephone calls – The dynamics of price elasticity'. Paper presented to ITS *Regional Meeting*, Leuven, August 1997
19 TAYLOR, L. D.: 'Telecommunications demand in theory and practice' (Kluwer Academic Publishers, Norwell, 1994)
20 MUNOZ, G. and AMARAL, P.: 'Demand for international call traffic in Spain, an econometric study using provincial panel data', *Inf. Econ. Policy*, 1996, **8**, pp.289–316
21 HACKL, P., and WESTLUND, A. H.: 'On price elasticities of international telecommunication demand', *Inf. Econ. Policy*, 1995, **7**, pp.27–36
22 LITTLECHILD, S. C.: 'Elements of telecommunications economics', (IEE Telecommunications Series 7, Peter Peregrinus, UK, 1979)
23 GATTO, J. P., LANGIN-HOOPER, J., ROBINSON, P. B., and TYAN, H. 1988: 'Interstate switched access demand analysis', *Inf. Econ. Policy*, 1988, **3**, (4), pp.333–358
24 BEESLEY, M. E.: 'Liberalisation of the use of British Telecommunications Network' (HMSO, London, 1981)
25 CANE, A.: 'Mobile phone market growth slows by half', *Financial Times*, 10 November 1997, London
26 OFTEL: 'Market information' (OFTEL, November 1995, London)
27 OFTEL: 'Economic evaluation of number portability in the UK mobile telephony market' (OFTEL, London, 1997)

Chapter 5
The economics of supply

5.1 The theory of the firm

Robinson Crusoe combined the functions of producer, consumer and regulator in one individual, at least until the arrival of Man Friday. The enterprises run by self-employed workers are still very numerous, but the majority of the national output in developed countries is produced collectively, with the firm or state enterprise as the basic unit of supply. In the theory of the firm it is profit maximisation which provides the dynamic that drives markets and results in efficient resource allocation. Private firms in competitive industries fit this model reasonably well. Empirical evidence shows that they are fundamentally profit maximisers, within whatever framework of institutional constraints they operate.

Profit maximisers with a large market share need some form of regulatory control over the prices that they charge, see Chapter 12. Enterprises that have a large market share but are not profit maximisers need perfomance tests to measure their efficiency. These are discussed in Chapter 15.

5.2 Agents and operators

5.2.1 Ownership and control

With separation of ownership from operation in the modern world, the board members of the average large company may hold few of its shares, allowing the possibility that they may not fully identify with the interests of the owners of the company which they control and on whose behalf they act as agents. This is a particular risk in monopolistic markets. It also affects companies not subject to the threat of takeover and those with inactive shareholders. A departure from profit-maximisation behaviour moves the business into an area of uncertain and arbitrary behaviour, where shareholder interests are neglected and wealth may be wasted.

The *theory of agents* is concerned with the economic behaviour of those acting on behalf of others. To take a small example, in large organisations the connection between the use of a resource and the bill for its use may be difficult to establish. The behaviour of business telephone users was mentioned in Chapter 4. Calls paid for corporately are difficult to control except by methods such as call barring. Internal systems of transfer charging place calls within the budgets of management units, but still leave the callers to balance their utility against a price which may be unknown.

5.2.2 Motivation of company boards and managers

Managers in operating companies have their own perspectives. These are important, given that they are the agents by which the supply side of the economic process is driven. Among the things which will affect their behaviour are:

- who the owners (public or private) are and what they want them to do,
- legal and environmental constraints,
- the need to satisfy customer and employee interests.
- their perception of what drives business growth and profitability.

The Cadbury Committee was set up to report on the financial aspects of corporate governance in the UK after a series of unexpected company failures and widespread unease about the way that the pay of directors was determined. Its report [1] was published in 1992 and contained a code of best practice. This set out standards for the protection of shareholder interest through good board organisation without any individual having unfettered powers of decision, an effective company secretary, independent nonexecutive directors and adequate financial reporting to the board and the shareholders.

Useem [2] argues that shareholder interests have been advanced in the US by the growing interest in management performance which has been shown by institutional investors, particularly the mutual funds. He says that the managers, who are experts in the exercise of power and see this interest as a threat, have responded with accusations that the mutual funds lack competence and with attempts to manage shareholder registers so as to replace hostile investors with those which are more docile.

5.2.3 Motivation of monopolists

Private and public monopolies may be driven by empire-building ambitions which have little to do with value for money for users. The managers may prefer to take part of their reward in terms of a quiet life and relax their control over efficiency. Privatisation is not enough to free monopolies from the suspicion among regulators that they may not be profit maximisers. Those running companies which are cash rich may be seen as wishing to find a safe home for excess profits, one which is out of the reach of the regulator and often abroad, regardless of the real merits of the investment. The companies, on the other hand, may feel that they are already too big at home and have to explore overseas for investment opportunities.

One safeguard for profit-maximisation motivation in monopolistic industries is the existence of competition in the capital markets. This is discussed further in Chapter 6.

5.2.4 Economic behaviour of state-controlled PTTs

State-owned PTTs operate under a variety of incentives. The UK nationalised industries were criticised for lacking clarity in their objectives. Faced with a host of alternatives, such as maximising service quality, business volume or market share, minimising political embarrassment to the minister and minimising prices to the user, the boards found that they had no ready way of setting stable priorities. This is an example of the type of problem which can be analysed and addressed by public-choice theory.

5.3 Conceptual framework for the production process

Adam Smith's example [3] of the reduction in unit costs brought about by the division of labour in a pin factory where '*One man draws out the wire, another straights it, a third cuts it, a fourth points it, a fifth grinds it at the top for receiving the head . . .*', resulting in a large increase in labour productivity, is an early exposition of the economics of a production process.

✳ In the terms of classical economic analysis the three basic factors of production needed for the supply of telecommunications services are land, labour and capital. Other more intangible elements include enterprise (sometimes regarded as a fourth factor of production), organisation and informational inputs. Information and the systems to provide it are vital inputs which may be bought in or provided by capital investment. Goods and services which are bought from other suppliers have the same basic composition.

Labour is paid for in wages. Land that is required in the form of physical sites for buildings can be regarded as part of the capital and commands a purchase price or rent. The most distinctive form of telecommunications land use is in terms of wayleaves and other rights to lay cables and fly lines across ground belonging to third parties. Payments for wayleaves, where they are made, have often been nominal. More recently, site rents for cellular mobile antennas have been much higher. Telecommunications and cable operators may have statutory powers to lay cables under public or private roads without payment, if they are essential for the provision of services. In such cases, the physical work is paid for by the operator but third party costs such as inconvenience are carried by the community. Radio spectrum is a special form of resource which, like land, is a finite natural resource with special allocation and pricing procedures. Capital inputs are those which, once bought, hold some value from year to year.

Enterprise is the catalyst in the market process. Profits accrue to the entrepreneur who provides the organisation and supply of capital and are classically described as the reward of enterprise. The profit made by a company, after paying for other factors, has to be at least as large as its cost of capital to retain external funds. In a fully competitive market this is all it would be, and is called *normal profit*. Short-term conditions and opportunities may create profits below or above this level. It is a fundamental feature of dynamic markets that these opportunities exist and that entrepreneurs look for them. Their possible existence determines the flow of investment one way rather than another.

5.4 Efficiency frontiers

There is usually more than one way of making a product, for example with different amounts of capital equipment per worker; Figure 5.1 illustrates this concept. Within a single proportionate mix (production method) some producers will use fewer resources than others, by

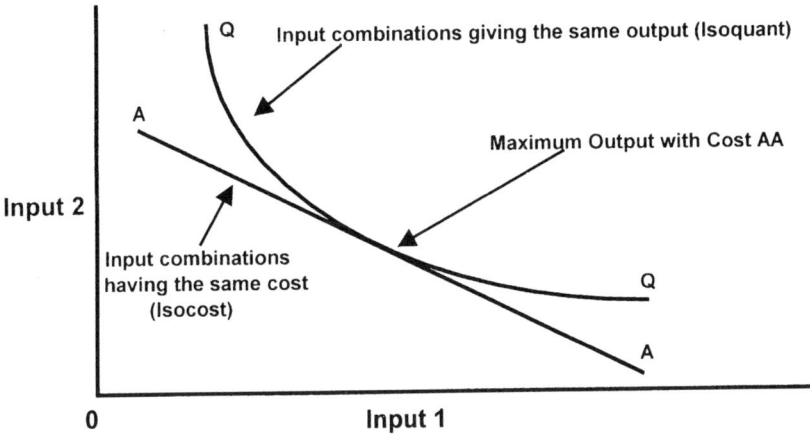

Figure 5.1 The efficiency frontier

being less wasteful. Isoquant QQ is a curve or *efficiency frontier* that, for the same total output Q, shows different combinations of inputs which produce it for the least cost for each proportionate mix of inputs. All the firms on QQ are *technically efficient*, that is they have minimum cost given the production method which they have chosen. The others use more of at least one input and are to the right of the curve. Isocost curve AA represents different combinations of two inputs which have the same overall cost, *C*.

There will usually be an optimum mix of inputs which minimises costs for a specified level of output among efficient producers. The one with the lowest cost *C* for output level *Q* is at the point where AA is a tangent to QQ and is said to be *efficient*. This firm is using the optimum mix and can be used as a benchmark against which to measure the rest. It has the highest level of output that can be obtained with expenditure *C*. The uniqueness of this point depends on the isoquant formula, but is likely to hold in practice.

Suppose that the two inputs are capital (K) and labour (L) and that *C* is the overall cost. The *marginal rate of substitution* (MRS) is dK/dL, the rate at which one input can be substituted for another to produce the same output. The chart shows two inputs and one output.

If a firm is operating with the optimum mix of inputs, the rate of change of output (Y) will be the same whichever one is increased, i.e. the marginal products, dY/dK and dY/dL, of the two inputs are equal. Output, capital and labour will have price components p_y, p_k, p_l and volume components v_y, v_k, v_l. It may be shown that under conditions of perfect competition the price of each input is determined by the marginal volume product of the factor valued at the selling price of the product.

In practical terms, if £10 is the market value of an extra unit of service, the cost will be £10 to produce it by extra labour and the same amount to produce it with extra capital. If production can be achieved with two hours of labour or one extra piece of equipment, labour will have to cost £5 per hour and the extra equipment will have to cost £10 for the equalities to hold. Management will then be indifferent as to which extra resource is used. The price ratio p_k/p_l is then $5/10 = 1/2$. The frontier concept generalises to the multi-input case and preserves uniqueness under similar theoretical conditions. Multiple outputs are considered later in the chapter.

5.4.1 Shifting frontiers

Empirical evidence suggests that efficiency frontiers are shifting, capital being substituted for labour for any fixed level of production and along the expansion path to greater output. Indicators of British Telecom's physical output, labour input and capital employed are in Table 5.1.

Table 5.1 British Telecom inputs and outputs

| | Year ending and year end | | change % |
	31.3.81	31.3.94	
Exchange Lines (000s)	20 552	26 640	29.6
Mobile phones (000s)	0	1 019	
Calls (millions)	20 281	32 784	61.6
Employees (000's)	251	156	−37.8
Net assets (£m)	6 302	15 719	149.4

(Sources: BT, Office of Telecommunications, estimates)

Table 5.2 shows how British Telecom's cost structure has shifted from labour to capital over the same 14 years. From 1981 to 1994 the percentage of BT operating cost spent on staff fell from 61.1 to 38.0, whereas depreciation rose from 13.7 to 20.2 per cent. Depreciation is here based on the historic cost of the assets, but the use of their replacement cost would not make much difference to the figures because of the fall in asset prices. Over the same period turnover trebled and the volume of output doubled.

Table 5.2 British Telecom labour and capital inputs 1981–94

Year ended March	Operating cost (£m)	Staff/ operating (%)	Depreciation/ operating (%)
1981	3 456	61.1	13.7
1982	4 271	56.1	15.1
1983	4 849	53.0	15.8
1984	5 345	50.8	17.0
1985	5 797	48.4	16.1
1986	5 729	52.0	18.6
1987	6 199	51.0	21.1
1988	6 990	50.0	21.8
1989	7 576	51.7	21.3
1990	8 264	50.5	21.7
1991	9 623	40.7	20.1
1992	9 922	39.9	20.7
1993	10 806	37.6	19.6
1994	10 660	38.0	20.2

(Source: BT Annual Reports)

The range of telecommunications statistics available in the UK is rather sparse compared with what can be had in North America, a reflection of the secretive nature of British regulatory and civil-service processes. In the US the same general trends emerge from the public data. There are numerous local and long-distance telephone companies. Figures for the largest, accounting for around 95 per cent of the total, are summarised in Table 5.3.

5.5 Economies of scale and scope

5.5.1 Economies of scale

Economies of scale were recognised in the classic economics of Marshall [4], and they are found in many sectors of economic activity. The level of output at which a *diseconomies of scale* set in, i.e. where average costs begin to rise as output expands, is of importance in the theories of perfect and imperfect competition, which are discussed in Chapter 6. Economies of scale are usually the result of large fixed costs, so that average costs fall as production increases, and network industries such as telecommunications, electricity, water and gas supply have them in strong measure. So do most manufacturing

Table 5.3 US telephone company summary 1988–94

Year ending and year to end	31.12.88	31.12.94	Change (%)
OUTPUTS			
Exchange lines (000s)	122 275	142 208	16.3
Local calls (millions)	379 036	456 206	22.7
Trunk calls (million minutes)	271 763	500 297	84.1
INPUTS			
Cable (000 miles)	3 535	3 857	9.1
Optical fibre in cable (000 miles)	2 290	10 075	339.9
Poles (000s)	19 534	19 836	1.5
Exchanges	15 095	23 151	53.4
Employees (000s)	688	553	−19.7
Plant assets ($m)	244 411	291 942	19.4

(Source: Federal Communications Commission)

industries, especially where massive integrated plant is used, as in chemicals or motor cars, or research and development costs are high (as in pharmaceuticals and telecommunications switching).

Telecommunications networks exhibit the potential for economies of scale in virtually all aspects of their operations and especially so in transmission. They arise from the queuing theory discussed in Section 2.7 and more importantly from fixed costs. The cost of digging up the road and laying ducts and cable is an example relating to transmission. The cost of the building, power plant and central processors are fixed elements in a telephone exchange. Convenient measures are the *asset-volume elasticity* (AVE), which is the proportionate response of asset volume to business volume, and the *cost-volume elasticity* (CVE), which is the proportionate response of overall cost volume to business volume. OFTEL [5] estimates of AVEs for BT are in Table 5.4.

CVEs can be translated into *average cost elasticity* (ACE). If a one per cent increase in traffic increases cost volume by 0.6 per cent, then the average cost per unit of traffic changes by $100 \times (100.6/101 - 1) = -0.4$ per cent. Figure 5.2 shows the relationship between the changes in cost volume and changes in business volume shown in Post Office annual reports for telecommunications in the UK. The scatter in the figures precludes a close estimate but theysuggest a downward

Table 5.4 Asset-volume elasticities for British Telecom

Asset type	AVE
Duct	0.0–0.1
Core transmission equipment	0.6–0.7
Transmission cable	0.1–0.3
Local switches	
traffic-sensitive parts	0.6–0.8
all equipment	0.15–0.2
Main switches	0.6–0.8
Average inland conveyance	0.2–0.3

trend of up to five per cent per annum in cost volume, offset by a CVE of between 0.5 and 1, giving an ACE of 0 to –0.5, including the trend. The downward trend comes from technical change and efficiency changes. It probably includes part of the scale economy effect as well.

The engineering evidence for economies of scale is strong, but empirical studies have to deal with price and mix changes which require large amounts of data to remove, and they may pick up technological change effects. Meyer *et al.* [6] reported an ACE of about –0.3 for the AT&T transmission network, based on work in by Mantell [7], and gave similar results for six other studies. Kiss [8] found scale economies to contribute an average of 2.8% per annum to annual productivity gains at

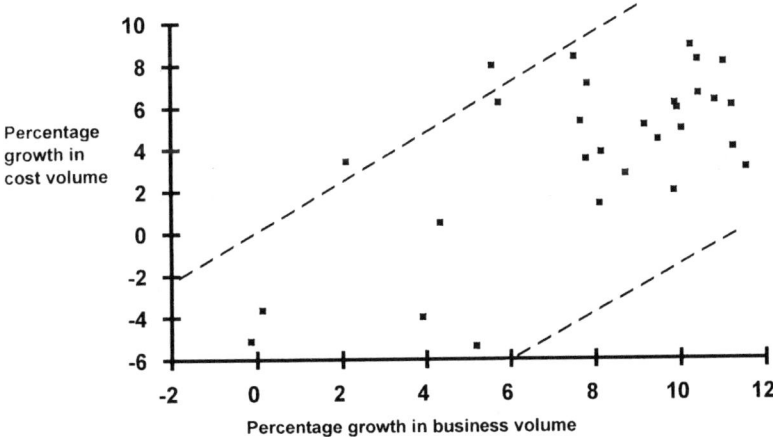

Figure 5.2 Scale economies in UK telecommunications

Bell Canada during 1953–79. McGowan [9], using 1968-83 data, estimated that output rose by 1.6% for each 1 per cent of extra input at Alberta Government Telephones (AGT), a CVE of 0.625 and ACE of –0.6. Bell and AGT produce multiple outputs. A mix of outputs which remains constant might be regarded as a single composite output. Where the mix changes, the methodology will pick up economies arising from changes in the product as well as those of scale.

The technical basis for the existence of economies of scale is *subadditivity*. (On this subject a full exposition is given in References 10 and 11 and there is a summary in Reference 12). Broadly speaking, subadditivity is found where the joint cost of two firms producing the same product is more than when a single firm produces the combined volume. Figure 5.3 illustrates this with a notional production surface for outputs produced by two different firms.

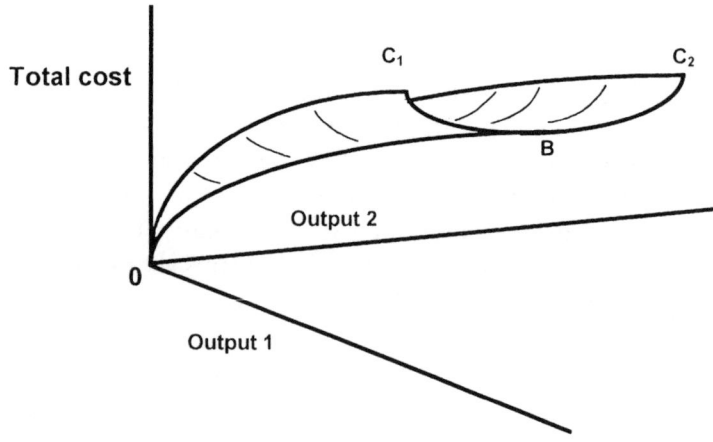

Figure 5.3 Subadditive cost function

The output of each firm exhibits economies of scale, so that average cost to the firm falls as output rises. Each point on the surface which links the two cost curves represents the cost of jointly producing a combination of the two outputs that is worth as much as each of the outputs on their own, for example £2 m jointly instead of two separate firms producing £1 m each. Point B marks the minimum cost of jointly producing outputs worth as much as those produced at C_1 and C_2.

Continued economies of scale exist if, as in the Figure, B is lower than C_1 and C_2. If it is higher, there are diseconomies and the function is *superadditive*. With an *additive* cost function the costs are the same for joint and separate production.

It should be noted that subadditivity ceases to hold where both firms have fully exhausted their economies of scale and may be unimportant if the residual scale economies are very small. Thus the theoretical cost advantages of monopoly may hold for low market demand levels but be easily outweighed by other factors where the level of demand is high. More generally, the number of firms that can operate within a market of given size depends on how many can enter without losing significant economies of scale from joint production. Thus, telecommunications competition that can be readily sustained in large, mature economies may not be sustainable in small, developing areas. Similarly, the number of manufacturers developing telephone exchanges for the PSTN is limited by the size of the world market for exchange equipment. This stems from the importance of the research and development costs, as identified in Section 3.3.

5.5.2 Economies of scope

Economies of scope exist where the joint costs of producing two different products are lower than if they had been produced separately. It may be illustrated by Figure 5.3 if the two outputs are taken to be different products, such as PBXs and local network services. Scope economies are found in most multiproduct industries and in telecommunications network services, where they may arise between similar services, such as the provision of facilities for delivering long-distance and local calls. There may theoretically be geographical scope economies arising from the aggregation of larger areas of territory.

The tendency of equipment suppliers to specialise in manufacturing technologies which was mentioned earlier in connection with BICC is as example of an economy of scope. Research and development costs and other overheads needed for the manufacture of metallic cable can be spread across the markets in telecommunications, broadcasting and the electricity industry. There will also be economies of scale, associated with high output volume.

There may be economies between industries if they use the same technology, for example in multimedia services. There the technical convergence is between the technologies of computing, broadcasting and telephony stemming from digitalisation and the use of computers, and has led telephone companies to enter or attempt to enter the field of TV distribution.

There has been a tendency to concentrate on the physical aspects of scale and scope economies, of which the technical examples are the most obvious. These examples can be basic, such as the use of a common trench into which to put different cables. Sharkey [11] distinguishes between these, which he calls *plant subadditivity* and the organisational aspects, which he calls *firm subadditivity*. As Sharkey points out, firm subadditivity is the more fundamental. It would be possible for a number of firms to share physical facilities and obtain the technological economies of scale (plant subadditivity) without actually merging, if they could agree on usage prices. Without firm subadditivity, management and marketing costs would be lower than if the whole job was done by just one company. This possibility is of mounting importance in the market structure of a variety of industries, including railways, gas, electricity and telecommunications, where rights to use a single network may be leased out to competing service suppliers. In telecommunications networks, traffic is commonly exchanged between networks or carried over lines leased from another operator.

A transeuropean trunk network would probably be able to operate more cheaply than a set of connected national networks, by being under a unified management, provided it could carry as much traffic as the individual networks. It would still need to connect customers to the nodes of its network, and this might be done through the PSTN if it were cheap enough. Alternatively, business access could be over lines owned by the trunk operator or leased from the local operator and residential access could be through cable networks. A number of the european collaborations are based on these ideas. Motivation may arise more from a desire to exploit price anomalies than from a reduction in underlying costs.

The extension of geographical range may create economies of scope, although these are often difficult to prove. There are, for example, small local telephone companies, even in the British Isles, with prices and costs which compare favourably with those of larger companies elsewhere. Geographical extension does not provide economies of scope that are as large as the economies of scale which come from increasing network usage, although joint headquarters costs may be reduced by geographical extension. In his study of nine european countries, Garonne [13] found some evidence of geographical diseconomies in local networks and thought that the breakup of the largest carriers would be likely to free some efficiency for the industry.

Marketing is important in modern industries, sometimes more so than technology. Examples include entertainment and information services, two industries where products are distributed over telecommunications networks. There could be diseconomies of scale and scope when combining disparate operations in a process which actually reduces the comparative advantage of the participants, and increases their joint costs rather than reducing them.

5.6 The supply curve

5.6.1 The supply curve of the firm

An individual firm will usually have been set up to produce efficiently at a certain minimum level of output, not necessarily much related to its current output levels. An entrepreneur establishing a new firm can choose the output level to cater for.

Whatever the history of the firm, the supply of an additional unit of output incurs an additional cost, which is the *short run marginal cost.* Marginal costs imply small, locally-uniform increments, and they may rise or fall as output increases. Typically, the marginal cost falls initially, especially when starting from a low level of production, then levels out or rises. There is an *average cost* associated with each level of output, which is the total cost divided by the volume of output. If the marginal cost curve is *U*-shaped, marginal and average costs are the same at the minimum point of the average cost curve, as shown in Figure 5.4.

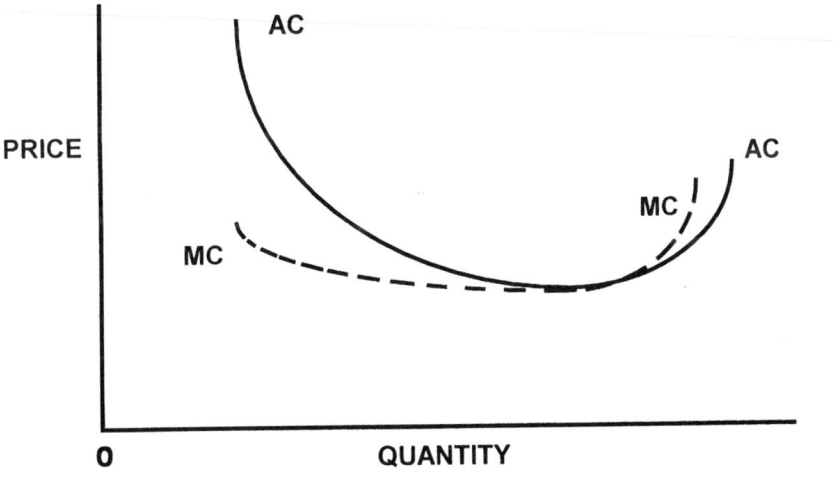

Figure 5.4 Average and marginal costs

The term *incremental cost* is used in three ways. One is for costs which can only be supplied in large increments, for example, an extra exchange may be needed when the capacity in an existing exchange is exhausted. A second is for the additional costs that a firm will incur to expand service in the future by, for example, adding new lines in an area already served. It is also used for extra costs incurred under a different scenario; such costs will be in lumpy increments. Incremental costs are averaged over the size of the increment to get an *average incremental cost*, which may approximate to marginal cost.

Since the firm could have been established to cater for other basic output levels, there are many possible short-run marginal cost curves, with shapes and positions which differ from one another. Figure 5.5 shows how these can form an envelope defining the possibilities for a newly established firm and an optimum size where average costs are at a minimum. The firm could have chosen to set up for a different level of production but, once established, it may not be able to evolve to minimum cost. A network built to provide one million lines within a conurbation, which it later expands to two million, may have somewhat higher average costs than one which was built to serve two million from the outset. In a world without perfect foresight, which is the one that we live in, there will also be firms which were established earlier using suboptimal technologies the early obsolescence of which was not foreseen. These firms may remain in the market in the short

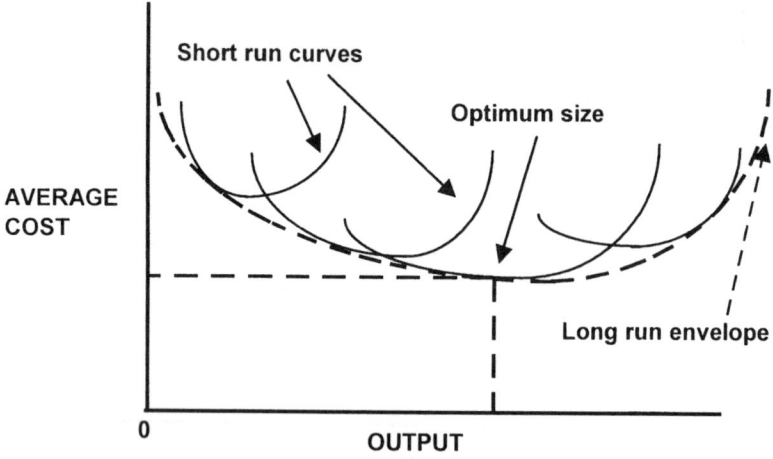

Figure 5.5 Short and long run supply curves: choices for the firm

run, although for any given level of output their costs will be higher than those of an efficient firm and they must eventually modernise their production method or go out of business.

The change in profit obtained from producing an additional unit of output is positive as long as the market price exceeds the marginal cost of production, but this is not a strong enough condition to keep a firm in business. Total profit will be negative unless the price equals or exceeds average cost. An individual firm will therefore expand production up to the point where the market price equals marginal cost provided that it also equals or exceeds average cost. If there is no level of production at which the market price equals average cost the firm must usually withdraw from the market because in the long run unsubsidised suppliers have to cover their average costs to avoid bankruptcy. The next Section considers how this is achieved under fully competitive conditions. Although such conditions are rare, especially in telecommunications, the implications for optimum resource allocations go wider. They are the foundation for most work on social welfare maximisation and regulation, as well as providing a reference point for the analysis of less competitive situations.

5.6.2 The market supply curve

Just as the demand curve presented in the market is an aggregation of the individual demand curves of all the potential buyers, there is a market supply curve which, as shown in Figure 5.6, aggregates the supply curves of all the suppliers. For any specified price there is a total supply which represents the sum of all the decisions made at the level of the firm concerning how much will be produced at that price. The *elasticity of supply* is the proportionate response of supply to a change in the market price.

It is easy to underestimate the elasticity of supply, and to suppose that a shortage in the market will result in permanently higher prices, or that a sharp rise in price will persist. As was seen in the oil market in the 1970s, the supply-side response to a rise in price is for marginal firms and resources (like North Sea Oil), which could not have entered the market at the old price, to increase the supply and ease the price back again.

Costs have a time dimension. Fixed costs are those ones generated by plant which has to be paid for whether it is in use or not, for example the cost of running a headquarters and all the other overheads. A little further out in time, some of these costs are variable.

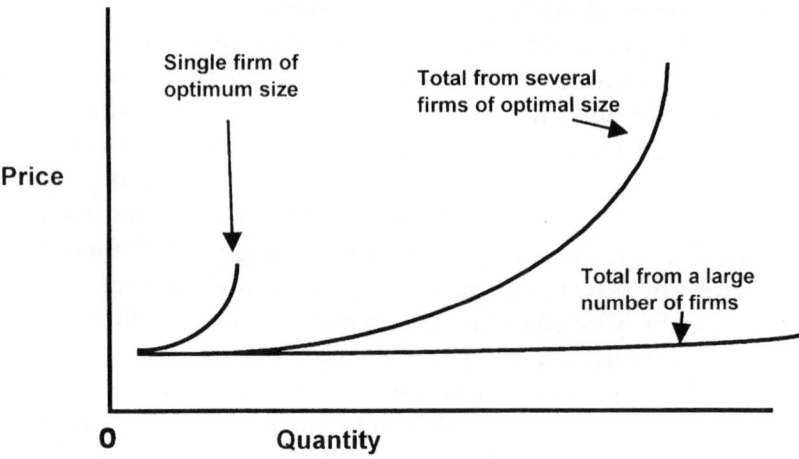

Figure 5.6 Market supply curve

Plant wears out and may be replaced by cheaper plant, or none at all, staff can be laid off and so on. For practical purposes, long-term effects creep in gradually over a period of time which is at least equal to the age of the oldest assets which are still providing a positive return in use. The economic definition of the long run is the period in which all costs are variable and all assets have to be replaced. It is also the period in which, as Keynes pointed out, we are all dead.

For each product there is a marginal cost, which is the cost of producing an extra unit. Marginal cost is likely to depend on the existing level of output of the product in question and possibly on production levels of other products of the enterprise. Costs have a time dimension. *Short run marginal cost* (SRMC) may be lower than *long run marginal cost* (LRMC) because there is less time in which to adjust the input volumes of long-lasting assets. In some circumstances SRMC may be higher because of a temporarily high opportunity cost for marginal resources.

5.7 References

1 Cadbury Committee: 'Report of the committee on the financial aspects of corporate governance' (Gee (Professional Publishing), London, 1992)
2 USEEM, M.: 'Investor capitalism: how money managers are changing the face of America' (Basic Books, 1996)
3 SMITH, A.: 'The wealth of nations': book I (Penguin Books Edn, Harmondsworth, 1970), Chap . I
4 MARSHALL, A.: 'Principles of economics' (Macmillan, Basingstoke, 1920, 8th edn.)
5 'Network charges from 1997, consulatative document' (OFTEL, London, 1997)
6 MEYER, J. R., WILSON R. W., BAUGHCUM, M. A., BURTON, E. and CAOUETTE, L.: 'The economics of competition in the telecommunications industry' (Oelgeschlager, Gunn and Hain, Cambridge MA, 1980)
7 MANTELL, L. H.: 'An econometric study of returns to scale in the Bell system'. Staff research paper, executive office of the President, Washington D.C., 1974
8 KISS, F.: 'Productivity gains in Bell Canada'. Conference paper presented at *Telecommunications in Canada: economic analysis of the industry*, March 4th–6th 1981, Canada
9 McGOWAN, F. S.: 'Allocative efficiency and total factor productivity'. Paper presented at *Sixth international symposium on forecasting*, June 15th–18th 1986, Paris
10 BAUMOL, W. J., PANZAR, J. C. and WILLIG, R. D.: 'Contestable markets and the theory of industry structure' (Harcourt Brace Jovanovich, New York, 1982)
11 SHARKEY, W. W.: 'The theory of natural monopoly (Cambridge University Press, London, 1982)
12 BEESLEY, M. E.: 'Liberalisation of the use of the British Telecommunications network' (HMSO, London, 1981)
13 GARONNE, P.: 'Network subscription and services usage in european telecommunications industries', *Inf. Econ. Policy*, 1996, **8**, (1) pp.25–50

Chapter 6
Market structure and dynamics

6.1 Firms and markets

6.1.1 How markets work

Market forces reflect the preferences of individual people or companies. When Adam Smith [1] wrote that each individual '*intends only his own security, only his own gain. And he is in this, led by an invisible hand to promote an end which was no part of his intention. By pursuing his own interest he frequently promotes that of society more effectually than when he really intends to promote it*', he recognised the efficiency of markets in allocating resources.

The way in which prices are determined in markets depends a good deal on the number and relative size of the participants, or market structure. The buyers in the market are on the *demand side,* with firms and other suppliers on the *supply side.* Telecommunications is not without its share of debates about the concepts of price and value. In the case of freely traded goods it may be hard to accept that the market price of a good or service can differ greatly from the notion of its intrinsic value or, in the terminology of the medieval theologian St Thomas Aquinas, its *just price.* The *labour theory of value,* a 19th century idea that price embodies the value of the labour used in production, still has its adherents and finds echoes in the efforts of some to relate prices to historic costs regardless of market conditions.

Modern economic theory is based on the market-clearing price, which is the price at which all the goods are sold and there are no unsatisfied buyers. Most of the buyers would have been prepared to pay more than they did, and for these the value of their purchases is in excess of the price they paid. In the long run, market-clearing prices must be at least as high as the average costs of all the firms supplying the market, otherwise one or more will go out of business. In all markets the short-run prices may leave shortages or unsold goods and there may be no price that would clear the market.

Competition tends to be a supply side concept, important to operators because it is a business opportunity and a threat. It is Darwinian, threatening all but the fittest. The viewpoint of the consumer is different. The ultimate purpose of production is consumption. From the consumer's point of view, choice is the key thing and competition is the mechanism which brings choice. Benefits include:

- Product diversity,
- Innovative product ranges,
- Lower prices for any given quality.

The first two benefits provide a better chance of getting the product type and quality closest to the optimum for the customer. Taken with the third, these benefits can be summarised as better value for money. There are libertarian aspects. Freedom depends on the ability to make economic as well as political choices.

Under competition, the short run is a period of turbulent market conditions in which market forces work towards a long-run equilibrium which they may never reach. Equilibrium here means a state where the participants in the market cannot better their own positions by changing volumes of the goods they buy or sell, or the prices at which they trade. Long run equilibrium can occur in competitive and monopolistic markets.

The process by which markets move towards equilibrium can be analysed as the pursuit of self interest by rational people with perfect knowledge of all relevant information. The extent to which this is realistic depends on the conditions of the industry and the times.

6.1.2 The market as a discovery process

The telecommunications industry is one in which large operators and equipment suppliers with considerable market power are in contest with smaller entrepreneurs who may be able to undermine the position of their rivals by discovering new techniques, products and markets. These discoveries give the market its dynamic and leave many, if not all, firms making profits which differ from the long-run norm. If they are lucky or clever they make a lot. Others make losses.

The existence of high profits attracts new entrants, creating extra competition which reduces abnormally high profits, bringing them nearer the norm.

Market dynamics of this type have been called a *discovery process* by economists of the Austrian school [2]. New knowledge, whether of how to make something, the meaning of the law, or the nature of demand, changes the horizon of the industry. Examples in telecommunications are:

(i)   The original Strowger switch (first automatic telephone exchange).
(ii) Legal challenges, such as those which permitted new entrants into the US terminal equipment supply markets and long-distance call markets from the 1960s onwards.
(iii) The exploitation of mobile telephony
(iv) Value added services, such as the audiotex information for birdwatchers and screen-based financial services data for stockbrokers.

There are many other examples, some developed by the big operators and some permitting new entrants into the market. Demand for Internet service has been dragging network evolution forward, rather than being stimulated by the introduction of new network features. It is perhaps a model for the discovery process and contrasts with what has happened, for example, with the development of ISDN. In the nature of the process, discoveries are surprises. Entrepreneurs are spurred to look for them by the expectation of extra profit more certainly than are public servants by following the rules and incentives that are set to motivate them.

Markets are an efficient way of solving most resource allocation problems and usually produce better results than any which could be foreseen by administrators. Markets may fail to get the best results where there are significant gains or losses to third parties, or where continuing economies of scale lead to monopoly. These are called *market failures* and may be used as a justification for intervention in the market process.

6.2 Arenas for competition

Telecommunications competition takes place in two arenas:

- The capital market, where companies seek funds and have their stocks bought and sold. There may be bond issues, privatisations, joint ventures, mergers, acquisitions,
- The product market, where there is competition between networks, service providers and equipment manufacturers.

The principal factors limiting effective competition in telecommunications have been the tendency for the industry to be dominated by a few large networks for technological reasons and, in many countries, a government belief that monopoly can best safeguard various national interests. In the US, monopoly resulted from patent protection.

6.3 Mergers

6.3.1 Capital market competition

Companies need to come to the capital markets for finance, and their shares are freely traded. This leads to pressure on their boards to improve their performance in terms of profits and dividends. A company which is seen to be operating inefficiently may be taken over and made to yield better value for the new shareholders. The pressure is only really effective, where the threat of takeover is real. Equipment suppliers are vulnerable. The boards of operating companies which are wholly or largely protected from foreign takeover may feel the threat rather less.

There are still many policy makers and economists who would deny that capital market competition is an effective force, at least when compared with competition in the product market. Two factors make it important:

(i)   Profit maximisation has, from the days of Adam Smith, been recognised as a creative force, the invisible hand which guides the market.

(ii) State-owned operators are essentially political rather than economic bodies. Their stock is not traded in capital markets, allowing them to pursue agendas other than profit maximisation if they wish.

Opportunities for capital market competition were set out by Littlechild [3] in the context of the UK water industry, where product market competition prospects were seen as more limited than those for telecommunications. His main points were:

(i)   The main incentive for efficiency must come from within the company, and be encouraged by regulator incentives.

(ii) The stock market sharpens efficiency by providing an immediate feedback on performance. Shareholders buy and sell on what they think of the company and its future profits.

(iii) The stock market influences control over resources. Efficient companies will command higher stock prices and be able to borrow on cheaper terms.

(iv) The stock market sharpens the market for managerial talent.

(v)   The takeover bid is the ultimate check on efficiency and provides the shareholder's major source of protection against bad management.

(vi) Takeover is a more effective form of capital market competition than is franchising.

6.3.2 The market for corporate control

Manne [4] wrote of capital markets in terms of the market for corporate control to emphasise the importance of potential takeover as a spur to management. Two of the seven regional operators created by demerger from AT&T in 1984 have since merged with other companies. The UK electricity industry was privatised as a set of regional monopolies and many of the companies have been taken over by other utility groups. Takeover has been a real threat (and opportunity) in Britain's fragmented water industry and several companies have been merged.

Bids and mergers call for clear thinking and an understanding of the odds. They are difficult to implement, requiring good management plans, an opportunity for synergy and the development of comparative advantage or a stronger market position if they are to succeed. Cultural differences between the companies can constitute a serious obstacle. A study of mergers in industries outside telecommunications has shown that these tend to be unsuccessful except where there is a clear basis of common interest. Takeovers usually have the following effects:

- Board members and senior managers of the firm taken over may lose their jobs;
- they cost jobs in both firms through rationalisation.
- too much is paid for the company being taken over;
- Shareholders of the firm taken over benefit from the excess payment;
- Shareholders of the firm taking over get little out of the deal.

There have been several unsuccessful takeovers in telecommunications. Examples are IBM/Rolm and AT&T/Olivetti, where mergers based on the supposed convergence of technology did not result in a market advantage. IBM's development of a telecommunications strategy included two attempts to get into PBX manufacture, first negotiating unsuccessfully with Mitel and then in 1983 buying a 19 per cent share in Rolm. IBM took an interest in the networking operations of MCI in 1987 and there were other attempts to link its computing strengths with the networking capability of a major operator. The Rolm and MCI interests were sold off in 1987 and little came of the other ventures, which failed for a mixture of technical, marketing and regulatory reasons. The history of the strategy is charted in Mansell [5]. IBM has built up a strong position in computer networking. In the case of cable TV and telephony mergers, there have been legal impediments arising from the market power which might accrue to those in the merger.

Unpromising though these precedents may be, capital markets still have a role to play in compensating for lack of competition in product markets. They do, however, indicate the importance of synergy and the creation of competitive advantage.

6.4 Perfect competition

A market in which there are so many buyers and sellers that no single firm or individual can influence the market price is known as a *perfect* market. Perfect markets are extremely rare. What makes them important is the that they result in optimal resource allocation, which is a concept widely used in regulatory analysis (Chapter 11) and welfare economics (Chapter 14). It is therefore useful to know how resources would be allocated in such markets. Perfect competition can only exist with four conditions fulfilled:

(i) Exhaustion of economies of scale

All firms have exhausted their economies of scale and face level or rising marginal costs as output is expanded. Marginal costs are then equal to or above average cost. Under such conditions goods will be sold at the average cost of the marginal firm, which will make normal profits.

(ii) Numerous suppliers

Demand is so large, in relation to the cost structure of the firm, that although every firm expands its output up to the point where its marginal costs are minimised (and equal to average cost), no company can supply more than a trivial part of the market. Demand is met by the combined output of many firms. Some of them will be more efficient or have natural advantages over others, so that at any particular level of supply, with its associated market-clearing price, there is at least one marginal firm which makes normal profits. These are just sufficient to meet the cost of servicing the capital employed by the firm. Marginal firms would eventually go out of business if the market price was any lower, taking profits below the normal level, and would make an abnormal profit if it was higher.

(iii) No suppliers can influence prices

All sellers are *price takers*, so called because whatever their own prices are, they can have no effect on market prices as a whole. They take their prices from the market because the demand curve faced by each individual supplier is horizontal, with infinite price elasticity. No goods offered at a price above the market price will be sold. An offer price at or below the market price will result in a demand larger than that which an individual firm can supply. Output cannot be expanded to meet this extra demand without raising average costs. No supplier has a market position strong enough to create a significant shortage and hence a rise in price, by restricting production.

(iv) Perfect knowledge

All the buyers have complete knowledge of all the prices and all sellers possess complete knowledge the production possibilities. Markets do not have to be perfect to be the most efficient available way of allocating resources, they have to be better than the other means which are available. The knowledge inherent in the market will usually be better than that available to outsiders, including the government.

The market demand and supply curves in a perfect market are illustrated in Figure 6.1, where equilibrium occurs at price P and quantity Q and the marginal firm just covers its average costs.

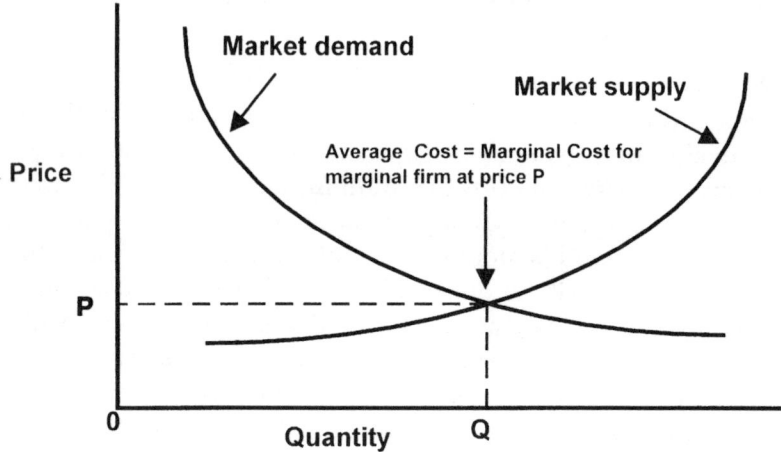

Figure 6.1 Perfect competition

No single individual is likely to have perfect knowledge. The major stock exchanges and foreign exchange markets are close to being perfect, which is why it is so hard to second guess them. Street markets are similar, but rather less perfect. Shoppers get the best prices by going round all the stalls if they have time. The price tags change during the day, moving towards market-clearing prices as the closing time approaches. Variations between stalls of fruit and vegetables provide the buyer with a range of alternative qualities and prices with no two offerings quite the same, a situation not unlike that faced by large business users when choosing a telecommunications supplier. In telecommunications, which has for many years undergone rapid market and technology change, the assumption of perfect knowledge is clearly inadequate.

In fully competitive product markets the prices that emerge from the market process have four important features:

(i)   The resources used to produce the current output are minimised, maximising *productive efficiency*, because competition drives down costs.

(ii) The prices equal marginal cost, maximising *allocative efficiency*. No other mix of outputs from the same resources is of greater value to consumers.

(iii) The output is distributed in such a way that consumers, given their incomes and the prevailing prices, would not wish to spend their incomes in any other way, maximising *distributional efficiency*.

(iv) No individual supplier or customer can change the prevailing market price.

Maximum economic efficiency requires maximum productive, allocative and distributional efficiencies. The importance of perfect market concepts in telecommunications is not that such markets exist, but that they define economic efficiency in a way which can be used as a test of the efficiency of other markets.

6.5 Imperfect competition

6.5.1 Market structures

Perfect competition will fail if marginal costs continue to fall up to and beyond the level of output which can satisfy the needs of the whole market. A single firm could then produce at an average cost lower than that of any combination of smaller firms and, sooner or later, cut prices sufficiently to drive them out of business. A market which has a single supplier is a monopoly. If there are two it is a duopoly. If there is more than one supplier but at least one has the ability influence prices it is called monopolistic, with the one having the most control over prices referred to as the dominant supplier. There may be several large suppliers and no others, in which case it is an oligopoly. A market with a single buyer is a monopsony. If there is a single supplier and a single buyer there is a bilateral monopoly.

All these cases are examples of imperfect competition. The basic theory of such markets goes back to a classic text by Joan Robinson [6]. Market prices are influenced by the behaviour of individual suppliers, none of which face a horizontal demand curve. They can raise the market price by restricting output, and it falls if more is offered for sale, as shown in Figure 6.2.

Suppliers in imperfectly competitive markets will therefore face declining marginal revenue (MR) as sales increase. For example, if a reduction in price from £100 to £99.99 increases demand from 1000 to

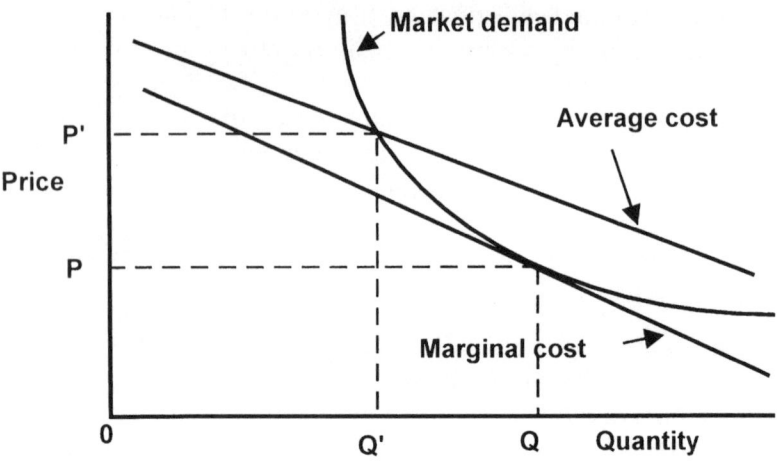

Figure 6.2 Imperfect competition

1001 then the marginal revenue is £(99.99 × 1001 − 100 × 1000) = £89.99. Profits are maximised when sales expand to the point where the marginal revenue from one more unit sold equals the marginal cost of producing that unit. This price will be above average cost if marginal costs are falling, the condition which usually gives rise to monopolistic supply. Abnormal profits are then made and output is restricted to a level lower than that which would match price with average cost, as shown in Figure 6.3. The monopolist would recover average costs by charging price P′ and selling quantity Q′, but make more profit by charging higher price P″ and selling the lower quantity Q″.

  The ability to influence market prices is called market power and may spring from regulation through the grant of exclusive rights or limited market entry. Large telecommunications operators with statutory monopolies experience inelastic demand in virtually all their markets. If they need more revenue they can get it by charging more, unless prevented by regulation. It is difficult to generalise about how many competitors are needed to reduce market power to trivial levels. In many product markets suppliers are able to strengthen their market position by product differentiation or brand imaging. Operators may attempt to reduce the substitutability of competing services by a variety of devices based on product differentiation, the encouragement of customer loyalty or the imposition of cost barriers such as excessive charges for interconnect. The need to change the telephone number when changing service provider is another example.

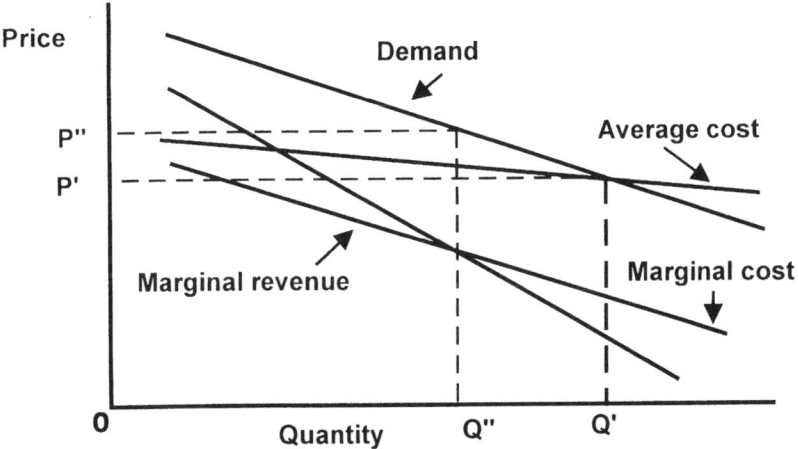

Figure 6.3 Profit maximisation under imperfect competition

Monopolies may arise out of market processes. As Littlechild [7] points out, market imperfection is the natural state of the world. The first entrant into any new market is a monopolist. A monopoly may also be established by legislation or decree in any market, whether or not it might arise naturally without the law. Markets with dominant suppliers usually exhibit leader–follower behaviour in their prices. The first move to a price change is set by the dominant firm, as price leader. Other firms follow suit, often maintaining the same discount against the price set by the dominant firm. They are price takers.

6.5.2 Duopoly

Duopolists may have roughly equal shares of the market and compete strongly for new business, yet still make high profits, as can be seen in some cellular networks. The existence of high profits does not necessarily stem from deliberate collusion (although they may do so tacitly, see Section 6.9). Each competitor may be able to edge market prices up by restricting supply, and may also be aware that competitors will have the same ability. There is no need for firms to get together and agree a course of action. Duopolies may not always remain in place naturally, they usually have to be protected by legal measures if they are to survive. Working models of duopolies include cellular mobile networks in several countries before wider liberalisation, international services from the UK during the period 1984–94 and, in the UK electricity industry, the generators PowerGen and National Power.

6.5.3 Monopsony

A single buyer can push down prices if faced with many suppliers, but faces a supply curve in which the price will rise if more is bought. The main cases in telecommunications are:

(a)   Operators facing effective competition in their own product markets, and buying labour or equipment specified in such a way that there are no other buyers for it.

(b)   Monopolists buying equipment specified in such a way that there are no other buyers for it.

(c)   Trade unions enforcing a closed shop which makes them the sole supplier of labour to an operator.

In case (a) the operator, as sole buyer, may be able to reduce the price paid for labour by hiring fewer people. The price paid for equipment may be reduced by reducing demand, if it is produced under increasing returns to scale. If competition prevents prices from being raised, there will be a reduction in sales without any offset from higher market prices. Profits are increased if the reduction in procurement and other production costs exceeds the lost sales, which cannot be guaranteed. A more likely outcome is that market forces lead to more flexible procurement.

6.5.4 Bilateral monopoly

Cases (b) and (c) above are examples of bilateral monopoly. In case (b) the monopolist will probably be able to raise prices by restricting supply. Profit is then maximised by restricting input purchases to depress their price, and by restricting sales and raising output prices to the point where marginal revenue equals marginal cost. Firms which are not the sole buyer may still have enough market power, as buyers, to impose tight terms of business on a market with many suppliers and few buyers.

Bilateral monopolies were common in telecommunications before market liberalisation. Monopoly PTTs would purchase from a single national equipment supplier or a cartel. Research and development contracts were placed in the same way. The arrangements for handling international traffic usually involved bilateral agreements between national monopolies. They covered such matters as the rate to be paid for handling inward traffic and market-sharing arrangements for outward traffic. A monopolistic buyer may share enhanced profits with its supplier by restricting output, raising the market price and agreeing an inflated transaction price for sales from one to the other. This is more or less what happened before the advent of international competition for services and equipment supply, except that some of the proceeds from higher market prices were lost in production inefficiencies.

In case (*c*) the most likely outcome is for wage rates to be pushed up with little effect on employment, the excess wage costs being passed on to the customer in the form of higher prices. The exact settlement price between the monopolists depends on their bargaining strengths. Samuelson [8] uses the theory to show that in bargaining between an employer and a fully unionised work force, the employer may wish to hold down wages, reducing the number of people seeking employment. Bargaining may then increase both employment and wages up to the market-clearing wage rate at the expense of profits, without increasing market prices. This has some resemblance to wage bargaining in telecommunications monopolies, but differs in that the monopoly employer can usually pass excess wage costs on to the customer.

6.6 Entry and exit barriers

6.6.1 Types of barrier

The ability of a supplier to influence market prices in the long or short run can be increased and maintained by barriers which stop new suppliers from entering the market or prevent those already in it from leaving. Some entry and exit barriers in telecommunications arise from technical and political considerations. These include:

- licensing;
- scale economies in network provision;
- scale economies in billing;
- inertia, reinforced by greater market awareness of the existence of the largest firm;
- reluctance of foreign operators to open correspondent relations for international traffic or to offer proportionate returns of profitable inward traffic;
- refusal to exchange traffic with competing networks on reasonable terms.

The telecommunications service markets in all countries have these features to some degree, limiting room for effective competition.

Entry opportunities change with the shifting legal and technological environment.

6.6.2 Barriers erected by dominant operators

Other barriers are erected by dominant operators to discourage market entry by competitors, most commonly involving a reluctance to exchange traffic with competing networks on reasonable and nondiscriminatory terms. Note, however, that it is not normal commercial practice to share facilities with competitors. Dominant carriers are acting no differently to companies in more competitive markets when they refuse.

Entry barriers which may be encountered by a long-distance carrier wanting to use a local monopolist for call delivery are grouped below by terms of business, volume economies and sunk costs:

(i)   Terms of business
  • interconnect refused, or available only at discriminatory rates, not based on cost.
  • interconnect elements bundled, i.e. the carrier has to buy the elements of local service that are not wanted.

(ii) Operating conditions
  • quality not as good as that enjoyed by the local operator for its own purposes, and interconnect provided at inconvenient points in the network;
  • local-loop transmission not possible by direct connection, but only through the local switch (the connection of two networks on the customer side of an exchange allows wider choice of routing, including the possibility of bypassing the local and trunk exchanges of the incumbent operator).

(iii) Sunk costs (unrecoverable entry costs)
  • nondiscriminatory access to poles, ducts and rights of way not available (the sharing of ducts and poles may allow a competitor to share in the scale economies of their owner, depending on what is charged for access);
  • colocation of competing plant within the same local exchange not allowed (this would enable more of the incumbent's network to be bypassed and sunk costs shared);
  • no non-discriminatory access to installation, repair, maintenance, billing and directory enquiry systems;
  • no nondiscriminatory access to emergency numbers (e.g. 999 in the UK, 911 in the US);

- no listing in the local telephone directories;
- discriminatory access to equipment needed for call routing and completion;
- no number portability (NP), so that the telephone number has to be changed when a customer moves from one local or mobile operator to another;
- Services not available for resale.

Most of these barriers have equivalents for a carrier wishing to enter the long-distance market, and there are others. The most important is dialling parity or *equal access*, whereby it is just as simple for a customer of the new entrant to dial into the competitor's long-distance network as it is for the competitor's own local customers. Early interconnect systems required special and sometimes lengthy prefixes to long-distance dialling codes, making manual dialling more time consuming and error prone. Tardiff [9] used consumer research to quantify some of these barriers for carriers in Japan, where NTT was the only carrier to provide local and long-distance services and the customers of new entrants had to dial an extra four digits. He concluded that:

- companies discounted against NTT by three to ten per cent because they lacked ubiquity and were less well known;
- dialling an extra four digits equates to a price disadvantage of about four per cent;
- delays in connecting new customers reduced the value of service by a further 2.4 per cent for each week of delay.

6.6.3 First-mover advantages in directory services

Subsidiary operations such as the production of directory enquiry (DQ) information have entry barriers which may be severe. The first entrant into a market such as advertising in classified business directories may quickly build up such a substantial product through brand image, scale economies in printing and attractiveness to advertisers that no competitor can get established. This is an example of a first-mover advantage.

Barriers which protect other forms of DQ service include the scale economies in building up a database and the printing of directories, the possible unwillingness of a dominant operator to pass information about customers to a competitor and distribution costs. Scope economies may be less powerful. Separate systems can and do exist for fixed, mobile and fax users. Distribution systems using books, CD-ROMs and online services to a large extent serve different markets, so it is access to the core database, rather than inherent economies, which may keep them together with a dominant operator.

6.7 Natural monopoly

An industry is a natural monopoly if a single supplier could meet all the demands of the market at a lower price than multiple suppliers could, without exhausting its economies of scope and scale. The technical basis is subadditivity, discussed in Chapter 5 and illustrated by Figure 5.3. The supplier would experience marginal costs which were below average costs on and beyond the production level at which market demand is satisfied at prices equal to average cost. Provision by a single supplier would then be cheaper than if the market was shared between several suppliers with the same supply curves. Although the mathematics of this can be demonstrated, there are difficulties in proving its existence with real firms. The potential of competition to improve, reduce costs, creating dynamic efficiency and moving the efficiency frontier, is outside the subadditivity formula but has to be taken into account.

Most utility monopolies are created by the grant of exclusive rights. They may be natural but they are evidently not so natural that legal support can be dispensed with. Fear of destructive competition, with several operators beggaring each other by competitive price cuts and being unable to fund their networks, so that prices to the consumer ultimately rise, is used to justify protection. The consumer's dilemma is that although a natural monopoly may theoretically have lower costs, these may not be experienced because:

- Monopolists tend to restrict output and raise prices to enhance profit;
- There is less pressure for cost control;
- The control of monopoly generates regulatory and management costs of its own.

Economies of scale are strong in local public and private distribution services. Gas, water, electricity, telecommunications, cable television and the delivery of such things as letters, newspapers and milk share the common characteristic that the average distance between delivery points is reduced when the number of delivery points is increased within a given area. Where physical visits are made, as with letters, the number of houses to be called at rises less quickly than the number of letters to be delivered, because some of the extra letters will be for houses which are already to be visited. In all these cases two operators working at the same level of productive efficiency will use more resources than a single operator.

The position changes when dynamic efficiency effects are included. There is practical experience available from the US. American towns are sometimes served by competing electricity or cable companies which have networks that overlap, so that customers have a choice between two suppliers, a situation virtually unknown in most of Europe. Studies have shown that in both electricity and cable TV distribution the prices to users tend to be lower in the cities with competing systems.

The entry of cable companies into telephony is now well established in the UK. Some of the companies have experienced difficulties with billing but, on the whole, they have provided a cheaper telephone service than the incumbent BT. To some extent this is a consequence of their being new, and forced to offer discounts against BT in order to get business. Levin and Meisel [10] give the results of a comparative study of 27 American areas with competitive cable TV service and 20 with local cable TV monopolies. It was found that customers of the competitive cable companies paid between $2.94 and $3.33 per month less for service and got more channels for their money. Levin cites other studies where it has been found that competition with radio-based (off-air) broadcasting lowers cable TV prices by up to 16 per cent. These results suggest that competition, rather than lower costs, may explain pricing of CATV telephony in the UK, but this is unimportant to the customers who are benefiting. It does not add credence to the idea that either the prices or the costs would be lower with a unified monopoly because the spur provided by competition would then be missing, offsetting the fuller exploitation of economies of scope. Further out, the interests of the customer would only be prejudiced if destructive competition drove all but one of the competitors out of the market and raised prices to a level higher than if competition had never existed, which seems unlikely.

There may be scope for a cable operator to offset losses on telephony by enhanced profits from TV operations, but the US experience suggests that competition from off-air systems, including direct broadcasting by satellite (DBS) services, would limit the scope for this.

Scale economies may be reached relatively early in long-distance transmission, leaving the residual cost advantage from a larger system as relatively small compared with possible dynamic efficiency gains. For terrestrial networks there are lifecycle effects which Noam [11] has analysed as *network tipping*. The network passes through three stages:

(i)   Initially it is a cost-sharing network in which the value of the network externality and the build-up of critical mass favour a single, unified structure.

(ii)   When the network continues to grow, internal coalitions begin to benefit some users at the expense of others, through cross-subsidy, and it loses optimality for some users.

(iii)   In the third stage, the internal dynamics, weakening scale economies and conflicts of interest cause it to tip over to a pluralistic structure, a network of interconnected but separately-owned networks.

Satellite networks have substantial fixed costs, but market entry has been so widespread for national and regional purposes, that the main benefit of scale economies must be reached well before market demand is saturated. The proliferation of global Sat-PCS systems (Section 3.7.3) suggests that here, too, there is room for competing networks in spite of the large fixed costs.

The evidence from empirical studies is summarised in Table 6.1, in which low-density, low-traffic networks are intended to include those in developing countries and the rural areas of more developed nations. In Noam's terms they would mostly be in stage 1.

Table 6.1 draws partly on the work of Kruse [12], who suggested a form of the classification and mentions Australian studies showing that, although mobile networks are subadditive within a technology, there is a net gain from competitive supply. The dynamic effects are those which arise from the competitive supply of infrastructure. The existence of a natural monopoly does not settle the question of who operates it. An inefficient competitor may need barriers to entry to prevent a more efficient competitor from taking over the market and the owners may need protection from a corporate takeover. The case for a natural monopoly of the infrastructure does not of itself imply a natural monopoly for the services which it carries. Competitive supply of services over a single infrastructure is possible by resale and other arrangements. This is examined further in later chapters.

Table 6.1 Natural monopoly and dynamic efficiency

Network type	Subadditivity	Dynamic efficiency effects stronger than subadditivity
Economies of scale		
Local, high density:		
urban residential	strong	about the same
urban business	weak	yes
Local, low density:		
rural	strong	possibly, with wireless
Long-distance:		
low traffic	moderate	unlikely
high traffic	weak	yes
Cellular mobile	weak	yes
Paging and		
other mobile	weak	yes
Satellite networks	weak	yes
Economies of scope		
Local telephony:		
+ adjacent areas	weak	yes, with high density
+ long-distance	weak	yes
+ cable TV	strong	yes in urban areas uncertain elsewhere
Network:		
+ equipment sale	none	yes
+ manufacture	weak	yes

6.8 Contestable markets

Work on the implications of market entry conditions initiated by Baumol *et al.* [13] and referred to generically as contestability theory has questioned the firmness of the link between monopoly and market power. The theory argues that, under certain cost and price conditions, a monopoly may be the most cost-effective way of providing services and that its potential inefficiencies and proneness to raise prices will be kept in check by the threat of entry rather than actual entry which breaks the monopoly. The essence of the theory is that, with certain types of cost structure, there is a price at which the monopolist can make a profit but no new entrant can do so. The monopoly is then sustainable. Market entry will occur if the price is raised, pushing the price back to the sustainable level. With a multiproduct firm, there are theoretical conditions where the prices required for sustainability are also those required for maximum economic efficiency, a remarkable result for the circumstances to which it applies, see Chapter 8.

In a contestable market there is a monopoly but are no barriers to entry or exit. Competitors can come and go without incurring significant costs by so doing and they all have access to the same technology. The monopolist cannot stop them from contesting the monopoly, or drive them out by temporarily cutting prices, because there are no exit costs. They will come back again if prices are raised. There may or may not exist economies of both scale (increasing sales volume) and scope (increasing product range), but there are no sunk costs.

Research was initially concerned with the conditions under which a natural monopoly could be sustained, with no entrant being able to undersell it and survive. At first it was hard to find any real industries where these conditions exist. A practical requirement was that the average cost curve should be substantially flat in the area where prices where likely to be set. This implied that output volume was large enough for scale economies not to be a powerful driver of cost, or at least that is was large enough for the major average cost reductions to have been achieved.

The first success claimed for the theory was derived from the analysis of city-pair airline markets by Bailey [14]. The key condition was that the major capital item, the aeroplane, could be readily switched between routes without significant entry, exit or sunk costs. Contestable market theory has some plausibility within telecommunications services because:

- scale economies are reached earlier by digital systems, reducing the scale economy advantage of incumbents;
- sunk costs can be avoided by the resale of capacity owned by the incumbent operator;
- technological change is making the equipment of incumbents obsolete.

Of these, resale is probably the most important. It applies even to the local loop, where it would be possible for many service providers to use the same infrastructure, as proposed by Harper [15]. The network would be run as an integrated monopoly but provide no services, and all services would be provided by operators competing freely over the same single network. This would allow plant subadditivity to be exploited in circumstances where there is no firm subadditivity (Chapter 5). Regulation capable of enforcing the provision of adequate service levels and fair prices would be needed. Enforcement might be assisted by franchising and the threat of market entry. The conflict of interest between the provision of service by a unified network operator and the desire of competitors to use the operator's network to deliver or forward their traffic by interconnection would be removed. The network operator would have an incentive to interconnect with any carrier willing to offer traffic. There would remain the question of line rentals, which might rise to an unacceptable level without competition from the cable operator.

In the UK, where the sharing of the local loop is already happening, or planned, in the retail distribution of gas and electricity, the operation may look familiar.

Contestability theory was developed by Baumol and Bailey at New York University and Panzar and Willig at Bell Laboratories [13] at a time when AT&T was trying to protect its monopoly, and the convenience of its use for this purpose has lead to some suspicion about the general case. There are, however, points at which the theory provides helpful insights. Real proof of its applicability would require it to be shown that, because of the threat of entry, monopolistic operators markets were behaving in the most cost-effective way. On this the evidence tends to be negative:

(a)   AT&T, BT and other operators faced with real competition have reacted to competition by cutting costs and prices. Input and output prices have been more closely aligned with cost.

(b)   The Swedish operator Televerket was unchallenged in its market until the 1990s despite the absence of legal entry restrictions. Most observers think that other barriers to entry may have been important. Market entry occurred in the mid 1990s.

The threat of entry has had observable effects on the behaviour of operators which previously did not face competition. There was a radical change to British Telecom's internal planning horizon. Long-term planning based on the expected trend of its own costs, prices and revenues was changed after the realisation that the costs of the new competitor, Mercury, were probably below its own. BT reached this conclusion after examining the likely costs of constructing a new, modern network and comparing them with the cost of the unmodernised network it was then operating. The planners used

Mercury's costs as a basis for future targeting and set about producing a more competitive cost structure using modernisation plans that were being developed as a framework.

6.9 Game theory

The behaviour of participants in imperfect markets has been explored as part of the theory of games. The duopoly case mentioned above can be presented as a 2 × 2 profit–payoff matrix, as shown in the example in Figure 6.4.

	Firm A's prices	
Firm B's prices	£2	£1
£2	Firm A profit = 6 Firm B profit = 6	Firm A profit = 9 Firm B profit = –2
£1	Firm A profit = –2 Firm B profit = 9	Firm A profit = 1 Firm A profit = 1

Figure 6.4 Duopoly

The duopolists each have a choice of the same two prices. The four cells in the figure show the profits made by each in the four cases. Both make profits of 6 at a common price of £2 but A knows that B could make a higher profit by a price cut to £1, pushing itself (A) into a loss of 2. B faces the same alternatives. The bottom right cell, where each charges £1 and makes a profit of 1 is stable, and may still result in A and B both making abnormal profits.

Although this example is oversimplified, it shows that duopolists with equal market power will realise that strategies to maintain a price differential will lead to a downward price spiral and are thus unstable. The downward progress would ultimately be halted if it hit the cost floor of the firm with the highest costs. This firm will be least able to support the price cuts and, with lower prices, would go out of business. Among telecommunications service operators this would usually be the smaller firm, because it has less fully exploited the possible economies of scale. The eventual result would be a monopoly. A more likely conclusion, in markets where government policy encourages the growth of competition, is that the dominant operator would conclude that an action which threatened to drive all competitors out of business would provoke a legal backlash preventing it from doing so. Where one duopolist has more market power than the other, there may be a stable position where the less powerful company sells at a discount which is acceptable to the other. For a further discussion see Reference 8.

One tactic would be to drive prices down to the level of the competitor's cost, leaving the competitor as the marginal operator and the dominant firm to enjoy lower costs (from scale economies) while still being able to charge higher prices (from the discount margin). The efficiency of the dominant firm is critical to the success of such a tactic. This form of market behaviour has been used to describe the rivalry of AT&T and MCI in the 1980s. It may not initially have fitted that between BT and Mercury, since BT's network was less modern and possibly of higher cost.

6.10 Measures of market power

6.10.1 Statistical measures

Market power exists where the market is a monopoly, a duopoly (two suppliers) or an oligopoly (few suppliers). It is difficult to generalise about how many competitors are needed to reduce market power to trivial levels. Two is certainly not enough. Analysts distinguish between measures of concentration among suppliers and measures of their ability to influence prices.

6.10.2 Measures of market concentration

The Herfindahl–Hirschman index (HHI) is a measure of market concentration; It uses the sum of the squares of the percentage market shares of the producers. A monopoly has the maximum HHI of $100 \times 100 = 10\,000$. The *entropy* index is defined as $\sum i(Si \times \log(1/Si))$ where Si is the market share of the ith firm. This is similar to the HHI but has different weights.

6.10.3 Measures of the power to influence prices

The Lerner index [16] is the proportionate difference between price and marginal cost, defined as $L = (P-MC)/P$, where P is the price and MC the marginal cost. The index has a value of zero under conditions of perfect competition, and it has no upper limit. A rough measure of the strength of market power is the price elasticity experienced by the firm. If demand faced by an individual firm is inelastic (i.e. elasticity less than unity in absolute value) the seller can always increase revenue by raising prices. Market power then exists. This can usually only happen when the seller has a large market share.

If the price elasticity faced by an individual firm is exactly −1, the revenue of the firm is unchanged by alterations to its selling price since the proportionate volume and price changes are the same. A price increase then reduces demand, and hence production costs, so that profit increases.

If the volume response experienced by the firm is larger than the price change, then the demand is said to be elastic. In the extreme case, where the firm loses all its revenue by raising prices, it has no market power.

6.11 Legal definitions of dominance

In Australia the Trade Practices Act prohibits mergers that would '*dominate a substantial market for goods or services*' or substantially strengthen a position that is already dominant; dominance itself is not defined in the Act. The Australian courts rejected a definition of dominance based on economics in the case of TPC *v* Arnotts Ltd & Ors, but developed a definition of dominance based on:

- the ability of a firm to behave to an appreciable extent independently of competitors, suppliers and customers;
- a high degree of market power;
- a commanding influence falling short of control;
- the ability to prevent effective competition or influence the conditions under which it takes place.

Background is given in Reference 16, where the Australian market for domestic and international services is analysed, paying particular attention to the weakening of market power brought about by the introduction of competition.

In the UK for the purposes of the Fair Trading Act 1973, a monopoly situation exists if at least 25 per cent of the supply is controlled by a single person or firm, or by a combination of firms (a complex monopoly). Such situations may be referred to the *Monopolies and Mergers Commission* (MMC) for investigation. A 25 per cent market-share test has a minimum HHI value of just over $25^2 = 625$. To achieve this minimum the remaining 75 per cent of the market has to be shared roughly equally between a very large number of other producers.

In the US the FCC has, since 1980, classified telecommunications operators as either dominant or nondominant, basing its definition on the presence or absence of market power. Stricter regulation is applied to those which are dominant. Control of a bottle-neck resource and the ability to make abnormal profits were among the tests applied. A bottle-neck resource is a facility such as a local line which is held by a monopolist and must be used by all those providing service to the premises. Those found to be dominant include AT&T (for both long distance and local operations), around 1500 other local service franchise holders and the domestic satellite operators. All foreign-owned carriers operating from the US were classed as dominant for all services on foreign routes. A foreign holding of 15 per cent or more was defined as ownership. Haring and Levitz [17] point out that market power does not only depend on control of production capacity. In 1988 AT&T owned only 40 per cent of PSTN long-distance capacity and on that basis was not thought to be dominant, but it was still considered to have enough market power to warrant control of its prices.

The US Department of Justice *Merger Guidelines*, issued by its Anti-trust Division in 1984 as an aid to merger policy under the Sherman Anti-trust Act of 1890 and the Clayton Act of 1914, contained two tests:

(i)   Can a '*small but significant non-transitory*' increase in price be imposed profitably?
(ii) Is the value of the Herfindahl-Hirshman Index for the industry unduly high?

The first requires that the price elasticity is below −1 in absolute value so that revenue rises as prices rise.

HHI values above 1000 were regarded as at least moderately concentrated in the *Merger Guidelines*, raising the possibility of a challenge to a merger which would increase the index further.

6.12 Cross-subsidy

6.12.1 Definition

If the operator sells a product at less than marginal cost, a loss will be incurred on the sale. This may be carried by the owners of the enterprise, who may wish to retain the product for noncommercial reasons, or it may be covered by overcharging for other products if market power permits. The latter is a cross-subsidy. Cross-subsidies from monopoly to competitive market sectors are anti-competitive.

They discourage efficient competitors from entering the subsidised market and they risk a misallocation of resources. State-owned enterprises often keep cross-subsidies in place for political reasons or to satisfy social welfare considerations, and they may wish to act anticompetitively. Profit-maximising operators may pursue several strategies. They may wish to cease making a loss as quickly as possible, while retaining the profitability of the other services. Operators that are subject to profit controls may see the cross-subsidy as something to be retained to satisfy the politicians in exchange for favours in other directions. Some may see a longer-term profit advantage from the discouragement of competitors. Regulators usually try to detect and eliminate cross-subsidies which are thought to operate against the public interest, but they may remain sympathetic to those believed to perform some social improvement.

6.12.2 Tests for the existence of cross-subsidy

Several tests for cross-subsidy have been proposed. The subsidy-free test was devised by Faulhaber [18]:

> 'A set of prices by a multi-product monopolist is subsidy-free (or free of cross-subsidies) if revenues at these prices cover total costs and if no subset of services produced by the firm could be produced at costs lower than the revenues generated for the subset by these prices'

No product or groups of products can produce profits large enough to support losses made by other products if this test is satisfied.

The stand-alone cost (SAC) test, is a special form of the subsidy-free test. It is the long-run cost of providing a service or subset of services on its own. Prices that generate revenue in excess of this are an indication of market power; those which are below it suggest predatory intent. A common situation is where the price covers marginal cost but makes a less than average contribution to overheads. This happens extensively within the market for local calls. It is not, strictly speaking, a cross-subsidy, although it may not earn a large enough margin to pay a proportionate share of the overheads. Any price which lies between the marginal cost of the operator (including the cost of capital) and the stand-alone cost of a new entrant will have no effect on the market structure of the industry and may have only second-order effects on resource allocation. This is because the potential new entrant will not be able start business and survive at a price less than the stand-alone cost and so will commit no fresh resources. The use, by an incumbent with economies of scope, of tariffs a little below SAC may generate abnormal profits without stimulating market entry, resulting a less than optimal resource allocation overall.

The incremental-cost test is satisfied where no subset of services produces revenue which is less than its incremental cost. The incremental cost is, here, the difference between total costs with and without the subset of services. If there are no diseconomies of scope the revenues from subsidy-free prices are never below incremental costs. If diseconomies of scope do exist they are an indication of inefficient production. It would be cheaper to produce the services on a stand-alone basis.

There are two problems associated with these tests. One is that individual services may pass the test even if the same services do not pass it in combination because of joint costs within the service group. These drop out only when the entire service group is taken out, creating incremental costs for the group which come to more than the sum of the incremental costs of the individual services. The other is that services may be substitutable, so that if one is withdrawn the demand for another goes up. These complications are addressed by the burden test (net revenue test) [19]:

> 'A price p_i for the product i constitutes no burden upon the consumers of other products supplied by the same firm if at that price the product's incremental cost is equalled or exceeded by its net incremental revenue (i.e. its [net] revenue after substraction of [net] revenue losses on other products j resulting from the cross elasticity of demand between i and j)'

The test has been used to decide whether prices are anticompetitive.

All these tests are in terms of the costs of services. Following Mitchell and Vogelsang [20], consumer subsidy-free prices can be defined as prevailing when:

> '... total revenues at demanded quantities cover total cost and ... no coalition of consumers could produce their demanded quantities at lower costs than what they pay under these prices'

This condition would be met, even with cross-subsidised service prices, e.g. access from call revenues, if all consumers bought the same mix of services. A stronger condition is that, while total cost equals total revenue, no consumption bundle of any single consumer or actual group of consumers can be produced for less than the revenue raised from the bundle. Tariffs based on the marginal cost of the bundles of individual customers have long been used in electricity pricing [21], although their introduction into telecommunications is more recent.

The various definitions require care in use in the context of telecommunications, but they have a place in the analysis of predatory price behaviour if the stand-alone cost is taken as the cost which a new entrant would face to provide the same service or services. For a further discussion of predatory prices see Reference 17.

The analysis of bypass problems, in which a competitor uses plant of its own to reach customers without using the local loop of the incumbent operator, makes use of the approach in these tests. Bypass is efficient if the incremental increase in cost to the bypasser is less than the reduction in cost to the operator bypassed. Inefficient bypass occurs when the competitor spends more on bypass equipment than the incumbent saves. Prices well above cost are likely to encourage inefficient bypass where entry is permitted.

6.13 Looking forward

This chapter has reviewed many features of the markets and suggested some ways in which they may be analysed. In terms of the Huber analysis of Section 1.9, there is much support for the view that networks are becoming increasingly pleuralistic, with competition, generally rather imperfect, becoming established in many areas of previous monopoly. There is no evidence that this is putting up costs. The arguments for natural monopoly have weakened, although they still retain some power in local networks and in developing countries. The separation of network operation from the services which are run over it, offers a possible way forward. Big operators still dominate most of the markets, however, and regulatory intervention is needed to keep them as open as possible. In terms of the discovery process, it is difficult for regulators to keep up with developments in the markets, which are highly innovative and growing rapidly, making deregulation the preferred alternative wherever practicable.

6.14 References

1   SMITH, A.: 'The wealth of nations, book IV' (1776)
2   KIRZNER, I. M.: 'How markets work'. IEA Hobart Paper 133, Institute of Economic Affairs, London, 1997
3   LITTLECHILD, S. C.: 'Economic regulation of privatised water authorities' (HMSO, London, 1986)
4   MANNE, H. G.: 'Mergers and the market for corporate control', *J. Political Economy*, 1965, **73**, pp.693–706 *in* FAIRBURN and KAY, (Eds.): 'Mergers and merger policy' (Oxford University Press, 1988)
5   MANSELL, R.: 'The new telecommunications' (Sage, London, 1993)
6   ROBINSON, J.: 'Economics of imperfect competition' (Macmillan, London, 1933)
7   LITTLECHILD, S. C.: 'Elements of telecommunications economics' (IEE Telecommunications Series 7, Peter Peregrinus, Stevenage, 1979)
8   SAMUELSON, P. A.: 'Economics' (McGraw-Hill, New York, 1976, 10th edn.)
9   TARDIFF, T. J.: 'Effects of presubscription and other attributes on long-distance carrier choice', *Inf. Econ. Policy*, 1995, **7**, (4), pp. 353–366
10 LEVIN, S. L., and MEISAL, J. B.: 'Cable television and competition: theory, evidence and policy', *Telecommun. Policy*, 1991, **15**, (6), pp.519–528
11 NOAM, E. M.: 'Network tipping and the tragedy of the common network' in ANTONLLI, A. (Ed.): 'The economics of information networks' (North-Holland, Amsterdam, 1992)
12 KRUSE, J.: 'Institutional options for east european telecommunications policy' in SCHENK K-E., KRUSE, J. and MÜLLER, J. (Eds.): 'Telecommunications take-off in transition countries' (Avebury, Aldershot, 1997)
13 BAUMOL, W. J., PANZAR, J. C. and WILLIG, R. D.: 'Contestable markets and the theory of industry structure' (Harcourt Brace Jovanovich, New York, 1982)
14 BAILEY, E. E., and PANZAR, J. C.: 'The contestability of airline markets during the transition to deregulation', *Journal of Law and Contemporary Problems*, Winter 1981, **44**, pp. 125–145
15 HARPER, J. M.: 'Telecommunications policy and management' (Pinter, London, 1989)
16 'The Australian telecommunications market, when does dominance cease?' (Bureau of Transport and Communications Economics, Canberra, 1992, working paper 6)
17 HARING, J. and LEVITZ, K.: 'What makes the dominant firm dominant?' OPP working paper 25, FCC, Washington D.C., 1989
19 BAUMOL, W. J.: 'Superfairness' (MIT Press, Cambridge MA. 1986)
20 FAULHABER, G. R.: 'Cross-subsidization: pricing in public enterprises', *American Economic Review*, 1975, **65**, pp. 966–77
21 MITCHELL, B. M. and VOGELSANG, I.: 'Telecommunications pricing. theory and practice' (Cambridge University Press, Cambridge, 1991)
22 CREW, M.: 'Electricity tariffs' in TURVEY, R. (Ed.): 'Public enterprise' (Penguin Books, Harmondsworth, 1968)

Chapter 7
Cost structures in telecommunications operations

7.1 Network optimality under conditions of change

If a telecommunications network were to be built from scratch at its ultimate size, the full benefit of economies of scale and scope could be obtained. In a dynamic world, however, demand grows over time and networks grow incrementally.

There is a second level of minimum cost which derives from the making of provision for future growth in a cost-minimising way. Although this, too, is impossible to achieve because a perfect knowledge of relevant parts of the future would be called for, it points the way to a type of optimality in which costs are minimised given the best available information about likely future events and it is this that network builders strive to achieve. Berrie [1] gives an account of a similar process in the planning of electricity generation capacity.

A factor limiting the cost minimisation of networks is the difficulty of forecasting demand and technological change. Telecommunications assets are not always long lasting, but there is ample room, in the fifteen-year life of a telephone exchange or the ten-year life of a satellite, for a cheaper technology to evolve. Additional calls are catered for mainly by the provision of extra plant or, as is often the case, the reconfiguration of existing plant to provide more capacity. For example, optical-fibre cables, which use digital transmission, make room for more lines in the duct network.

No operator will have perfect foresight, but some will make better guesses than others. City networks may be immensely complicated and far from minimum cost, reflecting the combined effects of incremental growth and the difficulty of forecasting. Telecommunications networks are always suboptimal compared with what might be built for a mature market with static technology.

Future-proofing may improve the cost characteristics of networks if they can allow for change. Analogue exchanges have been efficiently modernised by the installation of electronic control packages. New pieces of plant at either end of a cable can enable the same line to carry many times more information by increasing transmission capacity by using improved multiplexing methods. This can be done with optical fibre and with copper. The use of DSL technologies in the copper pairs of the local loop, described in Chapter 2, is an example. International and satellite services have been among the biggest beneficiaries of this type of improvement.

7.2 Operational cost determinants

7.2.1 Billing

Operational factors that are known to have a measurable and direct effect on costs are sometimes called cost drivers. One such factor is billing, which accounts for a significant element of total costs. Quigley [2] quotes a figure of five to ten per cent of infrastructure costs. For many years from the late 1970s, about 15 per cent of British Telecom's capital expenditure was on computing and office equipment of which much went on billing and customer support systems [3].

7.2.2 System density

The geographic density of the network has an effect on costs. Scattered rural areas are relatively expensive to serve because the average number of wire miles per line is higher and the economies of scale from the traffic volumes on individual routes are usually low. It is mainly capital costs that are affected. As Table 7.1 shows, labour productivity can be high even in scattered areas. In 1992 Austria, Spain, Sweden and Finland had fewer than 42 lines per square km but higher labour productivity than Belgium, West Germany and the UK with more than double the system density. Labour productivity figures are affected by the degree to which the operator uses labour as an alternative to buying in goods and services. Examples of activities which may be involved are catering, motor fleet management and the installation of plant.

Table 7.1 Density and labour productivity, Europe 1992

	Area in square km	Lines per square km	Lines per employee
Netherlands	41 863	177	212
West Germany	248 577	142	153
Belgium	30 518	140	159
UK	244 100	107	153
Switzerland	41 293	101	196
Italy	301 225	79	262
Denmark	43 077	70	168
France	547 026	55	193
Austria	83 853	41	189
Portugal	92 389	34	140
Spain	504 782	27	185
Ireland	70 284	16	86
Sweden	449 964	13	174
Finland	338 127	8	170
Norway	323 878	7	153

(Source: calculated from ITU statistics)

The countries in Table 7.1 are all in the top world income group. Looking across all income groups, low labour productivity is correlated with low penetration. There is a clear relationship which is shown rather broadly in Figure 7.1. Since wage rates are lower in the lower-income countries, labour costs per line will show more uniformity.

The ITU [4] has commented on substantial and not fully explained differences between the incremental costs of providing service in developing countries. Cost was measured as total investment divided by line growth, and as these are all countries with low penetration, most investment is for network expansion rather than for modernisation or replacement. Investment per extra line in 1992 varied from $1200 in Eastern Europe through a developing country average of $1500 to $2900 in Latin America and $6200 in Sub-Saharan Africa. The cost in India was more than twice as high as that in China.

Maitland [5] found large differences between urban and rural capital costs. The figures, which represent infrastructure alone, excluding labour and other running costs, are given in Table 7.2.

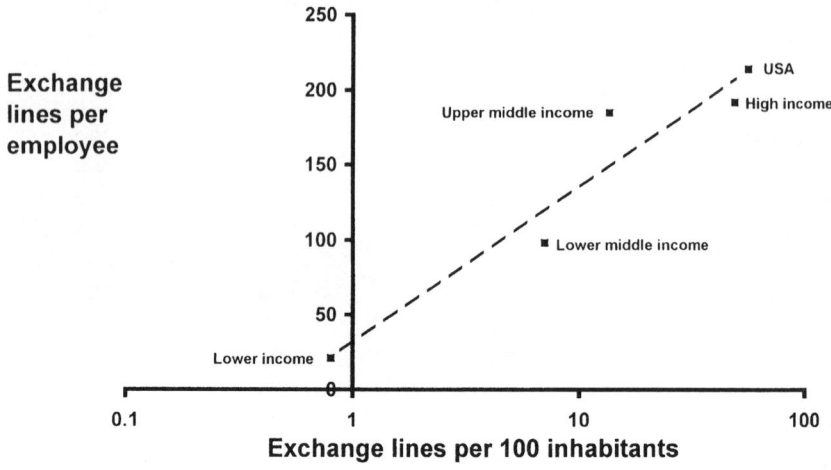

Figure 7.1 Labour productivity, penetration and income. World income groups and US

Table 7.2 Indicative cost per line

	Urban areas	Rural areas
Subscriber connection	40	400
Switching	30	60
Transmission	20	20
Land and buildings	10	20
Total	100	500

(Index: total urban = 100)

Rather different results were obtained for labour productivity in the regions of Sweden, shown in Table 7.3, in which the connection with density is less marked. It would appear, therefore, that it is mostly the capital costs which make scattered areas more expensive to serve.

Table 7.3 Labour productivity in Swedish regions

	Lines per sq km	Lines per employee
Stockholm	251.0	157.1
Göteborg	114.0	152.5
Malmo	83.5	175.2
Helsingborg	40.9	210.0
Vasteras	30.1	239.1
Uddevalla	27.1	177.8
Orebro	21.9	154.9
Uppsala	21.8	172.0
Nomkoping	21.6	192.4
Boras	21.4	199.2
Kristlanstad	20.1	229.9
Kalmar	18.1	180.7
Jonkoping	16.3	193.1
Gavle	10.9	167.4
Karistad	10.4	160.2
Sundsvall	7.2	134.7
Falun	6.3	171.1
Umea	2.3	161.0
Lulea	2.0	134.9
Ostersund	1.9	114.7
Total	13.9	169.2

(Source: Televerket)

7.2.3 Daily traffic profile

The number of telephone calls varies hour by hour in a systematic way, and there are very few at night. In business areas there is a daytime peak, probably with a dip at lunchtime, and the peak may occur in the evening in residential areas. Figure 7.2 gives an example for an exchange serving a financial district and shows how the amount of equipment needed is determined by the height of the daytime peak. Morning and afternoon peaks occur in business hours; dealings with the US create a substantial evening load. The Figure does not show how the peak load varies from day to day, but this has to be taken into account in system planning, as does the provision of a margin for expected future demand growth.

A different presentation of the same information is shown in Figure 7.3, where the load curve is sorted by frequency, with an indication of how demand is met. A major network operator would install plant to meet the peak load, within acceptable grade-of-service limits. An increase in the peak load requires more plant. A new operator may rely heavily on temporary capacity to meet peaks, for example lines leased from another operator. This is potentially more expensive, but may be the most cost-effective way of handling traffic which is not part of the base load. The highest part of the peak may be treated as overflow, handed over to another operator with spare capacity at a higher rate. Higher tariffs are usually charged for traffic in peak periods; this issue is discussed further in Chapter 9.

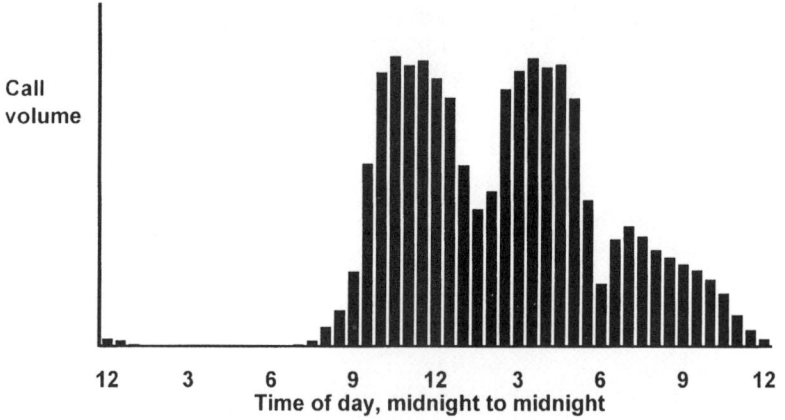

Figure 7.2 Business day call profile for a financial district

The modification to the load curve created by an extra customer or customer group is illustrated in the Figure 7.3 by a shaded area. In this example, the extra demand, like all demand below the peak, raises the load where there is spare capacity, thereby not increasing the capacity cost of the system above that which is already required to handle the calls from existing customers. Figure 7.3 does not include noncapital costs, such as maintenance and any fees paid to other operators for handling overflow traffic, so there may be an increase in costs overall.

7.2.4 Directories and directory enquiries

OFTEL [6] contains estimates of the cost of supplying 19 million telephone directories to BT's customers. The directories cost £1.18 each to print and another 30p each to deliver, including a contribution to overheads. Significant economies of scale were found in print runs up to 100 000 with small further economies up to two

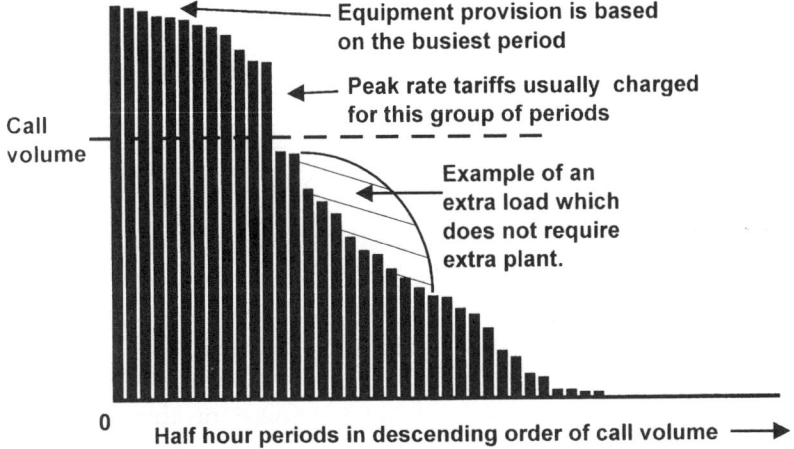

Equipment provision is based on the busiest period

Peak rate tariffs usually charged for this group of periods

Example of an extra load which does not require extra plant.

Call volume

0

Half hour periods in descending order of call volume ⟶

Figure 7.3 Daily load sorted by volume

million, beyond which scale economies appeared to be exhausted. Individual calls to international directory enquiry services cost an average of £1.33 to handle, although inland enquiries, being more fully automated, cost substantially less.

7.3 Cost and network design

7.3.1 Cost characteristics of mobile networks

Mobile networks are modular in form and are extended by the addition of extra base stations as demand requires. Marginal provision costs are not constant. A small increase in demand may require a large increase in the number of base stations in a congested area, as the customers may be anywhere.

Each base station can handle a fixed number of circuits, which are not dedicated to individual phones. The congestion level rises with the usage level in a way similar to the Erlang formula used to model congestion in telephone exchanges. The base stations and their associated network connections can be constructed as demand requires and without pronounced economies of scale. The number of customers per base station remains fairly stable.

Capital expenditure in the start-up phase for a British mobile operator in the 1990s was in the region of £250 000 per base station or £800 per connected customer. The average customer spent some £500 per annum on mobile telephony, compared with about £350 for the average PSTN user. A doubling of the number of base stations would require approximately twice as much capital expenditure. Marketing expenses are high — up to fifty per cent of revenues — and they are driven by the commissions paid to agents for signing up new customers and by promotional discounts on handsets. They are also generated by the high rate of churn experienced in the industry. This often approaches 30 per cent per annum, so that a mobile network operator has each year to recruit a cohort of new customers equal in strength to 30 per cent of its current customer base just to maintain the number of bill payers. The figures in this paragraph, drawn from a variety of industry estimates, are meant to be illustrative rather than precise.

The maturing of the network changes some of these features. Commissions to agents should represent a smaller proportion of expenses as the growth rate slows. The number of base stations should level off. Over 98 per cent of the UK population can be covered with 3000 stations, after which the need for extra base stations is determined mainly by infilling to relieve those which are overloaded, rather than by geographical expansion.

7.3.2 Radio links in the local loop

A wireless local loop (WLL) can be built up tower by tower, as demand requires. It is necessary to build the complete infrastructure in a new area to provide service, like a system based on ducts and cables. There is a considerable cash flow advantage against hard-wired systems because a system can operate profitably even with a low level of penetration. Duct and cable costs, including maintenance, are completely avoided and their absence makes the system physically more secure, a significant advantage for transmission over longer distances and in regions where theft or vandalism may be a problem. Terminal equipment in the home or office is more expensive because of the need to provide a base station or transceiver.

Systems being provided in the 1990s were estimated by industry sources as costing up to $1000 per line and sometimes more, a cost at least as great as that of copper but expected to fall to $500 or less quite quickly, at which point WLL would be fully competitive. A base station capable of serving 1000 lines might cost around $15,000. This is $15 per line if fully loaded, but it is more likely to have substantial spare capacity on it, especially in its early years. As much as 70 per cent of the capital cost may be in the transceivers and their handsets. None of these figures should be taken as precise, they are meant to illustrate the main features of the cost structure only. To arrive at end-to-end cost other elements have to be added, including local switching, long-distance conveyance and billing.

New entrants to the UK market have used radio to provide local services in cities. Radio has been used in the old East Germany, Hungary and the Czech Republic to provide a rapid extension of local service. Telefonica has used local radio links to extend universal service in rural Spain. British Telecom has replace fixed links with radio in remote parts of Scotland. China has used radio to provide local service. These examples are indications of the circumstances where the WLL is an appropriate economic choice.

7.3.3 Cable TV networks

Cable TV networks require a central facility known as the head end, cables to customers and a set-top-box to decode the signals on customer premises. The older cable TV networks of North America were mainly overground systems, with cable strung along poles or other overhead supports. More recent systems, notably those built in the UK following the issue of cable franchises in the 1980s, used buried cable for environmental reasons and were probably more expensive to construct. A 1982 report of the Information Technology Advisory Panel (ITAPS) [7] estimated the cost of providing cable service to a community of 100 000 people as in the region of £220 per house, made up of:

	£ per house
Head end and transmission	20
Cabling	50
Set-top box	150

The head end and street cabling represent fixed costs and the cost per house taking service depends on how many of the houses passed by the street cabling are connected. By the end of 1995 over £5 bn had been spent to build cable networks passing over six million houses [8] — a higher figure than might have been expected from the ITAPs estimate. In Italy, which was without cable TV at the time, a cost of Lire 13 trillion ($11.1 bn) was estimated in 1995 for the cost of networks to pass ten million homes [9]. This works out at about £700 ($1100) per home.

Given the costs and likely revenues, cable TV usually requires over 30 per cent of homes passed to take up service before it can be profitable. The high fixed costs encourage diversification to benefit from economies of scope, including pay TV, where customers pay extra for new films etc. and telephony.

Plant is not the largest current account cost. Depreciation accounted for only 20.7 % of operating expenses (excluding interest) in the Canadian cable industry in 1989 and the proportion was the same in 1994 [10]. In the UK, where the industry is much younger, depreciation was 30.9 % of operating expenses in the Telewest accounts for 1996. Although modernisation may have made capital costs more important in recent years, it is obvious that other areas of expense cannot be ignored.

The annual churn among cable customers is higher than that among the customers of PTOs. It is not uncommon for 25 per cent or more to leave during the course of a year, a similar rate to that in the cellular mobile market. In 1996, 33.4 % of the 528 000 UK residential customers taking cable TV from Telewest and 19.6 % the 627 000 who took residential telephony left their provider [11]. High churn rates will increase the labour-intensive marketing element of operating costs.

Things look rather different in cash flow terms. For example, Telewest's expenditure on fixed assets in 1996 was nearly double its total revenue and three times as large as gross trading profits [11]. By the end of 1995, Telewest had fixed assets of £1.1 bn, almost three times the mean revenue for 1995 and 1996. The company had 401 000 cable TV customers and 444 000 cable telephony customers, giving an average asset value of £1302 per customer if the two groups are counted separately (it is not known how far the two groups overlap). Rapid expansion of the network is the main reason for the high investment level. Cable operators in the UK have not expected a positive cash flow to appear before year five at the earliest and perhaps not until their networks are substantially complete, after up to ten years of operation.

7.3.4 Telephony via cable TV

The cost of providing a ubiquitous, interactive optical-fibre local network in the UK has been variously estimated at up to £20 bn (around £800 per line) and no single operator is willing to make the investment without being allowed to provide all the potential services that such a system permits.

Estimates of the value of the economies of scope derived from the joint provision of cable TV and telephony have usually shown them to be significant. Fletcher [12] assumed a 25 per cent cable TV penetration and put the cost of upgrading a cable TV network to telephony as $433 per line, compared with $1925 per line for a separate telephony network. The exact value of such economies depends on the specification of each system and for the present purpose is not particularly important.

Electrical integration, that is the use of the same cable to distribute telephone calls and cable TV, may have different, and possibly smaller, advantages than those from physical integration, represented by using the same duct for two separate cables. On the whole, British cable operators have stuck to noninteractive (one-way) coaxial cable for TV and laid separate copper pairs in the same duct for telephony, thereby gaining most of the economies of scope in the market sectors of interest to them, without much extra capital cost. Some of them may actually show diseconomies. One example is marketing, where the different natures of the product markets may create too large a learning task for the managers to absorb efficiently. In other directions, cable operators may exercise tighter cost control and they use a less massive infrastructure, for example in ducts.

The 1996 Telewest annual report shows that the company's telephony revenue (£159.6 m) exceeds that of cable television ((£121.2 m), so for the company telephony is hardly a marginal service. The average annual telephony revenue per residential customer was £243 (£22.95 per month), against £275 for the average residential cable television customer. The telephony revenue, at £68.85 per quarter, is well above the median BT residential figure of £45.42 quoted in Table 4.6 for 1993. A substantial part of the revenue (33 per cent in 1996) is paid to other operators for interconnect. The foregoing figures show that, if Telewest is taken as being representative of other UK cable telephony operations, the finances differ from those of the PTOs in costs, and particularly in the relative investment levels, more than they do in revenues. Telephony overheads are presumably similar to those for TV operation and inflated by the marketing efforts needed to cope with and offset churn. Fletcher's incremental cost figure of $433 per line only relates to capital costs.

These are all, at this stage, figures that relate to the early years of a market which has yet to mature, and in which a rise in penetration rates would increase capital utilisation. At the end of 1996 the Telewest residential penetration rates were 22.6 % of homes passed for TV and 27.5 % for telephony. The strategy of UK cable operators in taking a significant share of the residential telephony market suggests that if cable telephony is treated as an incremental cost, the operators think that telephony provision is profitable, perhaps more so than TV.

7.3.5 Satellites

Launch costs with their associated risks are a significant part of the cost of satellite communications. The cost of building and launching a single geostationary satellite in the mid 1990s was usually between $150 m and $350 m [13]. Press reports put the cost of an Ariane V launch failure in 1996 at $500 m, much of which represented the value of the scientific equipment on board.

Failures rates of one in every 15–20 launches, expected to fall to one in every 70 for a new generation of rockets, were quoted in the *Financial Times* (7 June 1996). Insurance rates for a major operator were quoted at around 18 per cent of insured value in 1995 and 14–15 per cent in 1996. The cost of providing $2 bn of cover for ten launches was reported to be about $185 m (*FT* 12 January 1995). Premium rates are volatile in a competitive insurance market. High insurance premiums may lead satellite operators to carry some of the launch risks themselves, especially where a noncommercial load is carried.

Satellite capital costs are large and lumpy but the capacity that they provide is remarkably cheap. With a working life of around ten years even a $300 m satellite will amortise at no more than $30 m a year. A modern satellite with digital-circuit multiplexing can provide at least 100 000 channels, bringing the cost per circuit down to below $1 a day. At a five per cent circuit occupancy, this is about one cent per minute. The cost of capital, back-up capacity, maintenance, management and other expenses has to be added but, even with this addition, it is clear that the cost of the satellite link is a minor part of the cost of an international phone call.

Many satellites combine PSTN and television broadcasting functions, suggesting that, to the extent that the two types of transmission are not fully interchangeable, there may be economies of scope.

7.3.6 Information services

Information services display large economies of scale because much of their costs go into building up and maintaining databases. The computing equipment needed to store and handle the information also shows economies of scale. Telecommunications networking is less important as a source of scale economies where, as is usually the case, access is over the PSTN or a leased line paid for by the customer. Where they do exist, economics may be a feature of the scale of operation of the customer, in having enough business to justify high-capacity access lines, rather than of the information provider. Geographical extension of the area served by the database, for example to another country, generates economies of scale within the database by increasing its usage. Economies of scope exist in the databases, but they are less marked than economies of scale. Alongside such large broadly-based companies as Compuserve there are many which specialise, serving the financial markets, law, journalism and other sectors and having an international clientele as Reuters, for example, does in providing financial information services.

An information service provider may act as a retail outlet for other organisations as well as developing databases of its own. Information service providers usually run at a loss for a lengthy period after entering the market because of the fixed costs and the need for heavy marketing in the early years. Some never get into profit before closing down. The typical industry structure is an oligopoly with a few large companies dominating the market, or market sector. Competition can be strong and with a constantly shifting technology and information content a big market share does not guarantee abnormal profit, or any profit at all.

7.3.7 Number portability

The introduction of number portability into the PSTN requires calls to be transferred from the operator supplying the original number to the operator currently providing service to the customer. This introduces extra costs. Following OFTEL [14] these are:

- network modification and training costs;
- the one-off cost of transferring a customer;
- additional traffic costs for the calls transferred, which have to be passed on to the new operator;
- customer costs for new or modified handsets (mobile only);
- caller costs, from set-up delays for transferred calls.

IN technology is not essential, but it might reduce cost. In 1995, BT estimated the cost of modification as £220 m for its fixed network [15]. Network modification and training costs were estimated at £8 m for UK mobile networks in Reference 14. Customer transfer costs were put at £19 per customer, becoming less with experience. Additional traffic cost per mobile users were put at 0.65p per minute.

7.4 Capital consumption

7.4.1 Asset valuation

The economic value of an asset is the discounted value of the future net income which can be earned from it. This may be some way removed from the cost of replacing the asset with a modern equivalent asset (MEA) and involves making guesses about:

- the cost of plant to be bought in future by itself or competitors;
- future maintenance costs for the asset;
- future volumes of services provided by it;
- future prices for those services.

There are strong elements of circularity about the calculation, notably in the price assumptions.

The MEA is a prudent basis for asset valuation. In a perfect market it would be the same as the economic value of the asset because:

(a) Customers will not buy if the price exceeds the value of the net income that the asset provides.

(b) If the price is less than the value of the net income, buyers will make an abnormal profit from using it until market forces drive down the price for the services derived from it.

Equipment markets are not perfect, but the MEA value can be a useful proxy for the economic value because it is less subjective and easier to determine. No assumptions about future asset or sales prices are required. Where asset prices are expected to fall, operators may use accelerated depreciation methods to recover more of an asset's cost in the early years of its life and give a rough adjustment in the right direction.

7.4.2 Depreciation: accounting for capital consumption

Capital is consumed through use, by wearing out or becoming obsolete. The loss in value during the year is capital consumption. Economic depreciation is the change in economic value of the asset during a period, usually also a year. It is the discounted value of future earnings from the asset, which is rarely known with any precision. Depreciation (amortisation) provisions are deductions from profit to represent the fall in asset value owing to capital consumption. Depreciation is also deducted from the book value of the assets. The simplest and commonest provision for capital consumption is straight line depreciation, in which equal amounts are allowed for each year of asset life, e.g. 20 % of the original purchase price each year for five years for an asset with a five-year life. Under conditions of competitive and technological uncertainty, straight-line depreciation may lead to an overstatement of asset values towards the end of the period. Two methods are commonly used to counter this: the easiest is to shorten the depreciation period, the second is to use some form of reducing-balance depreciation.

In its basic form, reducing balance depreciation is taken as a fixed proportion of the depreciated value of the asset at the beginning of the period. Thus, an asset costing £1000 and depreciated at a rate of 25 per cent would be depreciated by £250 in the first year, £(1000 –250) × 0.25 = £187.5 in the second year, etc. with some residual adjustment at the end of its working life. By way of comparison, for an asset with a ten-year life the cumulative total of depreciation with a 25 per cent rate amounts to 76 per cent of the original investment cost by the fifth year, compared with 50 per cent (five equal slices of ten per cent) for the straight-line method. By the tenth year 94 per cent is covered, compared with 100 per cent for straight-line depreciation. Although this is not full recovery, the front loading of the depreciation provisions with reducing balance would be worth paying for if the provisions could be converted into a cash flow by building them into prices. In this example it can be shown that, with an interest rate of 2.9 per cent, the two methods would provide equal value.

Average asset life is of interest for some purposes, e.g. in competition studies where new entrants may displace the plant of an incumbent, or to standardise the comparison of net assets in balance sheets when profits are being compared. An exact figure could be built up from the bottom as a weighted sum of the lives of all the assets but the information required is usually kept out of the public domain. The actual working life of an asset may differ from the asset life used for accounting, tax or regulatory purposes. Equipment may still be used long after its accounting life is ended, or it may be withdrawn from service prematurely to cut maintenance costs or improve service.

In the 1950s the average service life of many assets was shorter in the US than in the UK. Manual trunk exchanges, for example, had lives of 15 years against a UK figure of 30. Buildings were put at 43 years against a UK figure of 60. Operating companies will survey the history of plant in service to update estimates of average asset life. The combined effects of more rapid technological change and market restructuring have tended to shorten asset lives. Modern digital equipment is typically seen as likely to become obsolete relatively quickly compared with the earlier systems. Competitors accelerate the development of cheaper alternatives and it may be necessary for other operators to introduce these alternatives to control their own costs.

Depreciation provisions permit a rough estimate of average asset lives through the depreciation rate, which is the depreciation provision expressed as a percentage of the gross book value of the assets being depreciated, i.e. their original purchase price, before depreciation. The implied mean asset life can be calculated by assuming that the average asset is half way through its accounting life. Table 7.4 shows typical asset lives and depreciation rates used by european operators in 1995 and compares them with figures in use 20 and 40 years earlier where these are available. The figures are based on the accounting practice of operators in France, the UK and a major satellite group.

Table 7.4 Typical asset lives used for accounting, Europe

Asset class	1955	1975	1995
Life in years			
Freehold buildings	n/a	60	40
Cables and transmission	n/a	28	25
Exchange equipment	n/a	20	12
Computers and office equipment	n/a	7	4
Satellites – GE0	n/a	n/a	10–12
– LEO			5–6
BT depreciation rate (%)	6.2	8.8	13.3
Implied mean asset life (years)	32	23	15

Some operators are using shorter lives than those given in the table. Telefonica [17] used 5–15 years for transmission equipment and 8–25 years for domestic networks in 1955. North American monopoly operators used similar lives in the 1970s and have also tended to shorten them.

Undersea cables may have planned lives as long as 25 years, but their operational lives depend on a balance of maintenance against replacement costs. The first seven transatlantic voice cables were TAT–1, opened in 1956, up to TAT–7, laid in 1983. All had been taken out of service by 1996, replaced by optical-fibre cable with lower maintenance costs and greater capacity only thirteen years after the last was commissioned.

7.4.3 Current and historic cost accounting

The preparation of accounts which are based on actual cash flows in the money of the day is called historic cost accounting (HCA). Total depreciation provisions for an asset add up over time to the original cost of the investment, which is held in the asset base at a net book value, i.e. historic cost less the depreciation provisions. The annual return on capital is expressed as a percentage of the asset base at historic cost.

Correction for the effects of inflation may be made through current cost accounting (CCA). It is particularly important that depreciation provisions are maintained at a realistic level when there is general price inflation. If assets in the books are depreciated at their historic cost only, serious funding problems may arise when they are to be replaced. A better system is to value assets at their MEA replacement cost.

Depreciation provisions need to be augmented (or possibly reduced) by supplementary depreciation. This requires the original purchase price of each asset to be adjusted to its replacement cost and depreciated by straight line or other methods to a current value, from which depreciation in the current year can be calculated. The procedure may show losses or gains in real terms owing to the changes in real asset prices or uneconomic depreciation rates. These have to be taken into the balance sheet if the economic value of the assets is to be accurately shown. Backlog depreciation is a term sometimes used to describe this adjustment.

A full application of CCA requires many other adjustments, including changes to monetary asset values and liabilities, known as holding losses or gains, which affect operating profit. This is a large subject with its own literature, to which an introduction is given in Reference 18. CCA valuations provide the basis for a more realistic analysis of the true profitability of the enterprise. With full CCA a real rate of return on assets can be calculated, using cost profit and depreciation on CCA definitions and expressed as a percentage of current assets at replacement cost. Publicly-owned monopolies in Britain were, for many years, given real rate of return on asset figures as targets. The figures were related to the supposed cost of capital needed to finance their operations. The rationale was set out in two White Papers [19, 20]. Financial accounts, such as those published occasionally by BT and more regularly by British Gas, usually show a lower rate of return than the historic cost accounts generally published by companies.

CCA has theoretical attractions, but it is disliked by accountants because of the numerous subjective assumptions which it requires. Telephone companies have been known to write off assets which are still earning them an income and have experienced difficulty in finding a good method of revaluing assets to current value. As will be discussed later in this Chapter, investors are interested in the return on cash invested, rather than assets employed, and CCA may obscure this. The capital markets understand historic-cost accounts best, so this basis is the one which profit regulation often takes as a reference point. The case for some form of inflation accounting is strong but not over-whelming, and it is not widely used unless the rate of inflation is high.

7.5 Analysis of individual service costs

7.5.1 Methods of cost analysis

Cost analysis has proved to be important since privatisation, and the amount of effort put into getting good information about costs has increased. Operators often have poor information about the details of their costs on a product-by-product basis, far less than is needed in a privatised and commercialised environment. In a perfect market this would not matter much. Prices would be set by the market and operators with costs which were higher than the market prices could not survive. Those with lower costs could not affect prices, although they might make higher profits if they enjoyed special advantages which gave them economic rents. In the imperfect markets that characterise telecommunications, analysis of costs is the only effective way of finding a basis for prices, short of introducing fully effective competition. The prices which result have an effect on resource allocation and, where the law permits, on market entry by efficient competitors. The method of cost analysis is therefore important. The costing will usually be carried out using historic-cost accounting figures.

7.5.2 Principles of fully allocated costing for services

Where there are fixed costs, or economies of scope or scale, the incremental cost of adding an additional unit of output will be less than the average cost. If incremental costs are used to set prices they will be insufficient to recover the whole costs of the business and hence to permit the long-term survival of the firm. Accountants and regulators have sought ways of attributing the excess costs to individual products and service lines for the purpose of setting tariffs. Although all such systems are arbitrary in relation to fixed costs, in the sense that there is no real causal connection, some make more commercial sense to accountants and marketing managers and some seem more equitable to regulators. A marketing manager would probably want to mark up prices most in areas where the competition was weakest.

The result of these concerns has been the development of costing systems which allocate all costs, fixed or variable, to one or other of the product lines. Fully allocated cost (FAC) accounting usually distributes costs on a proportionate basis more influenced by ideas of fairness or equity than the way in which the competitive markets actually work. It may not be all that inaccurate. The name itself suggests its origin. Most industries talk about market opportunities for cost recovery; monopolists can allocate them. Fixed costs are often distributed in proportion to variable costs. The methodology has become highly developed in North America and latterly in the UK under regulatory pressure.

The five main elements of cost analysis are:

(i)   Distinguishing between fixed and variable costs. Fixed costs are treated as an overhead to be recovered by prices which incorporate a surcharge on variable costs and spread total costs over all the products, usually on the basis of FAC.

(ii) Analysing the operational structure of the business to identify activities for study. This is sometimes called activity-based costing and may be used as part of an operation to identify costs that can be eliminated.

(iii) Sample surveys to help split up large blocks of variable non-capital cost among individual activities,

(iv) The allocation of capital costs to a level as close as possible to final services. Capital costs such as local exchanges, which are shared between separately-priced elements of service, such as calls and access, are cascaded down to final services in a manner similar to that used for noncapital costs.

(vi) Assessment of capacity costs for use in peak/off peak charging schemes.

Fully allocated costs may be used in conjunction with a fully allocated asset base to produce financial results by service (FRBS) in which a return on capital is calculated for each service or product. A profit statement of this kind was published by British Telecom and its predecessors for many years. It is still sent to OFTEL annually and published by that Office from time to time.

7.5.3 Problems of cost allocation

The system can be controversial and even confusing to those who operate it. Billing is an example of a cost with large fixed elements. Basic billing costs are generated by the exchange line, which is the unit to which the bill is sent. If all calls were free, there would be no need to meter them for billing purposes and no billing cost attributable to them. It may be that a degree of metering would still be needed for network management and marketing purposes. The introduction of charging for calls is likely, however, to increase billing costs as a step increase followed by an marginal cost per call. There are substantial elements of short to medium term fixed costs, and these may be allocated by reference to variable costs or to relative revenues.

There are well recognised economic difficulties associated with the process of fully allocating costs to services, even where the practical problems involved with data collection can be solved:

(a)   The process is artificial where fixed costs exist.
(b)   Many costs are shared between services. Billing costs and the costs of the central processor in the exchange are examples of areas where there is an element of sharing.
(c)   Historic costs are unreliable where there is rapid technological change or inflation.
(d)   Cost allocation procedures are often complex and subjective.
(e)   Market distortions tend to be associated more with the costs of a new entrant than with the costs of the incumbent operator.
(f)   Telephone companies often have a poor understanding of their own costs, in terms of causation.

The stand-alone cost mentioned in the previous Chapter is estimated in relation to a notional company which provides a restricted range of services, using data drawn from the experience of real companies for its cost elements. Current or forward-looking methods are the most appropriate to use since SAC is used in the context of market entry.

7.5.4 Interconnection

Interconnection is not a simple service. The trunk operator may want interconnection points deep within the network, using large parts of the incumbent's capacity. Alternatively, interconnect points close to the customer may be required, bypassing more of the incumbent's switches and lines. The incumbent may have to install new capacity to handle traffic from the connected operator, and may need to know if the capacity is likely to be given up within a few years. Costing manuals have appeared and methodologies are becoming more formal. OFTEL [21] provides estimates of the cost to BT of the local conveyance elements in the year to March 1995. These are shown in Table 7.5. Depending on the point of interconnect, a call may pass through one or two tandem exchanges before reaching the local exchange.

Table 7.5 Incremental interconnection costs, BT 1994/5

Segment	Pence per minute
Local Switch	0.404
Conveyance to tandem	0.234
Single tandem segment	0.638 (0.404 + 0.234)
Conveyance between tandems	0.446
Two tandem segments	1.084 (0.638 + 0.446)

7.6 Cost trends

Trends in input costs from year to year have two components:

(i) Input prices per unit of electricity, optical fibre etc. purchased.
(ii) Technological change, reducing the volume of inputs needed or producing cheaper substitutes.

Real unit costs for two of the factors of production, land and labour, are on a static or rising trend. Building costs are relatively labour intensive with little scope for productivity improvement. These, and rents, also tend to rise in real terms, although erratically, and are offset by technological trends in equipment design. OFTEL [22] has forecast no change in the real UK prices of duct, land and buildings over the period 1996–2002, but has said that BT expects annual increases of up to two per cent. Most digital equipment is physically smaller than the equipment which it replaces, releasing space in buildings and ducts. Labour costs are one of the largest inputs for PTOs. Wage rates rise in real terms at an average rate which is close to the real growth in GDP per head. Prices for most other service inputs, such as audit, legal advice and property taxes also tend rise in real terms.

Significant real price-reduction trends are largely confined to the third factor, capital equipment. OFTEL [21] estimates of real UK equipment price trends are given in Table 7.6.

Table 7.6 Real average annual procurement price trends

Asset Type	1995/96–1999/2000 %	1999/2000–2001/2 %
Local Switches	–3.0	–10.0
Main Switches	–3.0	–10.0
Transmission cable	–4.5	–8.0
Transmission equipment	–2.0	–7.0
Network computers	–30.0	–30.0
All inland conveyance	–2.1	–5.4

(Source: OFTEL [21])

Within a particular technology, prices fall less quickly because the manufacturing inputs include labour and other elements which may be on a rising trend and scale economies in manufacture may have been exhausted. The downward trend expected is largely owing to step changes in technology.

The cost of a digital local exchange depends more on the number of local lines it serves and less on the number of calls it handles than does an analogue exchange, creating a higher proportion of non-traffic sensitive costs. OFTEL [21] estimates that only 27 per cent of the gross replacement cost (GRC) of digital local exchanges is traffic sensitive. Fault rates have fallen with the progress of modernisation. As a result, additional call traffic can be handled with fewer extra engineers. Digitalisation has reduced transmission costs; the longer the distance the stronger the reduction has been.

7.7 Cost of capital

7.7.1 Cost of capital and return on capital

Telecommunications networks use large amounts of capital. The annual debt service, dividend and depreciation charges may exceed the paybill, making a proper appreciation of their cost of great importance for the sound financing of operations.

The cost of the loans is the interest to be paid on them, the rate of interest depending on economic and market conditions and on the credit rating of the borrower. The existence of high or unstable inflation will increase the borrowing cost, as will actual or potential political instability. For companies which issue bonds, the cost of loan finance appears daily in the loan-stock columns of the financial press.

The cost of the issued share capital is the return to investors in the form of profits, including those kept within the company as well as those paid out in dividends. Retained profits enhance the value of shares and provide a capital gain to investors in lieu of a dividend. Dividends are paid from what is left of the profits after interest has been paid. The return expected by shareholders will be higher than the rate of interest paid on loans because it is the shareholder who carries the main risk.

The cost of capital is payment expected by the suppliers of the capital, in interest in the case of loans and dividends or capital growth for shareholders. The cost varies with the nature of the enterprise and is a weighted average of the separate costs of loan and equity capital. It is a forward-looking figure, reflecting the cost of new borrowing. A company will not engage in new investment unless its expected return on this investment exceeds the cost of the capital required. The expected return on any one decision is a function of the expected income stream and the original investment. In principle, the actual percentage rate of return from an individual asset could be worked out from the value of the stream of profits which it generates during its operational life. This information is almost never available because the asset is usually an extra piece of network equipment with no separately identifiable income stream. It is lumped into the corporate asset pool and the profits derived from it are pooled with all the other profits. The return actually achieved by the investment is as the internal rate of return (IRR) and almost certainly differs from the cost of capital.

Analysts may calculate an annual percentage return on the basis of total profits made and total capital employed as shown in the balance sheet each year. This is the return on capital (ROC) or rate of return, and it might differ from both the cost of capital and the IRR, because individual investment decisions involve an initial expenditure followed by a stream of benefits.

An ROC calculation takes a cross-section, at a single point in time, of all the cashflows resulting from past investment decisions. It is therefore in a different dimension to that of the original investment decisions that generated the cash flows, each of which was seen as creating a stream of income over future years. The ROC aggregates the tranches of benefit arising from all the past investments in one year and expresses the total as a percentage of the depreciated value of all past investments, usually on a historic cost basis.

The relationship between the ROC and the IRR is not precise and may be systematically different under conditions of growth or technological change. An alternative approach through CCA makes allowance for inflation. With this method of accounting the assets would be updated to a current replacement value. The rate of return obtained is a real return on capital, i.e. it is the ROC after allowing for inflation. The information in the ROC is now changed and some adjustments to profits may be required to allow for losses and gains from financial assets.

For example, suppose an investor wants to see an IRR, r % say, on a cash sum invested. This is sometimes called the required rate of return. The investor makes the same investment each year so that the new investment balances the depreciation, leaving the total assets unchanged at K. We suppose that a profit of P is made each year. With historic cost accounting the ROC would settle down at $100 \times (P/K) = r\%$. Suppose, though, that the assets were valued on an MEA basis and that the replacement cost falls by z % each year. This creates an annual holding loss of z % on the asset base. An achieved ROC on the revalued asset base would then need to be r % + z % if the rate of return on the cash sum invested is to remain at r %.

The annual rate of return is thus of uncertain value in monitoring the rate of return actually being made on past investments, in spite of its popularity with some analysts. This explains why financial markets make little use of the calculation, preferring to look at trends in the share price and the profits on a historic cost basis.

7.7.2 Market and specific risk

Risk can be analysed statistically as a variance round the expected rate of return, the larger the variance, the greater the risk. The market risk of an equity investment is measured as the covariance between movements in the investment return and movements in the stock market as a whole. If the two move together the covariance is positive. If, as may occasionally happen, they swing in opposite directions then the covariance is negative. Market risk cannot be avoided by diversification into more investments of a similar type. These will tend to swing in the same direction. The specific risk is the residual variance around the general market movement. Specific risk can, in principle, be reduced by diversification since it is uncorrelated with other risks. Some investments will do better and some will do worse; it does not affect the cost of capital.

7.7.3 Methods of estimating the cost of equity capital

For the stock market as a whole, the difference between the cost of risk-free borrowing and the average return on the market is the equity premium. In western countries it is usually between five and ten per cent.

There is extensive US literature on the cost of capital for public utilities, arising from the frequent use of ROC regulation for price control. OFTEL [22] is one of several OFTEL studies of the cost of capital which is faced by BT. It uses a weighted average of debt and shareholder capital and estimates the return required by shareholders on the basis of the market perception of BT's relative risk. OFWAT [23] presents a full exposition of the method and includes BT and US data. The cost of debt is observable from the market price of any loans or bonds which the operator has placed in the money market. The cost of equity is usually estimated with the capital asset pricing model (CAPM). With this model the cost is

$$\alpha + \beta e \qquad\qquad (7.1)$$

where

α is the cost of risk-free borrowing, usually taken as the percentage return on Treasury Bills

e is the equity premium

β is a sensitivity coefficient, known as the *beta* of the company

The equity premium is estimated from the percentage returns on stock and Treasury Bill data over extended time periods, up to 50 years in length. Over short periods its behaviour is erratic. The beta of a stock is the market risk, which can be estimated by regression analysis from observations of percentage changes in the company stock price and parallel percentage changes in the main stock market index.

Stable stocks which are not much affected by swings in share prices have low betas. Volatile stocks which swing more than the market does, have high betas. Telephone companies with monopolies are usually regarded as safe investments and their shareholders expect a comparatively low premium. Typical betas are:

0.25 to 0.75	dominant telecommunications operators
1	UK stock market
1.5 to 2	construction firms

If the model held exactly, the results for all firms would lie on a straight line and it would hold for the future. In practice, there is a lot of variation about the line, and the CAPM's accuracy is weakened by this and other factors. Studies of the cost of capital usually look backwards, albeit over long periods, whereas investors look forwards. The calculations do not allow for significant discontinuity between past and future.

There are several other methods of measuring the cost of capital. The arbitrage pricing theory (APT) is probably better but needs data which may be hard to find. It is an extended version of the CAPM, with provision for multiple sources of risk and return. The model for the cost of capital is of the form:

$$\alpha + \beta_1 x_1 + \ldots + \beta_n x_n \tag{7.2}$$

where

α is as in eqn. 7.1

x_i is the value of the ith factor

β_i is its coefficient

Coefficients are estimated, as before, by regression analysis.

APT tends to be used for utilities in the US and is described more fully in Reference 24. A fuller analysis of market and specific risk and the CAPM can be found in Reference 25.

7.7.4 Borrowing costs in imperfect markets

If capital markets were perfect, a firm would be indifferent about its capital structure, which is the mixture of various types of debt and equity in the balance sheet, for reasons given by Modigliani and Miller [26], discussed in Reference 25. The relative cheapness of a high proportion of the debt would be offset by an increase in the cost of equity capital (caused by the greater volatility and risk of profits after the payment of interest). Capital markets are not perfect, and prospective lenders take the debt/equity ratio into account when considering requests for loans. A company which is already carrying a high proportion of debt may not be able raise further loans at reasonable cost. The issue of further equity capital will be difficult if markets have doubts about the future financial viability of the enterprise.

What this means in practice is that companies tend to avoid raising external capital where they can, and finance a significant proportion of their new investment from depreciation provisions and retained earnings.

7.7.5 Internal and external finance

The proportion of investment financed internally is known as the self-financing ratio. The extent of self financing depends on the state of development of the business, especially the size of the investment programme in relation to turnover. In western countries it has been, during periods of rapid system growth, 50 per cent or less. In more mature systems, such as those in North America and the UK, the proportion of self financing may rise to 100 per cent.

The self-financing ratio appropriate to a particular operator depends on market conditions. In the imperfect capital markets which are usually found, the need to finance investment internally would be important in establishing a reasonable profit level, as well as the rate of return on investment. If capital markets were perfect, the rate of profit would reflect the cost of capital and new investment would reflect the future profit expected from it. The self financing would be of no importance because any extra funds that were needed could be borrowed at the established cost of capital, subject only to the caveat that a high level of borrowing might increase the cost of capital by increasing risk.

7.7.6 Dividend policy

Many companies believe that they can reduce uncertainty in capital markets, and hence the costs of borrowing, by giving reliable indications about their dividend policy, i.e. the dividends that can be expected from their operations. The belief is a reflection of the imperfection which exists in the markets and a common preference among investors for profits to be paid out rather than held within the company. Dividend policy would be irrelevant if capital markets were perfect, see Modigliani and Miller [26], summarised in Brealey and Myers [25].

Companies often try to keep the level of dividends on a relatively stable trend from year to year, where this can be done, rather than letting it reflect short-term fluctuations in profits. The intention is to indicate a dividend trend over the longer term, reducing perceived risk as long as all is well. A departure from the trend in dividend payments is meant to be taken by the markets as a signal of some underlying change and may be followed by a relatively sharp adjustment to the share price. This signalling works best for strong companies in monopolistic markets.

7.8 References

1   BERRIE, T. W.: 'The economics of system planning in bulk electricity supply', *Electr. Rev.*, **181**, September 1967 in TURVEY, R. (Ed.): 'Public enterprise' (Penguin Books, Harmondsworth, 1968)
2   QUIGLEY, P.: 'Permanence of patronage', *Communications International*, February 1995, p.46
3   British Telecom and Post Office: annual reports, financial results June 1989, GPO and British Telecom, London
4   ITU: 'World telecommunication report'. ITU, Geneva, 1994
5   MAITLAND, D. (Chairman): 'The missing link, report of the Independent Commission for WorldWide Telecommunications Development'. ITU, Geneva, 1984
6   'Provision of directory enquiry information services and products, consultative document'. (OFTEL, London, 1997)
7   Cabinet Office Information Technology Advisory Panel: 'Report on cable systems' (HMSO, London, 1982)
8   The Cable Companies Association: 'The case for cable', paper presented at *TMA conference* 28, Brighton, 1995
9   'Bell Atlantic picks partners in Italian new media minuet', *New Media Strategist*, June 1, 1995
10   Statistics Canada catalogue 56–205
11   Telewest annual report, 1996
12   FLETCHER, M.: 'Cable telephony: coming to the market', *Telecommunications*, September 1997
13   RIDDING, J.: 'Asian satellite market goes into orbit', *Financial Times*, 22 August 1997
14   'Economic evaluation of number portability in the UK mobile market'. (OFTEL, London, July 1997)
15   'Inquiry by the Monopolies and Mergers Commissions into number portability: explanatory note from the director general of telecommunications' (OFTEL, London, December 1995)
16   MORGAN, T. J.: 'Telecommunications economics', (Technicopy, Stonehouse, 1976, 2nd edn.)
17   Telefonica: 'Annual 1995 report', (Telefonica, Madrid)
18   CARSBERG, B. and LUMBY, S.: 'The evaluation of financial performance in the water industry: the role of current cost accounting' (Chartered Institute of Public Finance and Accountancy, London, 1983)
19   Treasury: 'Nationalised industries: a review of economic and financial objectives, Cmnd. 3437', (HMSO, London, 1967)
20   Treasury: 'The nationalised industries, Cmnd. 7131' (HMSO, London, 1978)
21   'Network charges from 1997, consultative document December 1996'. (OFTEL, London, 1997)
22   'BT's Cost of capital'. (OFTEL, London, 1992)
23   OFWAT: 'Cost of capital, a consultation paper' (Office of Water Services, Birmingham, 1991)
24   MUSSAVIAN, M.: 'An APT alternative to assessing risk', *Mastering Finance Part 10, Financial Times*, 27 May 1997, London
25   BREALEY, R. and MYERS, S.: 'Principles of corporate finance'. (McGraw Hill, London, 1984)
26   MODIGLIANI, F. and MILLER, M. H.: 'The cost of capital, corporation finance and the theory of investment', *American Economic Review*, 1958, **48**, pp. 261–297

Chapter 8
Welfare economics for telecommunications

8.1 Definitions and concepts

8.1.1 Welfare and the invisible hand

Adam Smith's 'invisible hand' was an early expression of the idea that resources are most efficiently allocated and the common wheal most effectively advanced when individuals are allowed to follow their own economic interests. Smith's use of the word 'frequently' does not imply 'always'. Welfare economics considers resource allocation at the level of society as a whole and investigates whether, at this level, a different allocation of resources would bring about a greater aggregate benefit. Since utility is a subjective measure, the total benefit itself cannot be measured. This does not matter as long as there are theoretical and practical ways of judging whether the total may be increased with alternative arrangements. Some approaches are more robust than others. What follows below is meant as an introduction to concepts that are widely used in the formulation of tele-communications policy. A more thorough presentation can be found in texts such as Doel and Velthoven [1]. Definitions are well covered in the ever useful *Penguin Dictionary of Economics* [2]. There are three basic approaches, Pareto optimality, the compensation principle and social welfare functions.

8.1.2 Public and private goods

A distinction is made between private goods, which are the purchases of private citizens, and public goods, which are purchased through taxation for public use. As shown in Chapter 4, the market demand curve for private goods is derived from those of individuals by adding them across on the horizontal axis, showing aggregate demand instead of individual demand at each price. If ten thousand people want to enjoy a private good that costs £1, they buy one each. This gives £1 on the price axis and 10 000 on the quantity axis. Public goods can be enjoyed by many users, each of whom would be willing to pay a different amount to use them, and who may actually pay a different amount through the tax system. Because the use is shared, more than one person derives utility from each purchase. Indeed, there may only be one purchase, shared by all. Doel and Velthoven [1] show a market demand curve for public goods which aggregates the total utility of all users on the vertical axis. If ten thousand people are willing to pay £1 each for a single shared facility, adding together the amounts that each individual would be prepared to pay gives £10 000 on the price axis for a quantity of one on the quantity axis.

The difference between private and public goods is less obvious in a dynamic market, or in one with practical resource constraints. The notion of aggregating the satisfaction of individuals through shared use implies that there is spare capacity for them to enjoy, which is often not the case. Many thousands of people can look at a single painting but not simultaneously. A bridge will eventually need replacement and long before that date it may, like the Severn Bridge, become so overloaded by a rising volume of traffic that an extra bridge has to be built. Telecommunications networks may have spare capacity, but all of their parts are eventually volume-sensitive.

There remains a difference in who pays. Public goods are paid for through the tax system and the purchasing decisions are taken by civil servants or politicians on behalf of consumers, rather than by consumers directly. The production of public goods may be undertaken by private or state-owned enterprises. The way in which allocative decisions are taken in the purchasing and production of goods and services by the state is a subject of public-choice theory (Section 5.2.4).

8.1.3 Pareto optimality

Pareto (1848–1923) was the first economist to show that market behaviour could be explained in terms of ordinal utility, or ranked preferences, without the need to attach absolute values to utilities obtained by individuals.

His definition of an increase in total welfare was a change which made at least one person better off without anyone being worse off. A situation where no person can be made better off without someone else being worse of is Pareto-optimal, and is a situation of maximum economic efficiency.

Pareto optimality is defined in terms of private consumption. It works by individuals each matching their own marginal utilities with market prices which, through the market process, are equal to marginal production costs. If public goods are to be included, and if they are paid for through taxation, there would have to be a way of matching collective marginal utilities against marginal costs, which is essentially a political process without market signals to guide it.

8.1.4 The compensation principle

A stronger approach is to define an increase in total welfare as occurring when, after the change, there are losers are as well as winners but the winners could collectively compensate the losers for their losses out of their winnings and still be better off than before the change. This is the compensation principle. It is not necessary for money transfers to actually take place. The compensation principle, unlike Pareto optimality, implies that the gains and losses of individuals can be measured on a common scale of value. It is therefore more demanding in its assumptions and less persuasive in its conclusions. As with the Pareto optimality from which it was derived, public goods are not readily included.

8.1.5 Social welfare functions

A social welfare function aggregates the preferences of individuals into one for society as a whole. Following Green [3] it can be written as

$$W = w(u_1 + u_2 \ldots + u_m) \tag{8.1}$$

where u_i is the total utility of the ith consumer out of the total of m consumers and w is some form of weighted average of the u_i.

Optimality can then be defined in productive and allocative terms. The problem with the welfare function, as with the compensation principle, is that is not obvious how to allow for distributional aspects. Public goods are included in the formulation. Total welfare depends on how goods and the incomes necessary to buy them are distributed between households.

8.2 Net social benefit

8.2.1 The components of net social benefit

The aggregate net benefit to society in terms of the excess of utility over production cost is the sum of three elements:

- Consumer's surplus
- Producer's surplus
- Net third-party benefits (externalities).

Net social benefit is a concept which derives from Pareto optimality with the addition of estimates of third-party benefits and costs.

8.2.2 Consumer's surplus

Most customers would be prepared to pay more than the actual price for something they buy. The difference between the value of the product to the user and the price paid for it is called the consumer's surplus. Only for the marginal customer is this zero; the concept is illustrated in Figure 8.1. The aggregate of consumer's surplus for all customers depends on price elasticity. Maximisation of customer value and maximisation of profit both depend on the way in which consumer's surplus and price elasticity vary between products.

Quantification of the consumer's surplus requires a knowledge of the demand curve all the way from the price at which demand is zero to the demand at the price level prevailing in the market. This knowledge never exists, although it may be estimated by extrapolation from known points on the curve. Economic analysis involving the

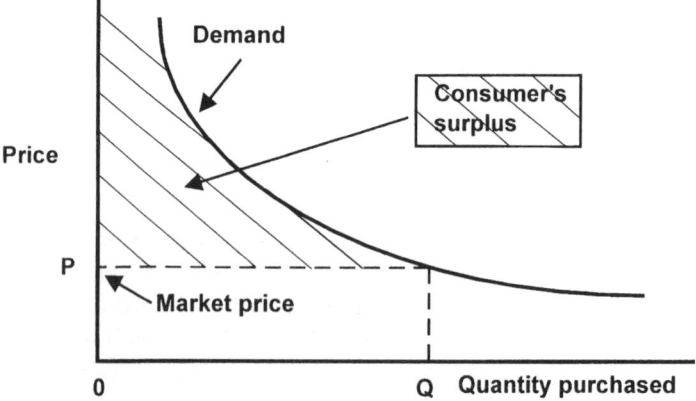

Figure 8.1 Consumer's surplus

consumer's surplus often concerns marginal changes in the surplus consequent on changes in price, rather than the absolute size of the surplus itself. Marginal analysis may gain robustness by requiring only a local knowledge of the shape and position of the demand curve over a price range where demand data exists, in the usual circumstances where demand continually rises or remains unchanged as price falls. This approach is based on revealed preference, a theory of demand developed by Samuelson [4] and reviewed in Green [3].

8.2.3 Producer's surplus

The difference between the cost of production and the gross revenue from sales is sometimes known as the producer's surplus. The cost includes the cost of capital. The marginal firm in a fully competitive market makes only normal profits and has no producer's surplus. Its trading profits are just enough to service debt and pay adequate returns to shareholders. More efficient firms, with lower costs but selling at the same price, will make an extra element of profit which is their producer's surplus. This surplus is a type of economic rent. Under conditions of imperfect competition all producers may be able to sell above their cost of production and generate a surplus which is the difference between the cost of capital and the abnormal profit actually made. Measurement of the surplus requires the calculation of the cost of capital to the firm, for which a methodology has been discussed, although there may be practical difficulties.

8.2.4 Externalities

There may be costs and benefits other than those directly contained in the production cost and the preference of the buyer. Examples of such costs are those suffered by third parties from transport congestion and atmospheric pollution. Congestion arising from high levels of usage and from the interference that can be caused by two radio networks using the same frequency in the same area are examples from telecommunications. Third-party losses of this kind are externalities, like the beneficial network and call externalities discussed in Chapter 4.

Externalities have been difficult to quantify in telecommunications. The network externality is important to business users. There may be relatively little further potential for growth in telephony but it operates powerfully in the rapid spread of fax machines, data networks and data services. When companies invest their own money in business networks they internalise part of the benefits. More may be internalised by collaborative arrangements with other companies and through co-operative sector network organisations such as SWIFT for banking and SITA for airlines. Even so, businesses with their own networks use the PSTN and benefit from the externality for many of their calls, often at least half. Only the on-net calls internalise it. The freephone service makes use of the call externality. The recipient of the call pays the cost on the strength of the value of sales which may made to the caller. Here the externality accrues to the caller, who makes a free call and derives value from it.

The amount, and to an extent the sociodemographic location of the additional consumer's surplus from a new line of an existing customer, depends on where the new line is. In principle, if an additional telephone line is opened in Clapham, Manhattan or Timbuktu, every other telephone user in the world gains a benefit by being able to make calls to it. However most calls are made within the local area, and the benefits are more local; relatively few cross national boundaries. It might appear from this that, for the most part, it is callers within a relatively short distance of the new line who benefit from the network externality. In practice other factors are at work. A new line in a remote area may attract as many calls as one in a town. It might attract more.

There is plenty of evidence that telephones attract calls, so that the total number of incoming calls, and hence the number of outgoing calls that are made, will rise if the network gets bigger. The calls attracted by a new line will not in most cases replace any that were previously made to other lines. Each new call will yield an element of consumer's surplus to the caller unless it is a marginal call with a utility which only just equals its cost. Measurements of a direct link with system size may get confounded with more general income effects which cause both system size and the calling rate to rise together, and analysts usually find that most models of the calling rate work well with macro-economic variables alone. System size has been used in international studies, where the explosive rate of traffic growth has been difficult to tie in with the more stable progression of the world economy.

The utility derived by the recipient of a telephone call is often taken as being subsumed in the price paid by the caller, and hence not an externality. This is a convenient, if imperfect, assumption which has been used to simplify the analysis.

8.2.5 Directory enquiry services

The value of the network externality is increased by the provision of telephone directories and online directory enquiry services (DQ) because these enable customers to make calls to people and organisations whose numbers they do not know. In 1984, French Minitel users alone made nearly 800 million enquiries, an average of 35 per line. Telemarketing has greatly increased the commercial demand for numbers. British Telecom was beginning to be swamped with calls before it introduced a charge for DQ in 1991. This, and the provision of information on CD-ROM to large users, reduced the demand, although it remained substantial. In the year to March 1996 the company, which had distributed 19 million directories, still received 605.5 million enquiries, an average of 22 per line [5].

The usefulness of these services is reduced by the reluctance of some customers to be included. In the UK the proportion of exdirectory residential customers rose from 24 per cent in 1991 to 37 per cent in 1997, and was as high as 56 per cent in London. Market research established that the avoidance of nuisance calls and telemarketing were the main reasons for wishing to be excluded (OFTEL [5]). In 1997, the directories available in the UK were far from complete in their coverage of other lines. Mobile telephones, fax lines and freephone, premium rate and helpline services were largely omitted. OFTEL research revealed a widespread demand for these numbers, although without reference to the extra cost which might be involved. The research did show the value attached to telephone directories by a sample of residential customers. Fifteen per cent of such customers said that they would choose not to have a phone book if their rental could be cut by £2 per annum without one. This is slightly more than the fully allocated cost of production and supply and suggests that directories are good value; there is, however, a case for unbundling them from the rental.

There is a kind of network externality associated with classified business directories such as BT's Yellow Pages. The more the directory is used by consumers, the more it becomes attractive to advertisers. This creates more entries, increasing the consumer interest [6].

8.2.6 Cost-benefit analysis

The quantitative measurement of externalities is usually attempted through cost-benefit analysis. Externalities are commonly found in the transport industry in the form of congestion and amenity costs. The case for a new road would include loss of amenity in the costs and set the total cost against the value of time, accident and other savings. Costs and benefits can be hard to measure even in terms of physical units. Translation into money requires subjective judgements about such factors as the value of a landscape, business and leisure time or a human life.

The surpluses of producers and consumers are measured in the same money and are implicitly assumed to be of equal weight in this calculation. There are circumstances in which this assumption is too simplistic, not making allowance for the role of firms as buyers of intermediate rather than final outputs. One of these circumstances is the optimisation of two-part tariffs, where large firms might obtain bigger discounts for the purchase of high volumes of telecom-munications services [7]. This might distort the market structure among firms competing in a perfectly competitive market, with possibly negative effects on the welfare of final buyers.

8.2.7 Commercial and personal discount rates

Founding assumptions of DCF calculations are that benefits are all re-investible and of equal discounted value to all parties. These may not hold up well in cost–benefit analysis. Harrison and Mackie [8] distinguish three cases:

(i)   Personal consumers have a personal time preference (PTP) interest rate for cash benefits, which is the minimum at which they will invest for future rather than present consumption

(ii) Government has a social time preference (STP) interest rate for cash benefits, reflecting its own priorities and borrowing costs.

(iii) There may not be investment opportunities available for benefits accruing to government or society as time savings or environmental improvements rather than cash benefits. A higher, social opportunity cost (SOC) interest rate may be needed to reflect this.

8.3 Pareto-optimal prices

8.3.1 Allocative efficiency

A price structure that encourages demand for services which are provided below cost and discourages demand for services priced well above cost is likely to misdirect demand and capital spending. Allocative efficiency requires that users pay the true cost of the resources which they use and that producers match demand and supply at the margin. The prices will be allocatively efficient in the sense that:

- The marginal cost of another unit of output is equal to the market price and hence to the value of that unit of output to the marginal buyer, that is the consumer who might enter or leave the market if there were a small change in price.
- The price does not encourage consumers to buy when the marginal production cost exceeds the value of the marginal output to the user
- The price does not deter marginal demand which could be satisfied at a cost lower than the price which the consumer is willing to pay.

Such prices are said to be efficient in the Pareto optimal sense. They are the product of market forces in fully competitive markets but not necessarily in other circumstances.

8.3.2 Compensation and social welfare considerations

Overall public welfare might be raised by prices which depart from Pareto optimality. Considerations include:

(a)   Industry dynamics: high initial profits may encourage more rapid service provision.
(b)   The existence of significant externalities.
(c)   Social engineering: income transfers between income groups may produce a net increase in wellbeing.
(d)   Marginal costs below average costs.

Evaluation of these factors can be difficult. Consideration *(a)* is partly an issue of the cost recovery pattern over the lifecycle of the product. The second and third are basically political issues. They may in some respects be more appropriately addressed through the tax and benefit system than through the manipulation of specific service prices.

As regards consideration *(b)*, the existence of the network externality has lead to claims that it justifies cross subsidy. For example, OFTEL [9] said that '*a case exists for holding exchange line rentals below the level at which they cover cost plus a reasonable return on capital*'. To sustain this argument, it would be necessary to prove that losses in allocative efficiency arising from overpricing calls and underpricing access in relation to cost is more than offset by the increase in third-party consumer's surplus arising from an extension of the network size that the underpricing of access creates. It would seem to require proof that more calls would be made within the framework of cross-subsidised prices, i.e. that the number of calls made to the new lines arising from the cross-subsidy plus the number of outgoing calls from the new lines would exceed the number of calls lost on existing lines because of the higher call prices. Resolution requires knowledge of the price elasticities of calls and access, as well as the incoming calling rate on the new lines. Complex problems of econometrics arise where, as is likely, cross-elasticities exist. In the UK at least, the demand for new lines appears to depend on the size of the total telephone bill and not just on the rental element.

8.3.3 Subsidised telephony for the poor

Consideration (*c*) above is redistributive and invokes the compensation principle. The extension of telephone access to people not well enough off to afford it at cost-related rates may be justified on the grounds of social equity and formulated into a policy. It might be that, in this sense, a poor person's pound is worth more than that of a rich person. In general terms, this would be consistent with the idea that the marginal utility of personal income falls as the income level rises. At this level, it is true of expenditure on virtually all goods and services, not just the telephone. There would need to be a stronger case to justify a subsidy for telecommunications alone. A case might be built on the special needs of elderly or handicapped people. This has been analysed by Sharkey and Sibley [10] in the context of a method of optimising optional two-part tariffs in which the welfare gains of different consumer groups are weighted according to income.

There is extensive literature about the welfare aspects of taxation. It is commonly used to provide public goods such as roads and schools but is unusual in telecommunications. A tax could be levied across the whole population or confined to telephone users. There are four main alternatives:

(i)   Government can subsidise the provision of specific goods and services, for example public transport or telecommunications network access in remote areas.
(ii)   A government subsidy can be given to individuals for the purchase of specific goods or services, such as telephone service.
(iii)   Cash allowances can be paid to needy individuals for the purchase of whatever goods and services they wish, including telephone service only if wanted.
(iv)   Direct income transfers between consumers through selective distortions of the tariff structure.

The most efficient of these alternatives is the third, since it provides a cash benefit without restricting the consumption choice of the pensioner. The fourth risks damage to the efficiency of the prices in determining resource allocation. Evaluation of net benefits of the various alternatives is a form of cost-benefit analysis and implementation is part of the general political process. These and other cases for subsidy may only happen with regulatory intervention. They are discussed more fully in Chapter 14, which covers universal service concepts.

8.3.4 Marginal costs and economic efficiency

The most common reason for a departure from Pareto optimality is consideration (*d*) from Section 8.3.2, i.e. the existence of cost functions which, even in the long run, feature marginal costs below average costs. Telecommunications operators usually have cost structures of this type. The economies of scale which commonly exist in telecommunications and other networks may lead to even the largest operators trading at business levels where they are still experiencing significant falling average costs as production volume grows. Marginal costs are then below average costs and setting prices equal to marginal cost would result in the enterprise running at a loss, so that it could not continue in business on the strength of its own resources. There are three ways in which this problem can be resolved:

(i) The government can provide a subsidy to the firm and finance it from a tax. Subsidy of the operator preserves Pareto prices and is likely to have a less distorting effect on the overall pattern of resource allocation than selective divergences from marginal cost for particular prices, but it raises political and managerial problems.

(ii) A variant of the first is for service provision to be put out to tender, with companies competing to provide the required services at minimum cost using prices based on marginal cost. The contracting out of loss-making or free services is widely practised in the provision of local-authority services such as refuse collection. Competition for the franchise gives an incentive to productive efficiency.

(iii) Prices can be raised. This is the commonest solution. An optimised *second-best* price or, for a multi-product firm, a set of prices, marked up to reflect overheads is used so that the operator can cover all costs, including the cost of capital.

There are many sets of prices which between them are high enough to produce revenues which cover all the costs if production is to be sustainable in the long term. All of them use fully allocated costs. Some are easier than others to justify in economic or equitable terms. Economics and equity may be in conflict. There is an extensive literature on second-best prices, of which a brief summary follows. Mitchell and Vogelsang [7] give a good general introduction.

8.4 Ramsey prices

The tariff structure can be optimised to maximise net social benefit subject to a break-even profit constraint. The optimum will usually be different if social welfare and externality considerations are included. A uniform percentage markup will not usually be optimal. A solution going back to the 1920s is Ramsey pricing, named after Frank Ramsey [11], who proposed it in the context of the optimum distribution of taxes. Broadly speaking, the product markups depend on price elasticities and cross-elasticities and are highest where the elasticity is lowest. Figure 8.2 illustrates the model for the ith product of a multiproduct monopolist.

In the figure, Q_iQ_i is the market demand curve, sloping downwards to the right because the firm is a monopoly. It is a function of p, a set of price and other variables, and written as $q_i = g_i(p)$, the ith market demand function, where p_i is the price set by the firm for the ith product, q_i is the demand at p_i and n is the number of products, $i = 1,2,..,n.$

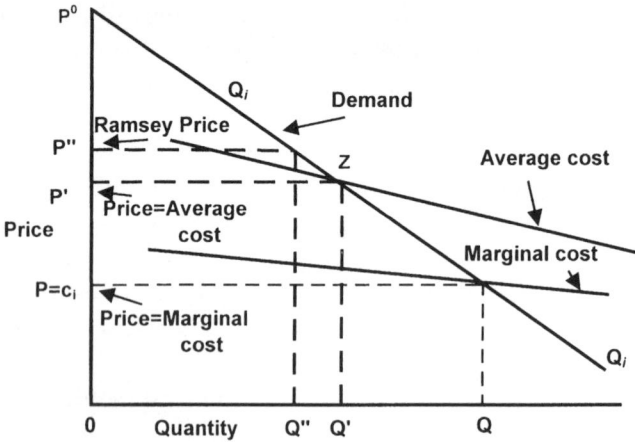

Figure 8.2 Price, demand and cost for the ith product

The Pareto-optimal price, $\mathbf{P}$, equals marginal cost c_i but the firm makes a loss because $\mathbf{P}$ is below average cost. The firm would have to charge $\mathbf{P}'$ to recover average cost, cutting demand back from $\mathbf{Q}$ to $\mathbf{Q}'$. Since we seek a set of prices for the n products for the maximisation of net social benefit subject to the profit constraint, individual prices may vary in their relationship to marginal cost. The Ramsey price for the *i*th product is $\mathbf{P}''$. It results in demand $\mathbf{Q}''$. The value of $\mathbf{P}''$ depends on the price, demand and cost of all n products.

The demand function can be rearranged to make price a function of demand, and is then called an inverse function of q. For example, with the demand function $q_i = a - bp_i$, the inverse function $p_i = h_i(q)$ is

$$p_i = -(q_i - a)/b \tag{8.1}$$

Total utility, U_i, is the area under the demand curve from its intersection with the price axis down to the market price. This is obtained from $h_i(q)$ by integrating for q and evaluating the integral from zero to q_i:

$$U_i = \int_0^{q_i} p_i dq_i \tag{8.2}$$

In this case the demand curve is a straight line and $U_i = q_i \,(a/b - q_i/2b)$. For example, with trial values of $a = 10$, $b = 0.25$ and $q_i = 6$, $U_i = 168$ and $p_i = 16$. On Figure 8.2, if $q_i = \mathbf{Q}'$ then U_i is the four-sided area $\mathbf{P}^0 0 \mathbf{Q}' \mathbf{Z}$. The firm's revenue is $p_i q_i$, which is then $\mathbf{P}' \mathbf{Q}'$ on Figure 8.2. Consumer's surplus is $aqp_i = U_i - p_i q_i$. With the trial figures, $U_i = 72$. On Figure 8.2 this is $\mathbf{P}^0 \mathbf{P}' \mathbf{Z}$.

The firm's marginal cost curve intersects the demand curve at price $\mathbf{P} = c_i$, where the demand $q_i = \mathbf{Q}$. The total cost of producing output q_i will be higher than $q_i c_i$ because marginal costs are falling and there may be fixed costs. Suppose, to simplify the exposition, that the *i*th product has constant marginal costs c_i (i.e. that the marginal cost curve in Figure 8.2 is horizontal at $\mathbf{P} = c_i$). Suppose also that it adds a fixed cost K_i. Producer's surplus S_i (here negative) is then

$$S_i = p_i q_i - q_i c_i - K_i \tag{8.3}$$

We seek a set of prices that maximises consumer's surplus plus producer's surplus subject to the net revenue constraint that

$$\Sigma_i [q_i(p_i - c_i) = K_0 + \Sigma K_i] = R \tag{8.4}$$

where K_0 is a central overhead cost not due to any individual product and R is difference between the revenue that would be obtained from marginal cost prices, and the total costs of the firm.

The constrained function to be maximised is then

$$\Sigma_i[U_i - p_iq_i + S_i] + L(\Sigma_i[q_i(p_i - c_i)] - R) \qquad (8.5)$$

where L is the Lagrangean multiplier *lamda*, a constant to be estimated. Differentiating for each q_i and setting the derivatives to zero gives solutions of the form

$$(p_i - c_i) + L[(p_i - c_i) + q_i(dp_i/dq_i)] = 0 \; for \; all \; i \qquad (8.6)$$

Defining the price elasticity for the ith product as

$$e_i = (dq_i/dp_i)(p_i/q_i) \qquad (8.7)$$

we can rearrange eqn. 8.6 as

$$(p_i - c_i)/p_i = L/(|e_i|(1+L)) \qquad (8.8)$$

where $|e_i|$ is the absolute (non-negative) value of e_i.

Prices are marked up from marginal costs by a factor inversely proportional to price elasticity; they are then the Ramsey prices. Profit maximisation under the same constraint, by the omission of consumer's surplus, reduces eqn. 8.8 to

$$(p_i - c_i)/p_i = 1/|e_i| \qquad (8.9)$$

So far it has been assumed that all cross-elasticities are zero. A generalisation to include cross-elasticities is due to Rohlfs [12]. The required vector presentation is in Reference 7. Most formulations of Ramsey pricing are based on the maximisation of the combined surpluses of consumers and producers, reflecting externalities only to the extent to which they affect either of these directly. For example, the network externality generates extra calls if the call price is lowered. This will show up in the calculations and there will be consumer's surplus transfers between customers within their existing budgets. Culham [13] uses cross-elasticities and incorporates externalities to model optimum call and rental prices for the UK. The Culham study found substantial welfare gains from raising rentals and the price of local calls at the expense of lower prices for trunk calls, where elasticities were highest. The results did not appear to be very sensitive to cross-elasticities.

One of the findings of Baumol *et al.* (see Chapter 6) was that a multiproduct monopoly with a subadditive cost function (Figure 5.3) could protect itself from the threat of entry by the adoption of Ramsey prices. Empirical evidence here is that they never do, preferring to seek legal protection. It has often been found that monopoly markups are in the opposite direction to those of Ramsey pricing, with the lowest profits sectors with low elasticity, notably access.

8.5 Telecommunications as a facilitator of economic growth

8.5.1 Conceptual framework

There have been many attempts to establish a link between telecommunications development and more general economic growth. The concept is vaguely defined but has four aspects:

(i)   Enhancing the value of total outputs obtained from a given set of inputs through more efficient pricing.
(ii)   Producing a higher rate of return for telecommunications investment than for other kinds of investment.
(iii) Meeting unsatisfied demand by shortening waiting lists and providing service to willing buyers.
(iv)   The acceleration of economic growth by the provision of an enabling infrastructure, service improvements and innovation.

These aspects are considered further below.

8.5.2 Enhancing the value of outputs

The widespread use of output prices which are economically inefficient has generated many studies showing the benefits to be obtained by a closer alignment of price with costs. The primary benefits come from the value of telecommunications outputs themselves through pricing for optimum resource allocation. The basic tariff imbalance that most telephone companies claim to experience is between calls and access. Profitability of calls is high and the excess is used to support losses made on the local loop, these losses arising mainly from a combination of high fixed costs and low charges for line rental.

Figures have been produced in support of this, but the methodology is contentious. Operators and regulators do have an agreed interpretation, either among themselves or with each other. The strongest evidence in its favour comes from the behaviour of new entrants to the market. Almost all of these have attacked the long-distance market first, as the quickest route to profits.

The World Bank has shown that telecommunications has obtained the biggest gains from sectoral restructuring and privatisation; the main source of these gains is more efficient pricing [14]. Welfare gains may amount to more than the total annual sales revenue. Culham's study for the UK was mentioned above. In 1993 the per-minute price in Germany for long-distance calls was at least seven times that of a local call, whereas costs differed only by a factor of three [15]. For an Australian study of international tariffs see Chapter 17. American studies have tended to show that this type of cross-subsidy makes users worse off, on balance, not better. A study by Perl in Reference 16 showed net welfare gains from a move to cost-based tariffs, summarised in Table 8.1. Within the total, groups with higher incomes would gain and those with lower incomes would lose.

Table 8.1 Distribution of welfare gains from cost-based output pricing [16]

Income group	Percent of households	Annual gain $
$6000 or less	11.35	−68.28
$6000–$12 500	19.15	−7.83
$12 501–$17 500	19.67	48.93
$17 501–$25 000	16.03	99.03
Above $25 000	33.80	181.95
Average		77.13

Input prices are a further source of potential gain from more efficient resource allocation. These may come from privatisation, the introduction of competition, or both. Peoples [17] found an increase in telecommunications wage flexibility, with more responsiveness to regional economic conditions, following the break-up of AT&T in 1984. Wage levels in the utility monopolies which were privatised in the UK during the 1980's became more responsive to market pressures. Moves to introduce greater competition into competitive and fuel-input procurement have accompanied privatisations in the British telecommunications and energy sectors.

8.5.3 Higher returns for telecommunications investment

The return on telecommunications projects supported by the World Bank has been above its average for infrastructure. Over the decade to 1992 the return was 19 per cent, against 15 per cent overall and it was three per cent ahead over the previous six years (World Bank, 1994). The biggest single source of benefit was the adoption of prices which were more efficient. Studies of the return obtained from investment in information technology and telecommunications have sometimes claimed a relatively high rate of return, over 500 per cent for some Internet applications. If such benefits are common, market forces should prompt private companies to obtain them by making the necessary investment, as many do, in private networks and other forms of communications technology.

The persistence of above-average returns over a long period implies an imperfection in the product or capital markets. These imperfections may be because of:

- under-investment by a government that will not allow private capital to be invested, or cannot borrow it,
- monopoly profits,
- rapid technological change, creating profit opportunities at a faster rate than they can be exploited,
- the inclusion of externalities.

8.5.4 Meeting unsatisfied demand

Long waiting lists can be an indication that more telecommunications investment is needed. The implication is conditional because the waiting list may be lengthened by low prices. Latvia had a waiting list equal to 26 per cent of system size in 1992, but average revenue per line was less than £8 per year (calculated from ITU statistics). Latent demand may also be masked by high connection charges or deposit requirements for suspected bad credit risks. The waiting list in India fell from 637000 to 250000 when an advance deposit scheme requiring a relatively large payment for registration on the waiting list was introduced in 1976 [18]. A truer test is the number of people willing to pay the resource cost of a line but unable to get one.

Long waiting lists are common in developing countries and have appeared elsewhere at times of public-sector capital shortage. They may reflect the high returns to be obtained from investment in other sectors, in which case the use of tariffs based on short-run marginal cost would be economically appropriate. Waiting lists arising from the way in which public expenditure is financed represent genuine unsatisfied demand and thus a market failure. The root of the problem here is often a combination of reluctance to adopt tariffs which are commercially adequate, a reluctance to provide public funds for a subsidy and a refusal to allow private investment.

Another measure of unsatisfied demand can be obtained by comparing the number of telephone lines per hundred inhabitants among countries with a similar level of GDP per inhabitant. A cross-section of the world's telecommunications markets shows a fairly strong relationship between income per head and the number of telephone lines per 100 population – see Figure 8.3. There is a good deal of local variation, but in broad terms each $1000 increase in per capita GDP is associated with an extra three lines per 100 people.

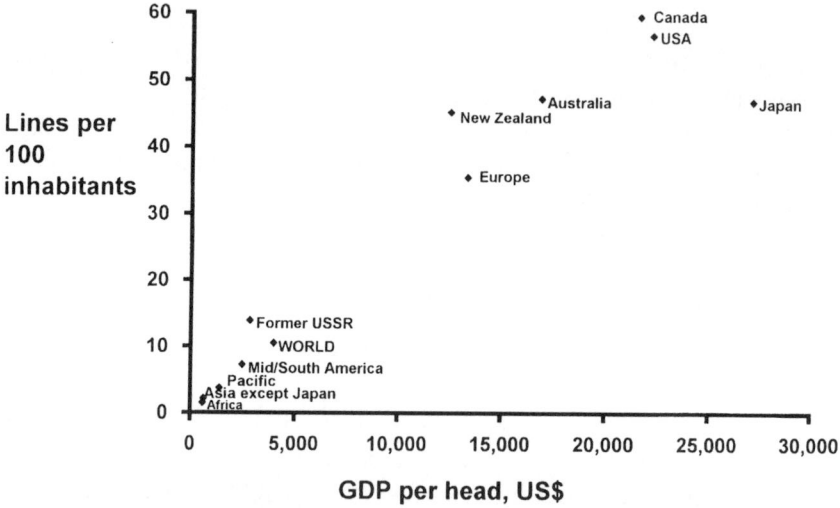

Figure 8.3 Telephone penetration and income per head, 1992

Some of the spread comes from the currency exchange rates used to reduce all the national statistics to a common currency. Tariff levels also play a part. Attempts to establish whether East European countries were underprovided in relation to national income before the collapse of the communist regimes in the early 1990s were made more difficult by differences in the systems of national income accounting and by the need to use arbitrary currency exchange rates for unconvertible currencies. Both factors probably led to a relative overstatement of East European GDPs. Some observers tried to correct for them.

The scale of variation is, even so, considerable, as the figures in Table 8.2 for countries with per capita GDPs of around $300 in 1992 show.

Some latent demand goes unmeasured in most developing countries. Requests for service from people living in areas where no service is provided at all may not be on official waiting lists.

Table 8.2 Income and telephone penetration in five developing countries in 1992

	GDP per capita $	Main lines per 100 inhabitants
Sudan	280	0.24
Niger	290	0.12
Nigeria	315	0.28
Kenya	320	0.77
China	320	0.98
Guyana	290	3.50

(Source: ITU World Telecommunications Report 1994)

Under monopoly conditions commercial viability and tariff levels are closely linked. Tariffs that are sustainable in a political sense must not appear excessive to the majority of individuals. Some governments have not allowed tariffs to be raised to pay for services which many individuals would willingly buy at a commercial price. Experience has been that, without these constraints, telecommunications development to meet unsatisfied demand shows an adequate rate of return on capital without subsidy [18].

8.5.5 Acceleration of economic growth

The use of economically efficient prices may create secondary benefits, for example where lower long-distance call prices enable value added services to be provided which would not otherwise be commercially viable.

High returns may be obtained from telecommunications investment where there are externalities, in the same way that they can from investment in roads and transport. Saunders *et al.* [18] quote a representative group of twelve programmes partly funded by the World Bank which were estimated to yield an average internal rate of return of 18 per cent, with externalities bringing the average up to 27 per cent.

Other studies have confirmed the high value of telecommunications in developing countries. Antonio and Canas [19] obtained indications of consumer surplus in Costa Rica. Average values attached to calls, expressed as a multiple of the call price, are given in Table 8.3:

Table 8.3 Average value of a call as a multiple of call price in Costa Rica [19]

Type of call	Multiple
Rural:	
emergency/health	11.2 – 23.6
productive/public/social	6.4 – 8.1
family/labour	1.6 – 2.8
Urban:	
business	50
residential	5

Consumer surplus exists in all countries, so these figures may reflect something that is rather general.

There have been other attempts to establish a link between the level of infrastructure spending and GDP growth. The 1994 World Bank development report [14] cites 14 such studies, all published between 1989 and 1993. Elasticity, measured as the percentage change in output with respect to a one per cent change in infrastructure spending, ranges from 0 to 0.44, averaging 0.17. Projects involving communications came out close to the average. Rates of return were calculated by taking the ratio of the discounted value of the increase in output to the discounted value of the infrastructure investment, and ranged up to 96 per cent.

Hardy [20] and Cronin *et al.* [21] found evidence of causality running in both directions. As an example US economic activity is a reliable predictor of future telecommunications investment *and* US telecommunications is a reliable predictor of economic activity. This recalls an unpublished UK Post Office study in which telecommunications investment was found to follow a rise in business rather than facilitating it. Feedback effects may be responsible, in a situation where there is neither cause nor effect, but a process involving both. In Latvia, which had a relatively high residential telephone penetration facilitated by low access prices in the early 1990s, there was little evidence of any spillover to accelerate growth in other parts of the economy.

For any of these findings to be useful in policy formulation, it would be necessary to show that telecommunications was not just effective, but more effective than other industries, as a growth accelerator. There may be other industries where a predictive relationship is found. Construction, for example, has often been used for pumping up a depressed economy.

In spite of the ambiguity of these studies there has been a widespread belief among industrial countries, and some of those in the developing world, that telecommunications investment can give a critical competitive edge. There have been attempts to force the pace, France's Minitel being perhaps the best known example. There has been extensive ISDN and broadband investment in France and Germany, the broadband being in separate cable networks, altough the financial results of these have tended to be disappointing. Investment has sometimes run ahead of the market, or even in a different direction. Singapore has invested heavily in information technology with the explicit aim of making the country a major hub in its region. Some commentators in the US watched Europe's development of ISDN in the 1980s and concluded that they were missing an opportunity.

A 1983 European Commission report [22] on European competitiveness identified telecommunications as being of strategic importance, but it fell short of suggesting that provision should be subsidised to accelerate development. A poor competitive performance was attributed to institutional factors such as national procurement policies in Europe and a better use of social investment in Japan and Sweden. There were no recommendations, but subsequent EEC action has been in the direction of institutional change.

A 1994 DTI paper [23] was one of several official studies on broadband policy in the UK. It concluded that reliable broadband communications were a necessary foundation for a full range of information and entertainment services and that, following market liberalisation in the UK, market forces needed little help from the government.

8.5.6 Summary of evidence

The overall position on the evidence so far is this:

(a)   There is a right level of telecommunications infrastructure spending for the level of GDP. It can be defined as the one where the marginal return from telecommunications investment is equal to that in other sectors. Inclusion of externalities in these benefits is appropriate, as it is in water, transport and other infrastructure projects.

(b)   In developed countries, market forces are likely to be the most efficient means of getting the best level of provision. The scope for institutional change and greater efficiency in pricing is large outside North America, Australia, New Zealand and the UK. There is no evidence that state ownership is necessary or helpful.

(c)   There are countries, for example, in eastern Europe, where service provision is behind the right level, or where all sectors, not just telecommunications, offer substantial returns from additional investment. Some of these countries are not large enough to support product market competition in all parts of the network, but all can benefit from privatisation and more efficient price structures. It has not been established that abnormally high levels of investment, beyond that needed to catch up on the sector norm for the size of GDP, would be the best way of allocating resources.

(d)   Institutional change and efficient pricing can bring benefits in less-developed countries, but they may not be enough on their own, especially where the scope for product market competition is limited. Telecommunications has a role to play in improving the efficiency of markets, by speeding up the spread of information concerning price and availability. Roads and transport can also improve market efficiency, and it is not obvious that telecommunications has a priority. There is a possible market failure in the supply of investment finance because low income levels are combined with high investment/revenue ratios among the operators. This is something that can be addressed.

8.6 Measuring potential welfare gains

8.6.1 Measurements and speculations

A possible measure of additional welfare is the ratio of aggregate additional welfare gains (in terms of profit, consumer's surplus and employee benefits) to total sales. The calculation could take into account estimates of the current value of future costs and benefits, using an appropriate rate of interest.

Mansell [24] compares market and managed solutions to the maximisation of public benefits from intelligent networks in two models. The idealist model relies on market forces to optimise public benefits from the new networks. An alternative analysis, rooted in a theory of imperfect competition, the strategic model, draws on Reference 25. Here market failure excludes many ordinary people from the potential benefits of the new networks. Policy makers and regulators pressure operators into providing universality for the benefits of access to electronic means of communication.

Much of the work is concerned with wealth-enhancement opportunities seen to arise for individuals and the smaller businesses. Although little is quantified, there remains an important issue of public interest. The PTOs, which usually have the strongest market position, may try to reinforce it by adopting IN strategies that lock users into proprietary configurations and segmented networks rather than creating a public network which is ubiquitous and offers equal access to all. If there is a task here for intervention on welfare grounds, it is to encourage open access through the encouragement of appropriate standardisation and the enforcement of non-discriminatory pricing. At a time of evolving technology and customer tastes there is a risk that regulators do not have enough knowledge to second guess even an imperfect market.

8.6.2 A case study: welfare effects of digital infrastructure

Greenstein and Spiller [26] have estimated the size of welfare gains arising from investment in fibre-optic cable, ISDN lines and common-channel signalling, three essential elements in the bringing of digital technology to local networks and the creation of information superhighways (Section 1.7.1). Consumer demand was found to be responsive to the provision of fibre-optic cable and ISDN (which uses common-channel signalling). An increase averaging 13 per cent was found. It was thought that this resulted from the effect of a whole range of new digital technologies, rather than the three specifically identified.

8.7 References

1 VAN DEN DOEL, H. and VAN VELTHOVEN, B.: 'Democracy and welfare economics' (Cambridge University Press, Cambridge, 1993)
2 BANNOCK, G., BAXTER, R. E., and DAVIS, E.: 'The Penguin dictionary of economics' (Penguin Books, London, 1992, 5th edn.)
3 GREEN, H. A. J.: 'Consumer theory', (Macmillan, London, 1976, revised edn.)
4 SAMUELSON, P. A.: 'Consumption patterns in terms of overcompensation rather than indifference comparisons', *Economica*, 1953, **20**, (77) pp. 1–9

5   'Provision of directory information services and products, consultative document' (OFTEL, London, 1997)
6   Monopolies and Mergers Commission: 'Report on BT Yellow Pages' (HMSO, London, 1995)
7   MITCHELL, B. M. and VOGELSANG, I.: 'Telecommunications pricing, theory and practice' (Cambidge University Press, Cambridge, 1991)
8   HARRISON, A. J. and MACKIE, P. J.: 'The comparability of cost benefit and financial rates of return'. Government Economic Service occasional paper 5, HMSO, London, 1973
9   'Review of British Telecom's tariff changes, (OFTEL, London, November, 1986)
10 SHARKEY, W. W. and SIBLEY, D. S.: 'Applications of public utility pricing theory to BOC pricing issue'. Economics discussion paper 11, Bell Communications Research, 1985 *quoted in* Reference 7
11 RAMSEY, F. 1927: 'A contribution to the theory of taxation', *Economic J.*, 1927, **37**, pp.47–61
12 ROHLFS, J. 1979: 'Economically efficient Bell-System pricing'. Bell Laboratory discussion paper 138', 1979, *cited in* Reference 13.
13 CULHAM, P. G.: 'A method for determining the optimal balance of prices for telephone services'. OFTEL working paper 1, London, 1987
14 World Bank: 'World development report 1994' (Oxford University Press, New York, 1994)
15 NEU, W. 1993: 'Tariff policy issues in Germany'. Paper presented at *Commed UK, European and International Tariffs Conference*, London, 1993
16 PERL, L.: 'The consequences of cost-based pricing', in MÜLLER, J. (Ed.) 'Telecommunications and equity' (North-Holland, 1986), *quoted by* GARNHAM, N. in proceedings of *Communications Policy Research* Conference, June 1988, Windsor (IOS, Amsterdam, 1989)
17 PEOPLES, J. H.: 'Wage outcomes following the divestiture of AT&T', *Inf. Econ. Policy*, 1989/90, **4**, (2), pp. 105–126
18 SAUNDERS, R. J., WARFORD, J. J. and WELLENIUS, B.: 'Telecommunications and economic development' (John Hopkins University Press, Baltimore, 1983)
19 ANTONIO, A. F. and CANAS, M.: 'The socioeconomic impact of telecommunications in Costa Rica'. Paper presented at 5 *World telecommunications* forum, Geneva, 1987
20 HARDY, A. P.: 'The role of the telephone in economic development', *Telecomm. Policy*, 1980, **4**, (4) pp. 278–286
21 CRONIN, F. J., PARKER, E. B., COLLERAN, E. K. and GOLD, M. A.: 'Telecommunications infrastructure and economic growth: an analysis of causality', *Telecomm. Policy*, 1991, **15**, (6), pp. 529–535
22 LOCKSLEY, G.: 'The EEC telecommunications industry, competition, concentration and competitiveness'. Commission of the European Communities, Brussels, 1983
23 Department of Trade and Industry: 'Creating the superhighways of the future: developing broadband communications in the UK'. Cmnd. 2734, HMSO. London, 1994
24 MANSELL, R.: 'The new telecommunications, a political economy of network evolution' (Sage Publications, London, 1993)
25 SMYTHE, D. W. and DIHN, T. V.: 'On critical and administrative research: a new critical snalysis', *Media, Culture and Society*, 1983, **33**, (3) pp. 117–27 *cited in* 24
26 GRENNSTEIN, S. M. and SPILLER, P. T.:'Estimating the welfare effects of digital infrastructure' (National Bureau of economic Research, Inc. Working Paper 5770, Cambridge, 1996)

Chapter 9

Telecommunications service prices

9.1 Background

9.1.1 Ideas about prices

If resource allocation is to be efficient, prices need to be high enough to sustain the supply and development of services at a level that customers are willing to pay for, but not so high as to generate excessive profits for the producer. Prices provide a vital signal to potential new entrants to the market, enabling them to judge whether their own costs will be low enough for them to trade profitably if they enter.

Telecommunications equipment prices are set in more or less competitive markets. Most others, especially those for basic domestic network services, are to a greater or lesser extent set by monopolists or by administrative means. However they are set, the prices will guide the resource allocation of consumers and, through their purchasing behaviour, that of producers.

Prices that are set administratively may be intended to emulate those which would prevail in a fully competitive market or, more realistically, a second-best set of prices which are intended to allocate resources efficiently within the constraint that costs are fully covered and the market is not fully competitive. They may have other objectives. The choice may result in a pattern of resource allocation which differs from that which would be set by market forces. For tariff theory, one of the implications is that the real focus of cost-based pricing should be the costs of a new entrant rather than those of the incumbent. Social aspects of tariffing are discussed in Chapter 14.

A price is said to be a linear price if a consumer's outlay is proportional to the quantity purchased. Linear prices are unusual, even in perfect markets. A price is a nonlinear price if the average amount paid for a unit depends on the quantity purchased. A street trader selling melons may offer them at 60p each, two for £1. Nonlinear prices are widespread in telecommunications services. One example is the basic two-part tariff for rentals and calls The maximisation of net social benefit can still be analysed through the Ramsey price approach of the preceding Chapter, but it now depends on the demand curves of individual consumers or groups of consumers rather than their aggregation in the market demand curve.

9.1.2 Perceptions of prices

Customers may be fully rational but they are not perfectly informed. They react to prices on the basis of what they believe them to be rather than what they are. Presentation and the selection of price points can assist with accurate price perception, which is often poor [1]. Call-charging structures in many countries are complex. Most are based on unfamiliar units of duration, rather than on minutes and seconds, and they may feature setup or minimum prices. Public awareness of the actual per-minute price of a call is often poor. Where time-of-day pricing exists, the break times may not be easily recalled. If they are changed, awareness of the change may take a long time to become widespread. This lack of awareness has sometimes led operators to adopt simpler tariff structures which may not reflect cost very closely but which are easier to market.

Producers and consumers use prices to make decisions on what to buy and where to buy it. Imperfect knowledge about prices reduces market efficiency because buyers base their decisions on perceived rather than actual prices. The improvement of knowledge about prices may itself incur a cost, through advertising and publicity. The simplification of tariffs, so that they are more easily remembered, may improve knowledge at the expense of suboptimal pricing. The balance of allocational advantage is open to doubt in the imperfect telecommunications services market since operators use inertia and imperfect knowledge of competitors to retain customer loyalty.

9.1.3 How an operating company sets its tariffs

The main thrust of the pricing policy of a commercially-oriented operator is to obtain as good a rate of return as possible, within service, quality and other constraints. There are typically five stages to the process of tariff setting for a large operator able to influence market prices and subject to profit control:

(i)   Analysis of costs
(ii)   Forecast of demand
(iii)   Calculating the revenue target
(iv)   Setting tariffs to meet the target
(v)   Negotiating them with the regulator

A minimum figure for the revenue target is the cost of capital; the maximum may be imposed externally. The self-financing ratio for investment is important. As discussed in Section 7.7.5, during a phase of rapid growth, at least 50 per cent of the new capital requirement may be borrowed. This is to avoid an excessive burden on tariffs. In mature systems, such as British Telecom in the 1980s, the external financing requirement may fall to zero.

Smaller companies and those in very competitive markets are free to try for as much profit as possible without price constraints. They take prices from the market or by discounting against the largest operator and possibly against each other. They still have to analyse costs, demand and revenue requirements and they use this information to formulate business strategy.

The basic tariff is normally in two parts, a fixed monthly charge and usage charges for different types of call. There will also be a connection charge to cover the costs of establishing a new connection. The usage charges are usually above marginal cost, so that they provide a further contribution to the fixed costs of the network. The advent of competition has put pressure on call prices, which are moving closer to marginal cost.

9.1.4 Criteria for regulated tariff structures

There are three different criteria for a tariff structure which in other respects brings in enough money to finance the service:

(i)   Fairness, often interpreted as favouring equal mark-ups for overheads, though equality and equity are not the same.
(ii)   Economic efficiency, favouring cost-based prices or, in imperfect markets, elasticity-based mark-ups (Ramsey prices, see Section 8.4).
(iii)   Social Welfare, which involves either or neither of these but maximises aggregate consumer satisfaction.

Fairness may result in a net loss in benefit across all consumers. It is to be distinguished from the maximisation of social welfare, which may not be achieved with equal mark-ups and may, in this sense, not appear fair.

Cost-based prices are easier to specify than to determine. There is a temptation to concentrate on the costs which are easiest to measure and to adopt a somewhat stylised approach which (as most regulators will admit) over simplifies the nature of the market. The criteria for optimum prices as described in Section 6.4 are the same for fixed and mobile networks, and for regulated and unregulated services. The high profile of cost-based prices in the literature to a large extent reflects the needs of regulators in the context of basic services on fixed networks. As will be argued in Chapter 12, this is second-best to making use of the knowledge in a competitive market, not the preferred way of setting prices or a reliable test of whether they are optimal. Cost-based pricing does not sit easily with marginal-cost or Ramsey pricing and its popularity is in large measure due to it being seen as better than most existing structures rather than in ideas of optimality. The method looks fair and less arbitrary than other methods. Firms operating in competitive markets are price takers, and the question is whether they can afford to stay in business given the prevailing price.

Ramsey prices are distrusted because the theory is based on the price elasticities experienced by the market as a whole. The price elasticities experienced by the individual firm will often be higher, reflecting the degree of market dominance in each of its product markets. The system would be difficult to regulate and easily abused. Elements of Ramsey pricing can be found in *value-of-service* pricing, which has been used at state level in the US since the 1920s. It imposes high line rentals on business users because the telephone is more valuable to them, and on charges for local calls at exchanges with large local-call areas, because there are more people to call (Vietor 2). The greater value of service to business users, associated with a willingness to pay higher prices, has been recognised in many countries. More examples are in Chapter 8 (Costa Rica) and Chapter 19 (Russia).

9.1.5 Short-run marginal-cost pricing

It is common in competitive markets for short and medium term shortages and surpluses to be met by special market-clearing prices based on short-run marginal cost. Telephone companies have used the connection charge as an effective regulator of demand when faced with long waiting lists. This may be a rationing device in a situation of chronic undersupply, it may be a response to temporary capacity shortages or it may reflect the cost of short-term expedients to match a surge in demand.

British Telecom abolished its peak-rate band in 1994 and gave network modernisation as the reason[3]. The decision came when capacity levels were high, and it might also have been an example of short run marginal-cost pricing.

9.2 The price for access

9.2.1 Option demand

Access to the network is a facility in its own right, for which there is an independent demand and an efficient price. The demand was called 'option demand' in Chapter 4 and it is the facility to make and receive calls. Somebody who made no calls at all might still want access so that they could be contacted in an emergency or by relatives for social purposes. Large businesses use some of their lines solely for incoming calls. Access also provides the facility to make calls, which is the more obvious part of its value.

Option demand on its own may not be as high as the rental typically charged in a two-part tariff. Implicitly, customers see the rental mainly as part of the package for making and receiving calls. The supply of an access facility can, however, be separated from call provision, for example under the Harper scheme above. A company can offer connections without calls, or calls without connections (by using another network to reach the customer), so that stand-alone access costs could be calculated.

The situation is not unique. There are plenty of other examples, such as records and record players, where two products are useless unless used together but are produced and sold separately. The allocation of costs between calls and connections is difficult and to some extent unnecessary except as an indication of entry costs. Market segmentation, which is discussed later in the Chapter, is already producing a range of access prices to suit different demand requirements.

9.2.2 Business and residential rentals

Many operators charge higher rentals to business users than to residential users. This is basically price discrimination. It arises from four main factors:

(i)   Residential customers tend to display a higher price elasticity.
(ii)   Political sensitivity of residential rentals, bolstered by welfare considerations.
(iii)   Higher maintenance costs on business lines.
(iv)   Tariff structures that have not caught up with the technology changes which have undermined call prices, and which have become too dependent on profits from calls.

Rentals ought to reflect cost unless significant externalities exist. A move to cost in one step could create problems over unpaid bills and differential income effects. The net loss of benefit among lower income groups in the Perl data (Chapter 8 [16]) comes from higher rentals, commonly believed to be subsidised from excess profits on calls. Groups with higher incomes have higher calling rates.

In some countries and at some times there have been significant debtors in the business sector so that many of these customers may also have been unable to afford to pay. This was the case in several excommunist countries during the restructuring of the 1990s. The de-averaging of rentals, i.e. the charging of rentals based on local or regional costs, is discussed in Chapter 14.

9.3 The price for calls

9.3.1 Setup costs

The cost of a call has three elements:

(i)   The call setup, which engages circuits and switches across the network and may result in an abortive attempt.
(ii)   The use of equipment during the call.
(iii)   Billing and administration.

Purely cost-based call prices would include a charge for calls which fail to get through. This is rare in fixed networks but it features for some calls to mobile phones. A constant per-minute or per-unit charge is the most frequent way of charging for long-distance calls. Local calls are sometimes free or have a fixed price regardless of duration. Some operators, especially in the US, charge more for the first minute than for subsequent minutes. This is also common with calls made on international phone cards.

9.3.2 Peak load pricing

The load curve discussed in Section 7.2.3 shows very uneven demand from hour to hour. It makes for efficient resource allocation if the callers who create peak equipment requirements pay the cost. Where there is no differentiation of call prices by time of day or day of the week, there is heavy congestion on parts of the network in business hours. At other times, particularly in the evening and at weekends, there may be unused capacity.

A price mechanism would be a better way of flattening demand peaks than allowing congestion to choke off the callers. Unused capacity in off-peak periods could be brought into use by reducing the price. Two and sometimes three different charging periods are commonly used. They cannot flatten peaks completely, but they improve allocative efficiency.

Off-peak prices should be high enough to recover marginal costs, but the prices could be lower than peak prices to reflect the use of the spare capacity of circuits provided to meet peaks. Time-of-day charging should reflect the cost of catering for peaks. Practical work is likely to reveal differences between peak and off-peak call price elasticities. There are likely to be cross-elasticities between peak and off-peak rates, shifting calls from peak to off-peak hours if the price differential between them is raised.

Operators usually make empirical adjustments with a view to reducing costs by extending the period of high load, and try ways of raising off-peak usage through marketing and the encouragement of suitable services. The existence of an upper limit to off-peak demand even if off-peak calls are free, which arises from other consumer costs such as time, complicates this approach. Large surpluses of trunk capacity have been created in the UK and the US with the progress of digitalisation, advances in transmission technology and competition.

9.3.3 Measured and unmeasured local service

The early electromechanical telephone exchanges had no metering equipment to count calls or measure their duration. If local calls are not to be supplied free, there has to be a way of measuring them. Early metering equipment only counted calls. The high number and low average value of local calls led to the development of metering equipment which measured the aggregate duration of calls rather than their number. Duration charging by periodic-pulse metering came to the UK with the introduction of STD in 1959. Pulses were fed into the meter at regular intervals during the call and were counted. In the US most operators did not install meters in electromechanical exchanges because of their cost. Customers were allowed free local calls and the rental was set at a level high enough to pay for average usage. Without metering, business users typically pay a higher rental than residential users, partly reflecting their higher calling rates. Free calls have been used as an incentive to join the service. They appeal to some customers because the bill size is capped.

Free local calls were available in the UK until 1921, when they were replaced by untimed calls with a fixed charge and a lower rental. The timing of local calls began in 1958. Many users in the US still enjoy free local calls, often as one of a range of service options, and they are offered as options by cellular mobile operators. The issues may be summarised thus:

(a)   Resource allocation would be suboptimal with free calling if there were no transaction costs, because the calls are provided below marginal cost. Metering equipment is part of the transaction cost.

(b)   Practical problems such as the unequal levels of peak-hour calling from day to day and the need for tariff schedules to be simple enough for customers to act rationally in relation to them, limit the extent to which the welfare benefits from cost-related pricing can be achieved. Park and Mitchell [4] found that the problems were large enough to make free local calling the optimum choice for voice traffic, under certain circumstances.

(c)   In exchanges where metering equipment is available, unmeasured service can enlarge customer choice by providing an alternative to measured service. As with other optional tariffs, this enlarges welfare, see below.

(d)   Digital exchanges are less sensitive to the level of calling and more sensitive to the number of lines served than were the electromechanical exchanges which they replaced. They also have lower metering costs. This makes the argument more marginal.

(e)   Free calling requires adequate capacity, or the system can be swamped with calls. This has happened in developing countries through gross underprovision and in developed countries from new services.

Free local calls are still common in unmodernised networks. Modern-isation has extended local call-charging, but a variety of other rental/call packages have been introduced. Many customers have a choice.

Calls connected through the operator, including all long-distance calls, could be counted and measured if paper records were maintained for each call. These paper records, or tickets, were the basis of long-distance charging in the UK until the replacement of operators by automatic metering equipment, which began in the 1960s. Meters had to be read physically, initially one by one and later photographically, using a camera which could record a block of meters in one image for later transcription. In Canada and the US, where there were often no meters, an automated system, based on paper tape carrying the billing data on each long-distance call, was introduced to replace manual ticketing.

Ticketing and meter reading were expensive processes, largely dependent on the technology used. Various types of metering equipment could be installed. Calls were basically counted or divided into units of time (called pulses in some countries) which could be counted. Exact timing of call durations was not possible, and the data provided by the system were poor. The British Post Office was unable to count the number of local calls, and the number was estimated by estimating the revenue from local calls as a residual after subtracting estimates of revenue from other calls, and then sampling to give the average revenue per local call. The introduction of electronic control to telephone exchanges enabled more precise measurements to be made, and the software in the latest digital exchanges can record and report the exact duration and destination of every call on each line.

The introduction of charging for local calls is likely to reduce their number. The calling rate in the UK fell substantially after free local calls were abolished in 1921. The average call duration may be lower in areas with free local calling, where customers may be more inclined to make brief calls for trivial reasons. Free local calling encourages long interactive calls with the Internet and other online data systems, as discussed above. System congestion would be reduced with a charging system and prices would be more efficient.

The cost of putting the necessary metering equipment into electromechanical exchanges could be substantial and outweigh the expected economic benefits from greater allocative efficiency. An alternative would be for charging to be introduced gradually, rolled out exchange by exchange as digitalisation proceeds. There may need to be an associated restructuring of line rentals.

9.3.4 Long-distance calls

Trunk calls cost more to handle than do local calls because they pass through more than one exchange. Trunk calls between adjacent regions have lower transmission costs than those between more distant points, although switching costs may be unaffected. The application of a common call charge for long-distance calls throughout a country may bring:

- Economies in the billing system (not large, however);
- tariffs that are easier to remember;
- possible externalities from greater social cohesion.

Whether these advantages will be large enough to outweigh economic losses arising from divergencies against cost depends on their size. Call costs are less dependent on distance than they were, but in a large country, distance-related rates can still be justified.

Calls on relatively heavily-used routes are cheaper to handle because of the economies of scale. A measure of geographical de-averaging of tariffs was introduced in the UK in 1982, when British Telecom brought out special tariffs for routes with lower costs. These were mostly high-volume routes where greater economies of scale could be expected. They were largely the routes on which competition was quickest to develop. BT's main competitor at the time was Mercury, which opened its switched service in 1986 and also offered lower tariffs on selected routes. The tariffs were subsequently averaged and in 1997 there were only three BT call price bands for the whole of the UK — local, regional and national. De-averaged prices have long existed in the US because of the local nature of tariff regulation and the spread of competition.

9.4 Prices for value-added and data services

9.4.1 Audiotex and other premium-rate services

The market for audiotex services (Section 2.4.9) has grown rapidly in recent years. In the basic audiotex operation an information provider sets up a recorded-message service which is accessed through the PSTN at special premium rates. The information may relate to personal, hobby or business interests. It may be updated occasionally or several times a day.

Some network operators provide a wider range of premium-rate charges than others. If the range is unduly narrow, the variety of information that the information provider can afford to provide is reduced. The establishment of network facilities to handle premium-rate services incurs cost and is innovative, justifying a charge.

9.4.2 Virtual private networks

Modern tariff schedules no longer offer a single-rate deal on rental and call price per minute. There is a range of choice, from low user schemes combining low rentals with high call charges, through to the largest bulk schemes with a high monthly charge and the cheapest calls. Virtual private networks (VPNs) are at the top end of the discount structure, occupying a place between PSTN calls and unlimited calling on private circuits.

VPNs do not dedicate capacity to any individual user. Circuits which are not fully used can be made available to other callers. Network planning has to allow for capacity sufficient to guarantee the VPN service level which has been agreed. The network efficiency of VPNs should be higher than that of leased lines because they keep traffic in a common flow instead of dividing it into separate streams, public and private. This enables some of the scale economies inherent in the use of high-capacity circuits to be extended to private circuits of lower capacity. The network-loading potential is increased, giving VPNs lower costs, and hence lower prices, than for leased lines. This advantage holds internationally as well as nationally, leading to the introduction of the international virtual private network (IVPN). An economically efficient tariff for VPN would take into account the usage profile, for data and equally for voice.

PTOs on both sides of the Atlantic and in the Asia/Pacific region have marketed virtual private networks vigourously, offering most of the advantages of private circuits. VPNs are particularly attractive to companies with insufficient levels of business to justify leased lines. The extent to which lower costs are reflected in lower prices depends on the effectiveness of competition. Some operators have tried to market IVPNs as part of a network-services management package which adds back an element of price.

9.4.3 Data network services

Companies using their own private networks for data transmission usually set these up with leased lines which are priced at a fixed rate however much or little they are used. Where the lines are physically dedicated to the user this is an efficient method of pricing, although the line rental should correspond to its cost. Within the company there may need to be a transfer-charging mechanism which ensures that company employees know the cost of the resources which they are using. As agents of the company, the extent to which they react to such price signals depends on the incentive structures that are laid down to motivate them.

Data transmitted over a public packet network or the PSTN locks up circuits for a time which depends on the volume of data sent and the transmission speed. The tariffs offered for this type of service may be call, volume or time related depending on which operator and price package is chosen. Volume charging has a superficial appearance of fairness, but is not always a good measure of the resources used because the duration of the transmission depends on the transmission speed and this may be controlled by the user, for example by the transmission speed of the modem on an analogue circuit. Where there is a value-added element to the charge, for example for the services of an information provider, an efficient pricing scheme would advertise the transmission and value-added elements separately and grade the value-added charge according to cost. Structures of this kind do exist. The call to the information provider may be at the phone company's regular rates, with the bill for the data coming from the information provider. Some parts of the database may be more expensive to access than others.

9.4.4 Call diversion, personal numbering and UPT

Call diversion and related services are more expensive to provide than telephony between fixed points. The cost structure for personal numbering was discussed in Chapter 7. The existence of setup, joining and usage costs justifies a multipart tariff. The numbers themselves may have a value, like those on car number plates, and there is a market in them even though no individual can *own* a number which is part of a national resource. At least eight UK companies were supplying personal numbering in 1997. Six of them raised connection charges and seven had annual charges. All had a usage charge for the basic service and most had other charges for supplementary features. Although some of the variety reflected the efforts of individual suppliers to identify their own costs and the nature of the market, most of it probably represented more enduring features of a new product for which market segmentation was already beginning to appear. Universal personal telecommunications has to be a premium rate service. The costs, especially for any international extension, would be significantly higher than for basic telephony. 'Universal' is, at least in the foreseeable future, something of a misnomer. Even true global coverage is rather far off.

9.4.5 Internet access and usage

Internet access pricing follows that of other online information services in having three potential elements, a time-based fee for a local call to the telephone company, a time-based fee to the Internet service provider and a value-added charge. Local-call charging is based on voice telephony and varies greatly from region to region. Where local calls are timed and priced at cost, Internet traffic can usually be provided for without undue operational problems, although like other forms of data traffic, it has a different profile and may impose a different marginal cost to voice traffic. Internet calls are, on average, longer and their temporal distribution may differ from voice traffic.

Operators providing free or untimed local calls in a tariff schedule designed for voice traffic can be swamped with demand from Internet users who keep the line open for long periods. This began to happen in some parts of the US in the mid 1990s, putting local-call options which had worked well with voice telephony under severe pressure from data users and strengthening the case for more efficient pricing, at least for Internet users where they can be identified.

The charges levied by Internet service providers (ISPs) are basically time based although with a series of flat rate options. Thus, a user may choose to pay £5 per month for up to four hours use, plus £2 per extra hour or may take another option such as £10 per month for ten hours plus £1 per extra hour or £20 per month for unlimited access. The marginal cost faced by the ISP for an additional hour is largely a matter of the transmission and switching cost, plus billing expenses. For inland transmission these can be low for data sent over the high-capacity leased lines which make up the backbone networks. Bearing in mind that packet switching is employed and that for some of the time on line there may be little transmitted in either direction, a payment of £1 (just under 2p per minute) can meet the costs incurred by the ISP. Plummer [5] reports that by 1995 there were some 12.5 million subscribers to online services (including the Internet) in the US. Three firms commanding over 80 per cent of the business, of which one had always run at a loss. Plummer notes the convergence of Internet access prices in the US, tentatively concluding that '*excess profits in the industry are impossible, and with decreasing barriers to entry, a long-run oligopolistic outcome is unlikely*'.

The low level of transmission costs and the improvement of systems for sending voice traffic over packet networks has led to the development of Internet voice telephony for long-distance and local traffic. This may be a useful service in its own right since it integrates telephony into the computer system on the desk, but interest has been stimulated by its apparent cheapness. The cheapness is a reflection of at least three factors:

- Overpricing of long-distance service on the PSTN;
- Internet pricing not fully adjusted to voice telephony;
- service quality differences.

The Internet is one of many competitive operations which will eventually move long-distance PSTN prices closer to cost as markets become liberalised so that, as with the untimed local call, Internet and PSTN prices are likely to converge. The service quality differences may remain. Long delays and blocked lines on the Internet at peak hours could be avoided by the provision of more transmission, switching and port capacity by its managers and those of the information providers. The balance between quality and cost which is preferred by users may remain different to that on the PSTN, but if not it will be a further factor leading to convergence of Internet and PSTN prices. In the long term the fate of Internet voice telephony may be similar to that of voice telephony by domestic resellers. The lines used, whether owned by the network service providers or leased from PTOs, cannot provide capacity any more cheaply than the capacity which PTOs have for their own use and the cost will often be higher through smaller economies of scale. For a while efficient Internet suppliers may have an advantage against inefficient PTOs. Competitive pressure is likely to wear this away. With a potentially higher cost base, money will then only be made out of Internet telephony if PSTN charges remain well in excess of cost, or if there is a market for a quality difference or for the convenience offered.

9.5 Market segmentation

9.5.1 Customer-class pricing

The identification of separate customer groups for tariff and marketing purposes is known as market segmentation and the use of separate price schedules for each is customer-class pricing. Both have become widespread following the advent of competition. In a social-welfare context it is necessary to distinguish between cases where net social benefit is increased and those where an operator uses market power to make abnormal profits from anticompetitive behaviour, selective price increases and the restriction of output, but where net social benefit is reduced. Special deals offering unlimited calls for a flat-rate fee, discounts and other marketing devices have enriched customer choice and increased allocative efficiency. These and other examples of nonlinear tariffs are now common and include:

- a choice between flat-rate and two-part tariffs;
- multipart tariffs, where the usage-related element is broken down into usage-level bands, with lower unit prices in the higher bands;
- block-of-time tariffs, where the customer buys a fixed amount of calling time for a fixed price and obtains a discount if it is fully used; this may be an option within a multipart structure;
- optional calling plans, where the customer is offered a range of two-part or multipart tariffs and can choose the one which seems to offer the best value at the usage level expected;
- leased lines, which offer unlimited usage for a fixed charge on unswitched circuits.

9.5.2 Two-part and multipart tariffs

Two-part tariffs are the commonest form of pricing. Discounts for large users are implicit in tariffs for leased lines, since these are not priced at usage-sensitive rates. The wide area telephone service (WATS) offerings in North America provided an initial extension of volume discounts to smaller users. A five-part WATS tariff offered by AT&T in Maryland in 1981 had a $30.40 per month access charge and separate charge bands with diminishing rates for the first 15 hours, over 15 to 40 hours, over 40 to 80 hours and over 80 hours of monthly use. Each of the bands had separate day, evening and weekend rates [6].

Today, a huge variety of discount schemes based on multipart tariffs is offered by the numerous US carriers. They are available for users of all sizes, at each step trading a higher rental for lower usage charges. The UK followed later, but operators now offer a range of discounts covering most users. The general effect, as illustrated in Figure 9.1, is to change the relationship between call usage and bill size from a straight line towards a curve or envelope.

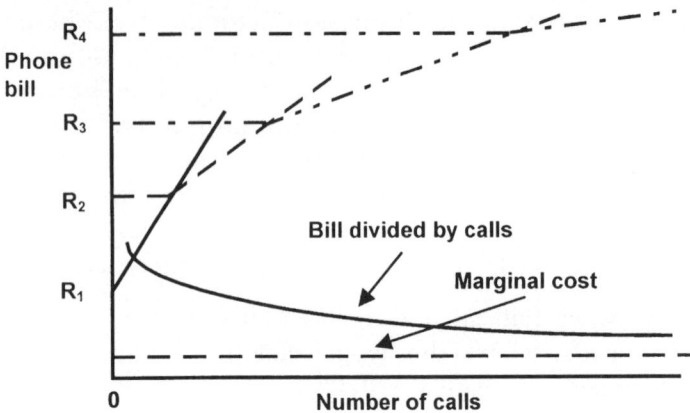

Figure 9.1 Multi-part tariff structure

The Figure shows a four-part structure with rentals rising from R_1 to R_1, and charges per extra call falling as rentals rise. All except R_1 have a free call allowance. Multipart tariffs can bring call prices nearer to marginal cost for high users and provide a cheaper telephone service for users making very few calls. This approach to tariffs has evolved through political pressures at the low-user end and competition for the business of larger customers.

9.5.3 Cellular mobile services

The tariffing of mobile services in the British market has been a long way removed from cost. A large subsidy from call revenues to handset prices keeps equipment costs low (handsets are sometimes free), but call charges are higher than for PSTN telephony and monthly bills are high. Since there are several networks and service is retailed by numerous intermediaries, the tariff structure possibly reflects the preferences of customers, although it is compounded by a lack of interoperability between mobile networks.

Mobile operators have complex multipart tariffs. Table 9.1 shows extracts from an eight-part tariff offered by ORANGE (one of the UK cellular mobile services) in 1997. The tariff allows usage on up to 50 lines to be pooled within a single option.

Table 9.1 Example of monthly mobile tariff options

Monthly charge for first phone	Minutes included in monthly charge	Extra minutes Peak	off-peak	Number of phones
(£)		(£)	(£)	
15	15	0.25	0.125	1
25	60	0.20	0.10	1–2
50	200	0.18	0.09	1–3
75	360	0.16	0.08	1–4
100	540	0.14	0.07	1–5
180	1000	0.14	0.07	1–10
425	2500	0.12	0.06	1–25
800	5000	0.12	0.06	1–50

Extra phones within an option cost £12.50 per month each.
No up-front charge for the handset.

As Figure 9.2 shows, the average and marginal charges per minute fall steeply with increasing usage, while presumably remaining above average and marginal cost at all usage levels.

Mobile operators are not selling into a homogeneous market but into one which is well segmented. Business users generally have their mobile phones paid for by their company and companies may be willing to pay high monthly bills. Selling mobile phones to residential customers who pay their own bills is a different marketing proposition.

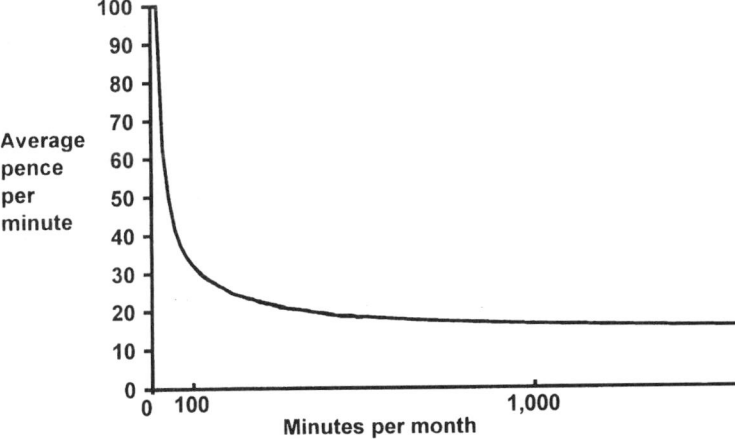

Figure 9.2 Volume discounts in the cellular mobile tariff of Table 9.3

9.5.4 Optional calling plans

Discount schemes will be unfairly discriminatory if they give benefits to customers in competitive sectors at the expensive of those who have no choice. Optional calling plans are examples of discount schemes which are not unfair.

Consumers may be offered a choice between a standard tariff and another tariff structure. This is called *self-selection.* Suppose that the standard tariff is itself optimal within the constraint based on the market demand curve and not on the demand curves of individuals. Suppose, further, that the exercise of the option does not reduce the profit of the operator. Rational and perfectly informed consumers will only choose the option if it makes them better off. Then, as may be shown more rigorously (e.g. Chapter 8 [7]), this must mean that the introduction of the option brings tariffs closer to Pareto optimality. The same is not necessarily true for sales to firms, since these are intermediate inputs and the prices affect competition between firms.

Optional calling plans (OCPs) were pioneered by AT&T, using arguments similar to those of Hausman and Mackie-Mason [7], which are summarised in Chapter 13. Discounted tariff packages were offered as alternatives to the standard tariffs. The lower prices resulted in more calls being made at low marginal cost. The FCC used Baumol's net revenue test (see Chapter 6) to decide whether AT&T was better or worse off after the introduction of an OCP. Schemes were not allowed through unless they passed the test and hence placed no burden on other customers. Options were exercised by the customer, not the operator. The early OCPs offered a block of calling time for a fixed price instead of the standard usage-based tariff and were *ex-ante,* i.e. the customer had to make the choice *before* the period in which the calls were made. The OCPs offered certainty about the bill size, a feature highly enough valued, by some customers, to persuade them to take up the plan even when it cost them more than measured service. Others took them through misjudgment. These factors, coupled with a high price elasticity, enables the OCPs to pass the net revenue test. Later offerings were more complex but worked in the same general way.

Wenders [8] discusses the case for local-service options, where the customer is given a choice between local calls that are free (flat-rate service), local calls paid for at a usage-related rate and local calls with a free allowance and a usage-related charge for higher usage. The rental is fixed at a different rate for each of the options. Rental and usage charges are at least enough to meet marginal costs. The operator can save measurement costs with flat-rate service. He concludes that there are circumstances where the combination of:

- Higher rentals for flat-rate service
- The loss of usage revenue
- Additional usage costs from call stimulation
- The saving of measurement costs

can make a flat-rate service option at least as attractive to the operator as a measured service.

9.6 Prices for radio and satellite-based service resources

9.6.1 Allocation of slots

Two nonphysical resources are required for satellite-based services. These are the celestial parking space, or slot, and the radio-spectrum frequency used for operation. The former has its terrestrial equivalent in sites for radio towers and wayleaves for cables. Each radio system, satellite or terrestrial, needs suitable frequencies for its operation.

With current technology there has to be a minimum spacing of two degrees between satellites on the geostationary orbit, and most are at least three degrees apart, permitting a maximum of 120 geostationary satellites. Spacings are likely to be reduced by technical improvements. The slots are allocated through a consultative process run by the ITU and based on the avoidance of interference between satellites by reference to technical parameters. For more details see Reference 9 and the proceedings of the periodic World Administrative Radio conferences (WARCs) at which allocations are discussed.

The allocation procedures will probably not result in an optimal allocation of resources because the parties involved may have an interest in excluding new operators and because there are no market signals to indicate the value of the slots. Unless sovereignty claims are conceded, the geostationary orbit has no owner and it is therefore not obvious who would be paid if the slots were sold by tender or auction. The trading of unwanted slots can partly rectify this, but it does not free up the initial allocation procedure.

9.6.2 Economics of spectrum allocation

Radio spectrum is a natural resource which is unusual in being spread evenly around the globe. At any point on earth the same range of frequencies is available.

Some argue that spectrum is such a vast resource that there is no shortage. Most would agree that this is not so, and that free provision will result in wasteful use and aggravate shortages. There is no cost attached to obtaining spectrum. The costs are associated with spectrum use and are of three kinds:

(i)   Those arising from the allocation process itself.
(ii)   The cost of equipment which enables more intensive use to be made of the basic finite resource.
(iii)   Externalities, notably the potential of signals to corrupt those of other users.

In an unregulated market there could be a cost associated with keeping other users out.

Spectrum users operating on the same frequency can seriously interfere with each others' signals if there is an overlap in the range of their signal propagation. Exclusive rights within an area and frequency band have a value and are needed to avoid chaos.

The value of spectrum derives from:

- The things that it can do for which there is no substitute, notably in mobile communications;
- Its ability to substitute other transmission media, such as land lines and undersea cables.

There are two basic ways of allocating it:

- Administratively;
- Through a market.

Spectrum is still mainly allocated by administrative means, giving operators the exclusive right to use particular frequencies in particular areas in exchange for a small charge to cover the costs of the allocation process. Charges are important, and the wrong ones can misallocate resources. Low cost encourages wasteful use which, if combined with a rationing system, can keep operators which are less wasteful out of the market. Overpricing discourages economically beneficial use. A market can be created by such measures as auctioning the rights to use spectrum and allowing spectrum usage rights to be traded. This will enable the users that are able to extract the most value out of exclusive usage rights to bid the most for them.

9.6.3 Spectrum auctions

The main difference between market and administrative solutions is in the treatment of economic rents. The value of spectrum may be enhanced by legal restrictions on market entry. The enhancement is a form of economic rent derived from the entry barrier. An operator will bid a positive sum for a license for spectrum, if there is the potential for making large profits from its use. Mobile licenses have been sold by governments for large sums e.g. in Greece. This appropriates some and perhaps all of the excess profit. A charge which extracts part of the monopoly rent will not affect resource allocation unless it leaves so little for the spectrum user that the project is not seen as profitable, given the risk and the cost of capital.

The New Zealand government began spectrum auctions in 1989 under the 1989 *Radiocommunications Act.* Within ten months 97 television, three cellular and 164 radio broadcasting licenses had been assigned to a wide variety of owners. Auctioning proved an effective way of resolving competition for channel assignments. A sealed-bid process was used, in which the license went to the highest tender but the price paid was that of the second highest tender. Prices paid were erratic. For the first round of television licences, bids ranged from NZ$ 0.685–2.371 million with second bids from NZ$ 0.1–0.401 million. Top tenders for the three cellular licences were NZ$ 101.2m, NZ$ 85.552 million and NZ$ 85.552 million. Second bids were NZ$ 11–25 million. The lowest three in each case did not exceed NZ$ 100 and five of them were zero. A full account of the auctions is given in Reference 10. The auctioning of Creman paging licenses is described in Reference 11. DM3.86 m (£1.25 m) was raised and there was little spread in the bids apart from two channels, for which no bids were received.

A wide spread of bids is not a problem unless the bids are very few. Since the bids are essentially valuations of the opportunity to make excess profits in a market with restricted entry, very low bids may only mean that the opportunity to make an excessive return on capital is seen as rather limited. Perhaps there should be more public concern about the high bids.

In the US, mobile communications operators and their licences are discussed in terms of the price per pop. This is the total cost of the bid divided by the total population in the area covered by the license. A 1995 round of personal communications services auctions raised $7.73 bn, about three quarters of what was expected. Most successful bidders paid $7 to $17 per pop and the highest was $31.9 [12].

Auctions of spectrum can also be difficult to manage. When the FCC auctioned capacity for mobile telephony licences in 1997 it accepted bids totalling $10.2 bn (£6.3 bn). The terms included a ten per cent down payment. Trouble began when it emerged that some of the successful bidders could not pay because the money markets decided that they had overbid, making fund-raising difficult. The FCC then agreed to softer terms, including the option to return some of the spectrum for re-auction in exchange for debt reduction [13]. Proposals to use auctions in the UK were announced in 1996. The first application planned was for the spectrum requirements of three third-generation mobile operators [14].

9.6.4 Spectrum management by users

So far, the discussion has been about the economics of spectrum distribution to operators. At least as important is the way in which spectrum is used once it has been obtained. Cellular mobile operators design their systems in a way that maximises the use of the spectrum which they have obtained. Frequencies can be re-used between cells. Interference is controlled by signal strength and cell size. Within the cell, the number of simultaneous calls that can be handled depends on the way in which frequencies are allocated to calls. This is largely a matter of technology and standards options.

The increasing demand for local services puts pressures on all parties concerned to develop systems which use spectrum efficiently. There would be no incentive to adopt such methods if an operator could get unlimited amounts of spectrum without charge. There would be an incentive under a rationing system, but it would not be an efficient way of allocating resources at the margin. Operators might have to spend money on unnecessary spectrum economy if there were spare capacity not made available to them, or not spend enough if allocated more than they needed. Price signals indicating the value of spectrum at the margin would assist in efficient resource allocation.

9.7 Political economy of tariffs

There is a basic problem of political economy in setting tariffs, illustrated in Figure 9.3, based on a chart in Wenders [8]. If calls are over-priced and access is underpriced, line profitability will rise with the number of calls per line. If, then, normal profits are made overall, the operator will make a loss on low users which is balanced by an extra profit from high users. This gives gains to the low users (mainly residential customers) at the expense of losses to the high users (mainly businesses), creating a cross-subsidy from shaded area B to shaded area A on the Figure.

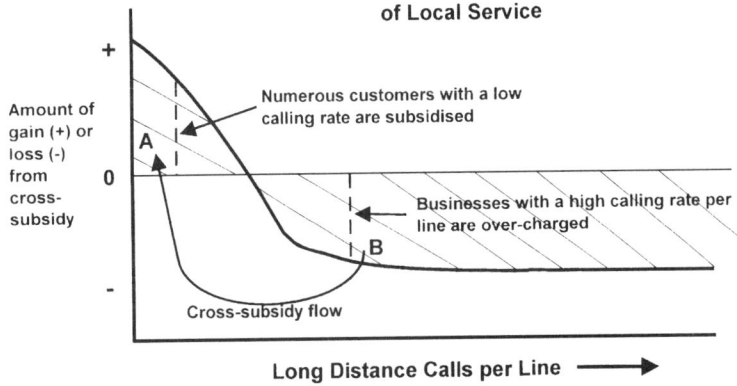

Figure 9.3 Gainers and losers from cross-subsidy of local service

The numerous residential small customers all have votes and can be a strong political lobby. Large users are far fewer and, as businesses, are without votes. This tends to entrench cross-subsidies from large users to small even in western economies, although liberalisation will reduce the problem.

The Perl figures in Chapter 8, in which the percentage of winners outnumbers the losers, are not at first sight consistent with the Wenders analysis, although the net overall gain found by Perl and his exclusion of business users makes them difficult to compare.

A full analysis of the winners, losers and income-transfer effects of a cross-subsidy from calls to access is rather more complicated. Among customers that would still have had a phone without the cross-subsidy, the cross-subsidisation is partly within the bill. The prices may not be economically efficient but they only result in income transfers to the extent that large users are providing more of the access subsidy than are small users. On the whole, this is likely to mean that businesses and private citizens who make a lot of calls subsidise people who have low incomes and customers who live in rural and other high-cost areas.

There is an income transfer for people who would only have a phone if access was subsidised from calls although, again, some of the subsidy is within the bill. There is no income transfer to or from shareholders because profits are unchanged.

Businesses form part of intermediate and not final demand. The income transfer effects of over-charging businessmen for the their use of the telephone depends on if or how the excess costs are passed on to customers. Competitive pressure will limit the extent to which prices can be adjusted but, at least in the long run, the excess costs cannot be carried by the shareholder. Government departments will find the required finance from taxation. Excess costs will be highest in those industrial sectors which use a lot of telecommunications, notably banking, insurance and finance. The output of these sectors will have the greatest appeal to those who are relatively well off and it is this group which contributes most to the subsidy. The products of the agriculture sector, which is a relatively low user of telecommunications, absorb a larger proportion of the income of the poor.

In countries where the penetration rate is low, the high proportion of the population who have no phone receive no benefit from the cross-subsidy, which only benefits those who can afford to join the network. However, they will contribute part of the cost, if they buy the goods and services of firms with excess telecommunications costs built into their prices.

Prices that are set by political or administrative action are subject to many noneconomic pressures and frequently depart from optimality, or from those prices which would prevail in a more competitive market. The cross-subsidy illustrated in Figure 9.3 stimulates demand for lines among low users, but deters people from making more calls even when they are willing to pay more than the additional cost for doing so.

9.8 Tariff trends

9.8.1 Current changes

Tariffs across the world are adjusting to structural changes in underlying costs and the development of greater competition in markets. The changes in cost structure have produced and exaggerated differences in profitability between services, to which the predominantly monopolistic suppliers initially adjusted rather slowly. Greater competition is accelerating the change.

International services are a driver of change. In developing countries they are a source of investment finance, tending to keep tariffs up and encouraging governments to delay competition. In developed countries they are an arena of competition for transit business and the business of multinationals. Dominant operators, new entrants, resellers, dial-back operators and card-based operations are exerting a downward pressure on international tariffs.

9.8.2 Reduction of cross-subsidy

Since 1980 there has been a shift in the balance of tariffs away from call charges and towards fixed charges in many countries. Much of it reflects changes in the underlying cost structure, keeping up with events but making no difference to the relative profitability of calls and line rentals. Residual shifts have been driven by actual or potential market pressures and an interest in cost-based tariffs. The process is known as rebalancing. The rate of rebalancing has shown little difference between liberalised and unliberalised countries for trunk calls. Rentals for access have risen in real terms and call prices have fallen in most countries. Some developing countries are becoming more rather than less dependent on call income, mainly owing to the buoyancy of international revenues. Until the early 1990s tariff rebalancing in developed countries did not do much more than keep pace with cost movements, leaving profit imbalances substantially unchanged. The position has changed since then, with increasing pressure on international tariffs.

9.8.3 Tariff trends in the UK and the US

British Telecom has made substantial changes to the balance of its tariffs and there has been an overall fall in real terms. Changes since 1983 are summarised in Table 9.2.

British Telecom continued to raise rentals and cut call charges for at least a decade, while still keeping a large imbalance in profits. International services remained highly profitable in spite of strong competition from other carriers while the local loop was claimed to lose money, see Table 9.3.

The introduction of subscriber line charges in the US was the main factor in pushing up local telephony charges over the 12 years from 1983, with a particularly sharp increase in the first six years, as shown in Table 9.4. Falling toll prices enabled the average residential phone bill to fall by over three per cent per annum over the period, excluding call volume changes. Indices are weighted averages of nominal prices in the service groups, expressed as a percentage of the average in 1983. Real indices are calculated by deflating the indices of nominal prices by the consumer price index.

Table 9.2 BT price changes 1983-96

Change of price of:	Nominal cumulative (%)	Real cumulative (%)	Real average annual (%)
Exchange line rentals:			
domestic	+82.3	+7.9	+0.6
business	+83.0	+8.3	+0.7
Local calls:			
standard	+15.2	−31.8	−3.1
cheap	+29.3	−23.5	−2.2
Long-distance group A:			
standard	+6.4	−37.0	−3.8
cheap	+50.7	−10.8	−1.0
Long-distance group B:			
standard	−51.5	−71.3	−9.9
cheap	−14.5	−49.4	−5.5
Long-distance group B1:			
standard	−40.4	−64.7	−8.3
cheap	4.6	−38.1	−3.9
International calls (from 1993)	−22.3	−27.4	−10.1
Directory enquiries (from 1993)	−41.0	−47.6	−12.1
Weighted average	−2.0	−42.0	−4.4
Retail prices (July)	+69.0		

Real average annual change is compound, calculated as eqn. 4.2 in Section 4.1.

(Source: calculated from OFTEL [15])

9.8.4 Services to the business community

The direct beneficiaries of tariff rebalancing in the US and the UK have been business users. Scope for similar moves exists elsewhere. Bergendorff [16] puts the Swedish subsidy from business to residential customers at SEK 1.7 bn (£160 m) per annum on a total telephony revenue of SEK 14 bn (£1,330 m). There is a tendency to overlook the way in which benefits to business should flow through to residential customers. Prices that are better related to costs should improve resource allocation in the economy. In this case residential customers, who are the final buyers of all goods and services, gain the benefits of greater economic efficiency in many products where the contribution from telecommunications is neither stated nor measurable.

Table 9.3 BT return on capital by service 1972-94

Service	1972 (%)	1978 (%)	1983 (%)	1991 (%)	1992 (%)	1993 (%)	1994 (%)
Rentals:							
business	8.0	−1.1	3.2				
residential	5.8	−0.5	−5.0				
All rentals	6.4	−0.6	−3.4	−10.6	−11.1	−9.0	−6.1
Inland calls	10.1	−0.5	10.5	54.6	48.9	44.7	46.9
International calls	24.8	25.7	18.8	72.8	51.4	53.0	53.9
Private circuits	11.7	0.5	7.8	11.8	12.3	9.7	17.0

Rentals first showed a negative return in 1978
Inland calls exclude call boxes before 1991
International calls include private circuits before 1991
Private circuits includes international from 1991

(Source: Post Office Annual Reports, OFTEL, calculations)

Table 9.4 Tariff rebalancing in the US

	Year			Average change	Annual %
	1983	1988	1994	1983-88	1989-94
Average monthly rental with					
free local calls $	11.58	16.57	19.00	7.4	2.8
Real rental $	11.58	13.94	12.87	3.8	−1.6
Real index (1983 = 100)	100	120.4	111.2	3.8	−1.6
Intrastate toll					
Index (1983 = 100)	100	97.1	88.2	−0.6	−1.9
Real index (1983 = 100)	100	81.7	59.8	−4.0	−6.1
Interstate toll					
Index (1983 = 100)	100	70.1	74.7	−6.9	1.3
Real index (1983 = 100)	100	59.0	50.6	−10.0	−3.0
Whole residential telephone bill					
Index (1983 = 100)	100	117.4	123.3	3.3	1.0
Real index (1983 = 100)	100	98.8	83.6	−0.2	−3.3
Consumer price index	100	118.8	147.6	3.5	4.4

(Source: calculated from FCC statistics)

There should also be more direct benefits in any but the most monopolistic markets. A business should be forced by the pressure of competition to pass on all the telecommunications cost savings to its customers in the form of lower prices, though again this is unlikely to be in a way which is explicit.

9.9 Tariff diversity

9.9.1 The effect of competition

The initial effect of the introduction of competition is to encourage the development of diversity in the price and quality of services. In a mature and perfectly competitive market, there would be one price for any particular service and quality and residual differences would reflect differences between product. Even early competitive pressures will start to drive prices towards cost, encourage pricing structures with finer graduations and bring in a wider range of quality choices, although perfect telecommunications service competition is not likely to appear.

The market segmentation described in Section 9.5.2, in which higher volume customers can obtain lower usage charges (bringing them down towards cost) by paying higher fixed charges, improves economic efficiency by bringing prices closer to marginal cost for the large users and is preferable to not doing it for anyone.

Table 9.5 shows a cross-section of 12 operators in the UK market of 1992. A range of usage prices had replaced the single rate that applied in the days of the BT monopoly. Most companies were offering alternative, volume-related, options. Fixed charges in respect of these options are not included.

Table 9.5 Dialled calls from the UK in 1992

| Route | Peak-rate pence per minute | |
	Lowest	Highest
Australia	55.57	70.00
Belgium	28.30	32.77
Canada	43.89	53.05
France	27.45	33.00
Germany	27.45	33.00
Hong Kong	66.35	86.01
Japan	82.94	109.57
Norway	38.14	43.30
US	31.85	53.05

The variety of tariffs in the US was at the time greater. Self's 1992 guide to AT&T long-distance services alone [17] ran to 48 closely-printed pages, with a choice of seven peak-rate schemes on each international route. The guide itself sold at ten different prices, depending on how many were bought.

9.9.2 Cost-based prices in developing countries

The elimination of cross-subsidies is a second-order consideration compared with the need to get the average tariff level correct. A more viable short to medium term strategy may be to rely on continued support from business users until the overall level is right, because they are more willing to pay. Many developing countries find a secure source of revenue for development funding in the profits which they make from international services, through high charges for outward calls. They can also make large profits on incoming calls (see Chapter 17). Most international calls from these countries are made by businessmen or tourists. Business users may also be willing to pay higher rentals, six times higher than residential customers in the case of Latvia in 1992 and 58 per cent higher than residential customers in the UK in 1997.

9.10 References

1 CRACKNELL, D. R.: 'The impact of structural changes in telephony prices'. Paper presented to ITS conference, Sydney, 1994
2 VIETOR, R. H. K.: 'AT&T and the public good: regulation and competition in telecommunications 1910-1987' *in* BRADLEY, S.P., and HAUSMAN, J. A. (Eds): 'Future competition in telecommunications' (Harvard Business School Press, Boston, 1989)
3 British Telecom: 'More big price cuts on the way', *BT Today*, September 1994 p. 1
4 PARK, E. R. and MITCHELL, B. M.: 'Optimal peak-load pricing for local telephone calls' (Rand Corporation, Santa Monica, 1986)
5 PLUMMER, A. C.: 'The second coming of videotex services: how is the U.S. faring with its on-line services industry?'. Michigan University paper presented at ITS *11th Biennial Conference* June 1996, Seville
6 PHILLIPS, C. F. Jr.: 'The Regulation of Public Utilities' (Public Utilities Reports Inc, Arlington, 1985)
7 HAUSMAN, J. A. and MACKIE-MASON, J. K.: 'Price discrimination and patent policy', paper presented at NERA Seminar, 23rd October 1986, London
8 WENDERS, J. T.: 'The economics of telecommunications' (Ballinger, Cambridge MA, 1987)
9 EWARD, R.: 'The deregulation of international telecommunications' (Artech House Inc, Dedham MA, 1985)
10 MUELLER, M.: 'New Zealand's revolution in spectrum management'. *Inf. Econ. Policy*, 1993 **5**, (2) pp.159–177

11 RUHLE, E. O., and STÜRMER, S.: 'A comparison of spectrum auction design in the US and in Germany – the ERMES example'. Paper presented at *Global networking 97* joint conference, Italy, 1997
12 'PCS bidders get their licences cheap' *Mobile Communications*, Issue 168, 23rd March 1995, London
13 SUZMAN, M.: 'Phone Licence sale shake-up', *Financial Times*, 26 September 1997.
14 DTI: 'Multimedia communications on the move'. Consultation Document' (Department of Trade and Industry, London, 1997)
15 OFTEL 'Pricing of telecommunications from 1977'. Consultative Document' (OFTEL, London, December 1995)
16 BERGENDORFF, H: 'Swedish policies towards universal service', SPRU/PICT conference paper 4, October 1990
17 SELF, R.: 'Long distance for less #13 AT&T' (Market Dynamics, New York, 1992)

Chapter 10
Industrial structure and ownership

10.1 Market liberalisation

10.1.1 World trends

Liberalisation is an umbrella term used for the introduction of competition into product and service markets by relaxing entry restrictions, and into capital markets by privatisation. The liberalisation of telecommunications markets began to spread round the world during the 1980s. It still has a lot further to go, but is unlikely to break the dominance of the largest operators. Equipment suppliers have found greater scope for export sales, while their home markets have been threatened by imports. Developments in switching, transmission and software made it possible for new operators to enter the market and bypass or retail the facilities of big networks with statutory monopolies. Plant became cheaper, with economies of scale accessible at lower traffic levels. With digitalisation, voice and data traffic could travel together and be largely indistinguishable.

The spread of competition has been uneven across telecommunications networks. Transmission, especially over long distances, has seen extensive market entry where this is permitted, facilitated by the existence of alternative transmission media, large traffic flows and the possibility of sharing economies of scale by retailing bulk capacity. Switches have tended to be a bottle-neck resource at local level, controlled by a single operator. So has transmission in the local loop from exchange to terminal apparatus.

The influence of technological change has sometimes been overstated. Liberalisation has less to do with technology than with a renewed interest in market solutions to supply-side problems. Competition could have been kept at bay by sufficiently restrictive regulation, but there has not been the will to do this. In the industrialised democracies, legislators and regulators were faced with a choice between permitting heavy restrictions on the use of leased lines, to prevent resale, or encouraging competition to reduce the cross-subsidies that made resale profitable. They opted for the latter.

Market liberalisation has followed different paths in the US and Europe. In the US, liberalisation was forced by litigation within an existing legal framework. This required creative thinking on the part of the courts and the FCC, but did not need any new laws. The 1996 Telecommunications Act came over twenty years after the process had begun. The spread of competition was slower within states than between them. State regulatory commissions tend to be protective towards local operators and their residential customers (who still, in some areas, elect the regulators). The FCC has been more energetic in encouraging interstate market entry. Intrastate long-distance rates have been less affected by competition than those between states.

Whereas the divestiture of AT&T was achieved through the courts, BT in the UK and NTT in Japan were privatised by statute. MCI won its network rights through a series of court actions, and Mercury got its right to enter the market under the 1981 British Telecommunications Act. Lawyers are becoming important in the British regulatory process, but they have yet to achieve the importance which American lawyers have.

In the UK, the customer-premises equipment market was liberalised without entry restrictions and quickly gained diversity and lower prices. Market entry into network services became a managed process in which, for the first five years, the Department of Industry tried to limit entry and create a quasiplanned approach to network development. The two operators that were allowed into the mobile cellular market were given roll-out targets to encourage the spread of mobile networks across the UK. A rapid relaxation followed in the 1990's and hundreds of operators now offer local, long-distance and international services. In 1980 there was only one.

Most states of the European Union did not initially show much interest in liberalisation and their PTT monopolies showed even less. Later, Brussels pushed often reluctant governments into accepting a more liberal regime, with competition permitted in telecom-munications service markets from 1998. More aggressive moves and a willingness to trust market forces that even exceeded the zeal of the UK came from Australia, New Zealand and some of the countries of eastern Europe. At one time it looked as if the collapsed regime in East

Germany would have privatised its telecommunications more quickly than the West German state-owned Deutsche Bundespost, with which it was merged after German reunification.

10.2 Privatisation

Privatisation and deregulation became widespread in many industries all round the world; telecommunications was among the privatisation leaders, but usually not the first. By the end of 1996, privatisations in around 40 countries had raised some $US 140 billion [1]. Many privatisations fell short of leaving a minority of the voting shares with the government, as Table 10.1 shows. Continental Europe and the Far east have been notably conservative whereas, in the Americas, voting control has always been relinquished.

The term privatisation covers a range of ten positions in the shift from full state to full private ownership. Ten stages can be distinguished, started with the unreformed PTT:

1   Government department without separate finances.
2   Production of separate accounts.
3   Separation of telecommunications from postal services.
4   Establishment of a separate corporation with employees who are not civil servants.
5   The removal of regulatory functions to an independent body.
6   The introduction of shares into the capital structure.
7   Sale of a minority of the shares to private interests.
8   Sale of a majority of the shares, with state retention of a Golden Share, or special share, giving it control over ownership.
9   All shares in private hands, but legal restrictions limiting foreign ownership.
10   Sale of all shares, no restriction on ownership.

This is more or less the path traced by the telecommunications arm of the UK Post Office as it evolved into British Telecommunications plc up to stage 10, the final step being the lapsing of the special share in the late 1990s. With few exceptions North American telephone companies have long been at stage 9. Few were ever state owned. The wave of privatisations that rippled round the world from the late 1980s onward often fell short of relinquishing a majority interest. In considering how far these privatisations made any economic difference we have to consider how they affected the behaviour of

Table 10.1 Privatisations 1984-97

Voting control retained	Voting control sold
Asia, Far East and Australasia:	
Australia (Telstra)	Australia (AUSSAT)
India (Bombay/Delhi)	
Indonesia	
Japan (NTT)	
Malaysia	
Pakistan	
Republic of Korea	
Singapore	
Europe:	
Belgacom	Hungary (Matav)
Czech Republic	Spain (Telefonica)
Denmark	United Kingdom
Deutsche Telekom	
Estonia	
France Telekom	
Greece (OTE)	
Irish Republic	
Italy	
Latvia	
Netherlands	
Portugal	
Slovenia	
Ukraine	
Yugoslavia	
Middle East and Africa:	
Israel	Guinea
South Africa	
North and South America:	
	Argentina
	Barbados
	Belize
	Canada (Teleglobe)
	Chile
	Guyana
	Jamaica
	Mexico
	Peru
	Venezuela

their managements. Some, even within the state sector, proved capable of behaving in an entrepreneurial and efficient way. Others, passing from state to private monopolistic status, were slow to display much change. Build/operate/transfer (BOT) schemes have been used in some developing countries, where a foreign operator upgrades and manages a network and then hands it back to the government after a period of franchised operation.

Governments have commonly limited the ability of capital markets to change the ownership of undertakings regarded as of strategic or vital interest to the national economy. Foreign control may be specifically illegal, or it may be ruled out indirectly by other regulations such as those in the 1934 Communications Act in the US, which restricted the use of radio frequencies by foreign companies.

In the UK privatisations of the gas, water, electricity and telecommunications services in the 1980's, the government held back a single special share. The owner of the share had the right to prevent certain of the founding articles of the company from being changed. These articles are the ones which specify key control factors such as the nationality of the chairman and the maximum shareholdings of individual owners, 15 per cent in the case of BT. The special share is retained by the government indefinitely or for a fixed term (initially, BT and C&W both had special shares). Some special shares had a limited life, for example five years.

The manner in which the shares of fixed-link and satellite operators have been sold in privatisations around the world has varied a good deal. The main possibilities were:

(i)   Public offering of shares, with special allocations for employees, customers and small investors. This method was used in most of the UK privatisations.

(ii) A trade sale (sale by tender), the shares going in large blocks to one or a small group of buyers in the industry. This was characteristic of eastern European privatisations.

(iii) Leasing or management contracts. Ownership was retained but management put into the hands of a commercial operator. This has long been used by small state monopolies in the Caribbean the Middle East and parts of East Africa. Cable & Wireless had a number of such contracts.

These privatisations transferred physical network assets to the new owners, often with an obligation to finance faster network growth.

Cellular mobile services have commonly been established in the private sector from the outset by the sale or auctioning of licenses. No network assets have been tranferred, the licenses being for the construction and operation of a new network.

The original logic for the privatisation of state monopolies, such as telecommunications in the UK, sprang partly from the belief that competition would bring benefits to the customer, as stated in Reference 2. It was believed that fair competition could not exist while a major participant was owned by the state, since the state operator was for practical purposes immune from the threat of bankruptcy. There was also a conflict of interest as long as the government was both owner and regulator. It was necessary to privatise in order to create conditions for fair and beneficial competition. It was also believed that privatisation would result in better asset management, greater operational efficiency, sound product innovation, an improved attitude towards customers and a greater ability to cope with change.

Some doubted whether privatisation alone would be sufficient to improve performance and the arguments have tended to follow several divergent tracks:

(*a*)   Competition is needed to get benefits from privatisation *or* competition would bring the benefits without privatisation *or* privatisation, under the right regulatory conditions, can bring all the benefits without competition.

(*b*)   State monopolists would act efficiently if they had the borrowing and diversification freedoms of private companies.

Empirical evidence in support or refutation of these views has been less prominent. Proposition (*b*), which echoes the original case made out in 1933 by Herbert Morrison [3] for nationalisation, is the failed UK model. The issues are not new ones, or confined to telecommunications, and have been argued over in North America for many years. It is sometimes not recognised that large organisations take time to adapt to a changed environment. The biggest privatised UK companies took at least five years to change significantly and continued to evolve for up to a decade.

Staranczak *et al.* [4] examined productivity data from ten OECD countries and concluded that although output growth was the main determinant of productivity growth, private ownership also increased productivity. No evidence of a connection between facilities-based long-distance competition and productivity growth was found, possibly because of the degree of competition in the countries examined.

ITU data [1, 5] shows that the privatisations in Argentina, Chile, Malaysia and Mexico were associated with a substantial investment increase. This is consistent with the expectation that privatisation will mobilise funds where governments cannot or will not do so, and that long waiting lists are an example of government failure in a commercial sector.

10.3 Service quality under privatisation

Private monopolies have sometimes been under greater pressure to improve standards, especially if the regulator becomes actively interested. By having power over the company, the regulator may become its most important customer for quality improvement. Private utilities in the US have tended to retreat onto the high ground of quality because of the regulatory appetite for service improvement. Under conditions of monopoly this may encourage excessively high standards, which do not represent the best trade off for the customer actually paying the bills. Privately-owned utilities in the US appear to have learnt long ago that a poor performance on quality could engender a powerful regulatory backlash.

Quality is subject to more commercial pressure where competition exists, although as long as regulation is in place, the operator has to balance the pressures coming from the customers who pay the bills and from the regulator who sets the rules. It has been the experience of both North America and the UK that regulated private companies cannot afford to let quality standards slip because of the regulatory risks. Businesses owned by the state may, on the other hand, ignore public complaints without a threat to their vital interests. Privatisation has made the utility industries more vulnerable to quality complaints and the same might be expected elsewhere, in both developed and developing countries.

Quality became a big issue for the newly privatised gas, water and electricity companies in the UK, as well as for telecommunications operators. There are already examples, especially in the water industry, where government quality requirements were imperfectly costed and may have been pushed too far.

The UK water industry provides a good example of the effect of privatisation on service quality. Water purity legislation originating in the European Commission required substantial expenditure by the British water companies, with a consequent rise in customer bills. In the ensuing debate, public expectations of water quality began to rise, but this followed rather than preceded legislative change. At the time of privatisation there were widely expressed fears that water quality was bound to suffer as water companies cut costs to drive up profits. In fact, nothing of the kind happened.

Most water companies soon saw that quality was potentially a damaging public and regulatory issue. They did their best to avoid criticism by persuading the regulator that they should be allowed the capital expenditure for quality improvement and the chance to make a profit on it, putting upward pressure on the price limits imposed by regulation. One water company underestimated the risks of poor service. It made large profits but inadequate provision for demand during a drought year. The waves of criticism and bad publicity which followed, from the customers, the press and the regulator, did great damage to its reputation and are likely to have been costly in commercial and regulatory terms.

British Telecom initially underestimated the importance of quality after privatisation and stopped publishing quality-of-service statistics on commercial grounds. A deterioration of service after its engineers went on strike and the poor state of its public payphones led to a storm of press criticism and the company suddenly found itself vulnerable. It gave up an allowable tariff increase worth around £100m as part of its efforts to save the situation. Statistics are again being published, the payphone network has been greatly improved and quality is no longer a significant issue.

The appearance of conferences devoted to quality topics confirms its new importance among operators. In the UK, the publication of quality indicators by OFTEL, e.g. Reference 6 and 7 and the analysis of quality issues in telecommunications by Oodan, Ward and Mullee [8] are further examples of the importance now attached to this aspect of service.

A study carried out into the effect of privatisation on service quality in the UK by National Economic Research Associates (NERA) [9], found a widespread improvement and confirmed these impressions. Privatisation resulted in an increase in data about service quality and ended self-regulation. The rather sparse information available from most industries before they were privatised made a comparison of trends difficult, but clear evidence of improvement was found in the water and energy sectors. Improvements in telecommunications service quality were substantial, partly arising from continued technological progress. NERA considered that BT was better able to finance new investment for quality improvement and agreed that the pressure from OFTEL raised the priority given to it.

10.4 Vertical and horizontal integration

10.4.1 Structural choices

A firm may wish to extend its operations away from the core business to related activities in order to become bigger and more profitable. First choices usually lie in one of two directions — vertical integration or horizontal integration. A risk to which both vertical and horizontal moves are prone is that the necessary management expertise is lacking. Manufacture of a product in a competitive market requires different management skills to those needed for a regulated service provider. Cable TV customers may not readily respond to marketing initiatives which work well for telephones.

Examples of vertical integration for a local telecommunications service provider are moves into:

- long-distance services;
- international services;
- customer equipment retailing;
- manufacture of telephone exchange equipment;
- installation of exchange equipment;
- construction of private networks;
- manufacture of telephone handsets.

Examples of horizontal integration for an operator are:

- geographical extension, providing local service in other areas, for example adding Wales and Scotland to England,
- providing cable TV over the local telephone network,
- adding mobile to fixed-link services.

There are three main factors encouraging integration:

(i) The exploitation of economies of scope.
(ii) Ensuring security of supply.
(iii) The reduction of risk.

Economies of scope may exist in research and development costs if they can be spread over a wider range of vertically or horizontally-integrated activities. Billing and marketing may be made cheaper by moving into related services. The main sources of advantage are likely to be:

- Efficient transfer of information about technology and market requirements.
- Synergy with other activities, particularly those using research

and development. For a network operator, these may be value-added network services. For a manufacturer, they include sales to other industries.

10.4.2 Switch manufacture

Security of supply has been a consideration for operators in the procurement of telephone exchange equipment, since a large quantity of expensive plant needs to be maintained and added to over a long period of time. For a manufacturer, the sale of an initial tranche of exchanges of a particular type is likely to encourage the operator into larger, later purchases for network extension and modernisation because:

- the initial costs involved in making the choice will already have been incurred;
- the operating characteristics of the equipment are well understood;
- customisation will already have been carried out;
- equipment prices can reflect scale economies.

Bell Canada [10] stresses uniqueness as a contributor to risk. A product for which one operator is the sole market has a high degree of uniqueness which may arise from the way in which the product is specified. The market is then controlled by the buyer. Switching equipment manufacturers with substantial investments in research and development and in need of secure markets have usually welcomed long-term agreements with their customers for similar reasons, whether or not the two are vertically integrated. There are examples of manufacturers vertically integrating into network construction for third parties and, particularly in mobile communications, the operation of networks themselves.

As with the practice of buying only from domestic manufacturers, vertical integration reduces competition between external suppliers and is likely to result in higher procurement costs if market and regulatory conditions are such that the extra costs can be passed on to the customer. The prices at which goods are exchanged between vertically or horizontally integrated units are *transfer prices*.

There are three resource-allocation matters affecting transfer prices for the supply of equipment:

(i)   Manufacturing costs may be higher than those of efficient competitors. The manufacturing unit charges transfer prices which show a fair profit but the operating arm could have bought the equipment from competitors for less. The return to the shareholder is less than it might have been (not sustainable in the long run), or network service prices are higher than they could have been. Either way, there is a misallocation of resources in manufacturing.

(ii)   If manufacturing costs are at least as low as those of efficient competitors but internal transfer prices are too high, this will result in a high manufacturing profit. If the high transfer prices are then passed on to customers in the form of high network-service prices the high profit will not show in the network-service accounts. Here, the overpricing of network services is likely to result in a misallocation of resources in the supply of network services.

(iii) Whatever the costs are, the transfer prices may be too low to cover them and the loss is recouped from excessive prices on network services in monopolistic markets. This is a cross-subsidy. It misallocates resources in the supply of both the equipment and services markets. In the equipment market, it is an unfair competitive practice.

The introduction of stronger network competition puts greater pressure on the in-house manufacturing unit to control its costs by making it more difficult to build excess equipment costs into network-service prices. With lax regulation, the excess profits on the manufacturing operation generated by excessive transfer prices will reduce the apparent profit from price-regulated network operations. External equipment sales will be at risk if excessive costs are reflected in prices to outsiders, so such sales will either not be made at all, be made at marginal cost or show a loss.

Operators competing with a vertically integrated operator may choose to buy network equipment from a third party for several reasons:

(*a*) They do not need to reveal potentially sensitive information about the design and local capability of the network to a competitor.

(*b*) The purchases do not add to the profits of a firm with which they are competing.

(*c*) They do not need to suspect any conflict of interest.

Rao [11] points to the rising importance of software in equipment design and concludes that '*The emergence of a competitive, stand-alone software industry, combined with a trend towards open operating systems and customer demand for flexibility, may have undermined the case for vertical integration in the telecommunications industry*'. Rao found that the separation of local networks from manufacture in the US after 1984 led to a reduction in research funded within operating companies, but that R&D intensiveness, defined as R&D as a percentage of sales revenue, had risen in all the relevant manufacturing sectors. There was no break in the upward telecommunications productivity trend. Vertical integration of a switch manufacturer with a dominant operator is now confined to Bell Canada Enterprises (BCE). Even here, the relationship is at armslength. BCE has separate operating divisions for domestic network operations (Canadian Telecommunications) and manufacture (Nortel), and its accounts identify them separately. The procurement of switching and transmission equipment is an important element in corporate strategy and is discussed further in Chapter 18.

10.5 References

1   BESANÇON, L and KELLY, T.: 'Telecom privatisations: the new realism', *ITU Speeches and Discussion Papers*, ITU website, 1998
2   MOORE, J. 1983: 'Why privatise?' in KAY, J., MAYER, C. and THOMPSON, D. (Eds.) 'Privatisation and regulation – the UK experience' (Oxford University Press, Oxford, 1983)
3   MORRISON, H.: 'Socialisation of transport' (1933), *quoted in* 1
4   STARANCZAK, G. A., SEPULVEDA, E. R., DILWORTH P. A. and SHAIKH, S. A.: 'Industry structure, productivity and international competitiveness: the case of telecommunications', *Info Econ. Policy*, 1994, **6**, (2), pp. 121–142
5   'World telecommunications development report 1994' (ITU, Geneva, 1994)
6   'Telecommunications companies: comparable performance indicators January–June 1996 business customers' (OFTEL, London, 1996)
7   'Telecommunications companies: comparable performance indicators January–June 1996 residential customers' (OFTEL, London, 1996)
8   OODAN, A. P., WARD, K. E. and MULLEE, A. W.: 'Quality of service in telecommunications' (The Institution of Electrical Engineers, London, 1997)
9   'The performance of privatised industries: A report by NERA for the centre for policy studies, volume 4: Prices and service quality'. National economic research associates, London, 1997
10 Bell Canada: 'On the need for large size and vertical integration in the research and development and manufacturing of telecommunications products', Bell Canada Special Task Force paper presented in Paris, June 1980
11 RAO, P. M.: 'R&D and innovation in U.S. telecommunications: recent structural changes and their implications'. Paper presented at ITS *11th biennial conference*, June 1996, Seville

Chapter 11
Forward planning

11.1 The planning process in telecommunications

11.1.1 The role of forecasting

Action is in the present and the future. Most management decisions have to be based on forecasts, for which history is an imperfect guide. Engineers want the figures for system planning, and regulators, especially in the UK and the US, want them as an aid to setting profits and prices. Operators use them extensively for business planning, budgeting, estimating the effects of regulatory constraints and other purposes.

11.1.2 The dominance of capital investment

Since assets such as ducts and buildings are long lived and expensive to establish, investment decisions, good and bad, have effects on the profits for many years into the future. Equipment embodies heavy research and development costs which have to be recovered over a long period of time. The management of investment programmes requires a great deal of effort which itself needs careful planning. New capital investment is large in relation to turnover and annual depreciation provisions are typically similar in size to the pay bill. Planning for growth and change is therefore a major activity within all telecommunications operators.

Figure 11.1 shows how British Telecom's investment has moved in relation to turnover over the thirty five years to 1997. The peak in the 1960s came at a period of maximum network growth. Spending later moderated with the move towards saturation of the domestic market and cost savings from use of digital technologies. The business was run by the Post Office until it was split off as a separate corporation, British Telecom, in 1981.

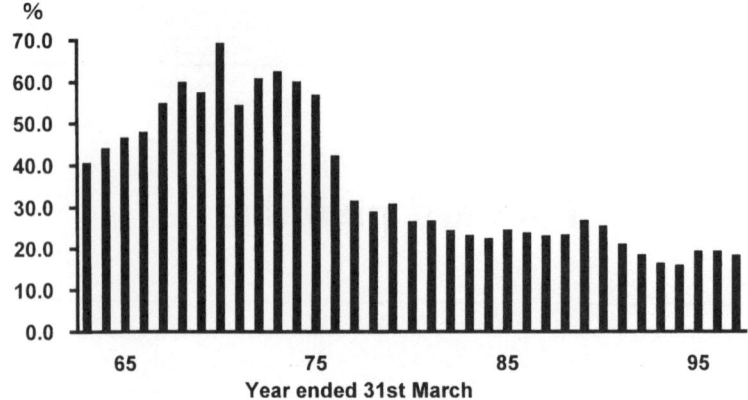

Figure 11.1 BT investment as % of turnover 1963–97
Source: Post Office and British Telecom Annual Reports

British Telecom's level of investment is lower than that of some other operators in developed countries, as is shown by Table 11.1.

Table 11.1 Comparison of investment as a percentage of turnover

Company	Year	%
Equipment manufacture:		
Alcatel, France	1994	4.1
Nokia, Finland	1995	11.4
Northern Telecom, Canada	1992	6.8
Long-distance networks		
AT&T (Plant additions)	1994	6.5
Telecom Denmark	1992	8.5
Networks with local services:		
Belgacom, Belgium	1995	33.0
Bell Canada	1992	32.8
British Telecom	1997	18.9
France Telecom	1994	22.1
NTT, Japan	1989	30.2
Telecom Australia	1990	34.9
Telefonica, Spain	1995	37.5
Seven RBOCs, US	1994	20.1
Networks in world income groups:		
Low income	1992	60.8
Lower middle income	1992	36.4
Upper middle income	1992	47.9
High income	1992	28.1
World	1992	31.0

(Source: Company Annual Reports, ITU [1])

In the monopoly conditions which commonly prevailed until the 1970s, the market requirement for network growth was reasonably predictable, at least in comparison with the forecasting problems of most manufacturing companies, even for long periods ahead. The forecasting of profits, prices and the sources of new capital could be relatively undemanding in a regulated monopoly, whether it was owned privately or by the state. In the UK, the government gave Post Office Telecommunications (as it then was) a financial target for the return on assets, and tariffs were usually set to achieve this – which was done with considerable precision where political factors permitted, as Table 11.2 shows.

Table 11.2 Return on sssets for telecommunications comparison with target, UK 1971–80

Year ended March	Target %	Achieved %
1971	9.6	9.8
1972	10.0	9.8
1973	10.0	8.6
1974	none set	6.9
1975	none set	6.4
1976	none set	6.2
1977	6.0	7.6
1978	6.0	6.1
1979	6.0	6.9
1980	5.0	4.6

(Source: Post Office accounts and report, 1979-80)

Targets were in nominal terms before 1975 and in real terms from 1977. Targeting was suspended during a government-imposed price freeze in the middle years. The stability of the return occurred in spite of large telecommunications service price movements. Average tariffs rose by 148 per cent nominally and fell by 26 per cent in real terms over the period.

The most demanding work was in the area of technological forecasting, where the need was to find an optimum time path for the modernisation of the network. For example, should the operator wait until a fully digital exchange design was ready or would it be more cost effective to upgrade older exchanges and defer full digitalisation? Mostly, this kind of planning was concerned with cost minimisation, using the lower maintenance of digital equipment to justify the replacement of older plant that was still in working order. Later, the revenues from new services that digitalisation would make possible became more important. Management of the relationship with equipment suppliers is a vital part of the process and can take up a great deal of management time. There is a good description of the capital planning process at British Telecom in Harper [2].

11.1.3 Aggregating risks

A telecommunications operator carrying out projects in different markets and countries will encounter a range of different risks. Work on portfolio analysis in Reference 3 first drew attention to the different effects of market and specific risks (Section 7.7.2) on portfolio returns and, by extension, on the expected return from a set of investments. Some risks move independently of all the others; others do not. The total variance is the sum of the variances of the individual risks plus the sum of the covariances between the risks. Choice of project mix can be helped by operational-research methods which optimise the mix of investments to obtain the highest return for any given overall risk factor. There is a fuller analysis of portfolio theory in Reference 4.

The conclusions that follow from the theory are at first sight surprising, especially as regards overseas diversification. First, although there may be a significant element of country risk which increases the base on which risk premiums are built, the covariance with domestic risks may be low because the world's economies do not all swing together. Overseas diversification may therefore reduce the average covariance of performance within the investment portfolio and hence reduce the cost of capital to the business.

Secondly, although diversification may reduce the overall risks of the firm, it does not necessarily benefit the shareholders. They could have made their own choices of investment outside the company and optimised their portfolios according to their own preferences. Investments will only add to shareholder value if they achieve a higher rate of return than the shareholder could have done with the same resources. There are ways in which a higher return might be achieved, mainly arising out of imperfections in the capital markets. One example is the opportunity to buy shares in foreign telecommunications operators through a trade sale rather than a stock market flotation.

11.1.4 Planning for uncertainty

Competition has greatly increased the difficulty of forecasting revenue and investment needs. Uncertainty about all parts of the process is greater and it extends even to the future existence of the operating company itself. Plant lives have shortened with modern technology (Section 7.4.1), but the fundamental consequences of the greater risk have not changed. An operator will not sink money into expensive, long-lived assets without the confidence that market and regulatory conditions will enable their cost to be recovered through revenues in later years.

Even under monopoly conditions there can be a great deal of uncertainty. Planning processes in the public sector have the potential to underestimate risk, particularly with projects which have a low return realised over a long period and perhaps financed by borrowings at low interest rates. Telecommunications has not been especially vulnerable to this. However, the experience of some other industries, where assumptions about technological change and future factor prices have been seriously wrong, provide a warning. Power generation investments provide several examples, including expectations of future oil and gas prices and the cost of decommissioning nuclear power stations.

In a world which is changing at what seems to be the fastest rate in its history, strategies for reducing risk and increasing flexibility by deferring spending decisions until they really need to be made, have become important. The outsourcing of plant and services to contractors, replacing internal by external resources, is a typical response. Outsourcing first came to prominence in the computer services industry. Companies that had built up large departments to provide internal computer services were offered management and service contracts which used the greater experience of the service companies to provide equivalent services at lower cost. The service companies could build up a comparative advantage through scale economies and by using their wider experience of corporate requirements. Potential savings are seen as coming from:

- lower costs from specialist suppliers;
- a reduction in fixed overheads because external contracts can be changed more quickly than internal manning levels, giving greater flexibility.

Network operators may be significant outsourcers, cutting staff numbers and closing in-house vehicle maintenance, catering and other ancillary services in favour of external suppliers.

11.2 The financing of investment

11.2.1 Funding for a commercialised telecommunications operator

In principle, an investment that promises a return higher than the cost of the capital needed to carry it out should be financeable in its own right by new borrowing. In practice, the imperfections of capital markets, which have already been discussed make the method of raising finance a matter of some importance. The essentials of the funding equation for a commercial operating company are summed up in a cash-flow statement, showing the sources and uses (often called applications) of funds, which is common in company accounts. This is a simplified example:

Sources (inflows)	Uses (outflows)
retained profits (after depreciation provisions)	additions to reserves and working capital
depreciation provisions	new investment
new borrowing	interest
sale of new shares	dividends
state grants and subsidies (if any)	taxes
Total of sources equals	Total of uses

Trading profits and depreciation provisions on earlier investments are a source of finance. Profits that are retained to reduce the amount of new borrowing needed to finance new investment cannot also be paid out in dividends. Retained earnings are disguised borrowing. If internal sources cannot provide all the required finance, the company will need to raise money on the capital markets. Telephone operators are usually regarded as low risk, but investors will not lend to a company which appears to be overstretched in its existing borrowing. The financial management of the company requires investor perceptions to be taken into account in the financing plans.

For a regulated monopolist there is a third possible source of finance, which is to obtain permission to generate more profit by raising prices. For example, in periods of rapid expansion, as has happened in some developing and East European countries, tariffs may have this role in investment finance. This is justifiable if the increase draws them towards, rather than away from, short-run marginal costs.

11.3 Forecasting horizons

Telecommunications business planners work to three horizons — the short, medium and long term. The short term runs through the current financial year and, as it draws to its end, takes a preliminary look at the next year. There will be a forecast of the main features of the accounts and probably a budget and considerable extra detail. Forecasts of revenue will usually be split between volume and price.

The medium term runs out to about five years and is mainly concerned with investment requirements and how to finance them. The term needs to be long enough to cover lead times for planning, ordering equipment and negotiating procurement and the time needed for regulatory negotiations. For the less capital-intensive parts of the business, five years may be a long enough period, but long-term planning often goes further ahead. Projections running twenty-five years ahead are not uncommon for examining issues in technology, demand and network evolution.

There are three main forecasting tools. These are:

- Point forecasting, aimed at the closed possible quantified prediction.
- Sensitivity analysis, to find the assumptions most critical to the accuracy of the point forecast.
- Scenario analysis, to sketch out broad themes in alternative futures and their implications for strategy.

In the short term, fullest attention is given to the point forecast. The emphasis shifts to the scenario as the time horizon extends. Whatever the horizon, forecasters use mathematical projections of income and expenditure and many assumptions, or postulates in the case of sensitivity analysis and scenarios about external events. Projection methods are discussed in Chapter 16.

11.4 Investment appraisal techniques

11.4.1 The valuation of cash flows

Small companies, and others with severe cash-flow difficulties, may use crude methods such as the period over which the cost of the original investment will be recouped (the payback period) to adjudicate between alternative investments. Even large companies with sophisticated methods of investment appraisal may pay attention to the payback period, or variants such as the number of years before the yearly cash flow becomes positive. In all these cases, there is a perception of risks which are somehow inadequately allowed for in other parts of the appraisal process.

The standard method in use among large companies is discounted cash flow (DCF). The method has sound economic foundations and is described in textbooks such as Reference 4. In essence, a net cash flow, usually annual, is built up from estimates of the investment cost and all related income and cost each year. A discount rate is chosen to obtain the present value of flow in each of the future periods. The present value of a sum in year n is the sum of money which, if held in the base year, and invested at a rate of interest equal to the discount rate, would accumulate to the net flow in year n. For example, with a discount rate of ten per cent, £100 accumulates to £110 in year 1 and £121 in year 2 at the chosen discount rate. £100 is therefore the present value of £121 in year 2. The net present value (NPV) is the total of the present values, positive and negative, over the whole period and including the original investment. A positive NPV indicates that, at the chosen discount rate, the investment yields a higher return than the discount rate itself whereas, if the NPV is negative, it yields something lower. It is possible, by iterative or other means, to obtain the IRR (Section 7.7.1) for the cash flow. This is the actual rate of return provided on the investment, and is the discount rate which produces a zero NPV for the cash flow. Modern computer spreadsheets will handle these calculations with no difficulty and can be useful for sensitivity analysis and other exploratory desk work, as well as being a convenient way to perform the general analysis.

Choice of discount rate will be critical for marginal projects. In general, the rate chosen should be at least equal to the cost of capital for the business carrying out the investment. The cost of capital may not be known very precisely and there are several reasons why a different, usually higher, rate may be chosen. If there is a capital shortage (possibly owing to a capital market imperfection) a higher hurdle rate may be chosen to filter out the most profitable of the projects or to allow for optimism. Hurdle rates do not discriminate by project size and are usually only one test. Some investments, such as those abroad in areas where the country risk is high, may be appraised at higher rates. Risk-related discount rates are justifiable to the extent that they reflect different foreign risk-free rates for undiversifiable risk, see Brealey and Myers [4].

A problem frequently encountered with telecommunications projects is that they have long or open-ended cash flows. For example, in considering the optimum long-term path for modernising a network, provision may be made for the replacement of long-lived plant which has not yet been installed. Plant values may be such that the present value of distant events in the cash flow may still be significant in the overall calculation. Another example is the projection of cash flows for an unquoted business which is a potential purchase. The business to be taken over has an indefinite life but no market price. It is customary in such cases to perform the DCF over a fixed period, perhaps 20 years in the network case or ten for a takeover. Cash flows beyond the final year are estimated as a *terminal value*. Terminal values are typically taken as a multiple of the cash flow in the last year of the discounting period, perhaps ten times in a takeover case. They can be large in size and, like other large figures towards the end of the discount period, may be an important determinant of the estimated return. This feature arises from the nature of the problem rather than the methodology and is another indicator of the long time scale of many telecommunications planning decisions.

11.4.2 Valuation of existing networks

The creation of a market in the shares of previously unquoted network operators has produced methods of valuing them. Company valuation techniques used in other industries have been developed to suit the circumstances of telecommunications operations. Cash flow projections derived from a model of the costs and revenues of the operator can be discounted to provide a net present value. The method requires reasonably accurate estimates of the relationships between tariffs, business-volume costs and the national economy. Future tariffs are among the most critical of the assumptions and are dependent on judgements about future price regulation and the strength of competition.

The results are often encapsulated in a calculated valuation per exchange line. This has some appearance of robustness, given that many costs depend on the number of lines and the calling rate per line tends to be stable, but it is exposed on the assumption about future tariff levels and does not take strategic aspects into account. Valuations per exchange line can be directly compared with those for quoted operating companies by using data on market capitalisation and exchange lines.

Besançon and Kelly [5] have drawn attention to the wide range of valuations which have emerged from privatisations. Although those for some, including Telmex (Mexico) and NTT (Japan) were sold for under $5000 per line, Singapore Telecom was sold for just under $20 000 and then appreciated in the stock market. Valuations will include a premium for the extra profits to be obtained from the exercise of monopoly rights. This also affects cable networks. For a further discussion, see *Tobin's q* in Section 15.4.10.

11.4.3 Valuation of licences for new networks

Valuation of licenses to operate new networks is dominated by the monopoly rent which is expected, as was discussed in relation to the auctioning of spectrum, in Section 9.6.3.

For cellular mobile services, the ultimate penetration is seen as a function of demographic and income factors. Calculations of cost, revenue and cash flow are carried out as for a fixed network, but the results are typically expressed as a price per pop (see Section 9.6.3).

11.5 The management of financial risk

11.5.1 Country and currency risk

The treasury departments of telecommunications operators and their suppliers, like those of other large enterprises, manage substantial balances of short-term funds in domestic and foreign currencies. They are responsible for organising the finance needed for investment in plant and corporate acquisitions at home and abroad. In all of these operations there is an element of financial risk which can be minimised, usually at some cost, but not completely avoided. Country risk exists for all foreign operations and is the subject of substantial research. It has various dimensions, reflecting the state of the economy, political stability and the trustworthiness of institutions.

Telephone company treasury departments usually avoid getting drawn too far into forecasting exchange rates because they do not have the resources or comparative advantage to improve over the activities of external financial institutions. Large in-house financial management operations have been run down in recent years and have been replaced by increased reliance on external sources of advice.

Imperfections in the capital markets result in varying discount rates between market sectors. Private individuals may face borrowing costs which are higher than those faced by companies. The existence of such differences may give rise to arbitrage opportunities. For example, a company may give a telephone customer credit through extended payment terms at a lower cost to itself than the savings are to the customer. Sometimes factors run the other way. A discount for prompt payment may benefit the company less than it saves the customer if the customer's alternative is to pay off high-cost borrowing.

11.5.2 Financial markets and the random-walk hypothesis

The prediction of future prices of financial assets has been extensively studied, especially in stock markets. Share prices usually follow a random walk, where today's price contains no useful information about tomorrow. If it were easy to predict price movements it would be easy to make money with this knowledge. Few manage to do so. Financial markets are, compared with most investors, close to being infinitely rich. Rules which involve deepening exposures to counter increasing losses may bring ruin.

Insider trading allows profit opportunities by using relevant information which is not available to the market as a whole. It is illegal in most developed countries but occurs in many.

11.5.3 Hedging risk

A future liability or asset in a foreign currency can be bought or sold on the futures market. There are well developed markets in most major currencies and many smaller ones. Rates are quoted in the financial press. The quoted rates are *not* independent forecasts of what future exchange rates might be. They are derived exactly from interest rate differentials for the same time period in different currencies. If this were not so, it would be possible to make money without any risk by arbitraging the spot and forward rates.

Financial institutions are able to offer a wide variety of operations to cope with uncertainty. A loan raised in a foreign currency and converted to sterling may, for example, be covered with a back-to-back arrangement whereby simultaneous foreign exchange cover is obtained for the future repayment of the loan. It may be both with a contract denominated in, for example, US dollars with a Third World country. The cost of risk reduction is significant and treasury departments need to assess how much of it the business is willing to carry.

11.5.4 Forecasting interest and currency exchange rates

Financial projections may require forecasts of interest and currency exchange rates. This type of forecasting is a major activity in financial institutions, but all concerned agree that it has limited accuracy. Theoretical considerations suggest that interest rates combine:

- a view of future inflation;
- a view of the future productivity of capital.

The 'rational expectations' hypothesis postulates that interest rates incorporate a view of the rate of inflation over the term of the loan. Index-linked treasury stocks, which pay a real interest rate plus the rate of inflation, provide a benchmark for what this expectation is. When forecasts of interest rates are required for financial planning they may be most robustly expressed in real terms. If other assumptions are to be made, this is best done explicitly to avoid implied but unrecognised arbitrage assumptions.

In the longer term, especially over periods where there may be significant differences in inflation rates between countries, the forecast of unchanged rates may not be adequate. This holds even where currencies are locked together in formal arrangements, since they may be pushed apart under the pressure of market forces. Links can be broken, even with powerful currencies such as Sterling, which was forced out of the Exchange Rate Mechanism of the European Union in 1992. The CFA Franc, a common currency used in the 14 states of the Communauté Financière Africaine Zone of West Africa and which was tied firmly to the French Franc, had the link broken with a 50 per cent devaluation in January 1994. Links with the US dollar held by several countries in South East Asia were broken in 1997.

The volume of currency flows through the financial markets dwarfs the total value of world trade or the reserves held by central banks. In 1989 the flow through the City of London was over £64 000 billion, compared with UK import and export trade totalling £265 billion and central bank reserves of £24 billion. The flows should not be attributed wholly to active speculation on arbitrage margins. A great deal of it is probably defensive, carried out by pension and other investment funds attempting to minimise risk across their portfolios. The pension funds are richer than central banks. For all of these reasons, operators making investments abroad try to avoid taking uncovered risks, especially in Third World countries. Where they have to be carried, such risks are offset by a high discount rate in project appraisal.

11.6 References

1 ITU: 'World telecommunications development report'. ITU, Geneva, 1994
2 HARPER, J. M.: 'Telecommunications policy and management' (Pinter, London, 1989)
3 MARKOWITZ, H. M.: 'Portfolio selection', *J. of Finance*, 1952, **7**, pp.77–91
4 BREALEY, R. and MYERS, S.: 'Principles of corporate finance' (McGraw Hill, London, 1984)
5 BESANÇON, L. and KELLY, T.: 'Telecom privatisations: the new realism'. *ITU Speeches and Discussion Papers* (ITU website, 1998)

Chapter 12
Telecommunications policy and regulation

12.1 Ideas about telecommunications policy

12.1.1 Stakeholder perspectives

The dominant telecommunications operator is usually one of the largest enterprises in a developed country, with one of the largest customer bases. Everything done can be highly visible. It is therefore essential to satisfy the reasonable ambitions of the main interest groups, sometimes called stakeholders, who are:

(a) The owners:
   - The providers of capital, who want an adequate return.
(b) Customers and suppliers:
   - the customers, who want reasonable prices, adequate service quality, useful innovation and choice;
   - the workers, who want reasonable pay;
   - equipment suppliers, wanting reasonable terms of business;
   - society at large, requiring responsible behaviour and attention to broad quality aims.

Stakeholders must feel that they are getting a fair deal. Failure to satisfy them spells trouble for the operator, through the capital markets or from regulatory intervention. For a privately-owned operator it is primarily the suppliers of capital that have to be satisfied. The management problem is how to maximise shareholder value within the constraints of the operating environment, of which pressures from other stakeholders are a part. Suppliers may have an interest in the survival and profitability of the firm, and will benefit most if it is a monopoly able to pass excess costs on to the customer.

From the point of view of the consumer and regulator the problem is rather different — how to maximise consumer benefits within the constraints imposed by the capital markets. This does not necessarily involve the profitability or even survival of any particular operator, it only requires that there is at least one operator able to provide service.

Competition in telecommunications will be at best imperfect because of the large capital requirement and the prevalence of sunk costs. This makes regulation inevitable for the foreseeable future. As with contestable markets, the threat of regulation may be (almost) as good as the real thing. The preference for competitive over regulatory solutions owes something to the celebrated dictum of Lord Acton [1] that '*Power tends to corrupt, and absolute power corrupts absolutely*', as well as to the belief in competition as a creative force.

One of the features which observers from genuinely competitive industries must surely notice in telecommunications is the discovery of truths which are well known in other industries, as competition extends into monopoly areas. Sometimes the wrong conclusion is drawn. For example, it is often said that prices should be based on costs. This is true, in the sense that such a proceeding would result in the optimal allocation of resources in a perfect market (Section 6.4), but are these the costs we observe in everyday accounting or the imperfect markets of telecommunications? A quote from Hayek [2] gives this a context:

> 'The misconception that costs determined prices prevented economists for a long time from seeing that it was prices which operated as the indispensable signals telling producers what costs it was worth expending on the production of the various commodities and services, and not the other way round.'
> 'It was this crucial insight which finally broke through and established itself about a hundred years ago through the so-called marginal revolution in economics.'
> 'The chief insight gained by modern economists is that the market is essentially an ordering mechanism, growing up without anybody wholly understanding it, that enables us to utilize widely dispersed information about the significance of circumstances of which we are mostly ignorant. However, the various planners (and not only the planners in the socialist camp) and dirigistes have still not yet grasped this.'

The fact is that most costs are not known at all precisely by operators or regulators. There are two ways in which this view of the nature of the market is relevant to telecommunications policy:

(i)   The role of costs in regulatory profit controls, especially as regards the choice between rate of return limits and price caps.
(ii)   In market liberalisation, where new entrants probably have lower costs than those already in the market and use market prices as signals.

12.1.2 Short-term disequilibrium as an incentive

Equilibrium is a long-term concept, whereas market entry happens in the short term. Few firms make normal profits, or make fresh investments to achieve them. The chance of making abnormally large profits drives most entrepreneurial investment. The *creative destruction* described by Schumpeter [3] is part of the market process. It can be seen working itself out in the reconstruction of telecommunications networks in the collapsed Soviet empire. In established markets, new telecommunications entrants may need to be encouraged by the chance of making extra profits. In developing countries, the chance of super-profits can attract foreign capital.

12.1.3 Financing services

The regulation of privately-owned operators in Europe and North America has not hindered investment finance and may have helped it. Publicly-owned operators in these regions will not generally have had their investment constrained where (as in Germany and the UK) they are politically powerful or ministerial oversight is weak, especially if they have a strong cash flow. In command economies, public control has often seriously impeded financing. Even after the collapse of communist regimes in eastern Europe, the low level of tariffs which were passed on to the new governments inhibited access to foreign funds as well as limiting the scope for self financing. A principal requirement for loans from international lending agencies has often been a government guarantee that tariffs should keep pace with inflation.

12.2 Policy objectives and their implementation

Most, if not all, countries have policy objectives which shape telecommunications services and tariffs. These policies are formulated through institutions which in varying degrees reflect the economic interests of society at large. In the most robust of the state monopolies they are more or less determined by the PTT itself, possibly within statutory guidelines which it may itself have helped to draft. The economic efficiency of prevailing policy then depends to a large degree on the inclinations of the monopolist. In more liberalised markets the formulation of policy has a greater exposure to economic considerations. However formulated, the policy elements commonly included are:

- the nature of the responsibility, if any, to provide some form of universal or affordable basic service potential for residential users and an efficient, modern network, especially for business users;
- the continued viability of the supply side of the industry, including the funding of investment and the discharge of its statutory obligations;
- the degree of competition allowed;
- incentives for the operator to improve efficiency and service quality;
- the way in which prices are set;
- the protection, if any, given to customers from the abuse of market power.

These elements can come into conflict, particularly where there is a trade-off with cost. Policy will be specified by a law, which may stick to a few principles or include detailed provisions for implementation. There will also be lower levels of legal control, through government regulations and licensing.

The implementation of policy may, in extreme cases, be kept with the monopoly operator run by a Ministry of Posts and Telecommunications or something similar. Separate regulatory and operational roles begin to develop when the operator is set up as a separate entity, state owned but with its own finances. None of these arrangements provides a guarantee of economic efficiency. They leave tariffs to be settled by political and administrative processes and management priorities to the whims of individuals, but they are sustainable as long as the only complaints come from customers.

The introduction of competition puts such arrangements under strain as rival firms begin to dispute the fairness of trading practices and regulatory decisions. Regulation then has to be put into the hands of a body designated to implement regulatory policy. The body may be embedded in the controlling ministry, which can leave it subject to heavy political pressure and create a conflict of interest for the minister, or it may be fully separated and answerable only to a separate or higher layer of the government. The Office of Telecommunications in the UK is an example of a regulatory body which is fully separated from the operator and the responsible ministry, as is the Federal Communications Commission in the US.

The terms of reference of an independent regulatory body are usually to monitor the sector and ensure the implementation of a telecommunications policy determined elsewhere by the processes of law. The granting of licences will probably be kept as a ministerial responsibility, with the regulatory body having the job of making sure that the terms of the licences are kept. It may be possible, as in the UK, for the regulator to persuade the regulated operator to agree to changes in licence conditions by threatening an appeal to the courts or to an antitrust body such as, in the UK, the Monopolies and Mergers Commission.

12.3 Economic regulation

Economic regulation of telecommunications is an American idea exported to Europe. The intention is to apply controls and incentives to a monopolistic operator that, as far as possible, make it behave as if it was in a fully competitive market. In its early history the regulation of telecommunications was mainly a practical matter concerned with the allocation of commercial rights and the limitation of prices to what was *fair* or *affordable* or something similar. Economic regulation emerged as a policy objective in the US and is now the foundation of market liberalisation laws across the world. The objective of economic price regulation is to achieve prices which simulate those expected in a competitive market. Regulatory bodies frequently pursue it as part of their implementation task, encouraging imperfect markets to simulate the efficiency of those which are perfect. The legal framework has to be at least consistent with the concept before a regulatory body can implement such a scheme. The main elements of economic regulation are:

- market structure;
- price and/or profit control;
- operational efficiency;
- quality;
- market efficiency and consumer protection;
- improvement of welfare through externalities.

12.4 Regulation of market structure

12.4.1 Licensing to control market entry

There are no controls over entry in perfect markets, and any proposal for limiting it needs justification in terms of potential customer benefits. In those UK markets where entry was restricted the supply side behaved less dynamically than in those, such as equipment supply and value added services, where it was not. The customer premises equipment market functions best without any entry restrictions. Customer choice increased in the UK after liberalisation (Section 10.1.1) and prices fell. In more managed markets, price reductions came more slowly and more subject to price leadership by the dominant operator.

The sale of monopoly rights has a long history as a device for raising money for the monarch or the state. Of itself it implies no particular interest in price controls or any other aspect of economic regulation. The sale of monopoly rights without associated price controls is still an important element in the auctions of mobile operating licences carried out in countries needing development finance.

The control of market entry to counter the potential effects of market failure may produce benefits in cases such as:

- a natural monopoly, where costs may be lower;
- spectrum use, where there may be interference between signals with unrestricted use;
- public service obligations, where the licence provides a means of enforcing service conditions which would otherwise not certainly emerge from market processes;
- the imposition of quality and other standards;
- the definition and control of potential anticompetitive practices of a dominant operator;
- stimulation of the entry of one or two strong firms willing to develop a new market in exchange for good medium-term profits e.g. in mobile communications;
- easing the transition from a protected monopoly to full competition.

The creation of a licensing regime is an important step in regulation. It provides, or at least should provide, a clear statement of the rights and obligations of the operator. Licenses with such provisions are valuable economic documents, especially for a privately-owned company given the right to enter a market in which entry is restricted.

They need appropriate care in drafting. Even for publicly-owned operators, the provision of a licence should be a useful step, given the confusion which may otherwise surround their priorities. None of the UK nationalised industries held licences. Their responsibilities were described, often rather vaguely, in the Acts of Parliament which founded them.

Licences do not necessarily restrict market entry. If they do, they may be selective. A new entrant may be free to provide one service, for example value-added or mobile, while being prohibited from providing others, such as long-distance or international fixed-link telephony. A general licence may be issued under which anyone can operate provided that they do so in the way prescribed by the licence, with or without payment of a fee. There may or may not be a requirement to register, and such a requirement may only affect large firms. Licences of this type have no economic value to the operator except as an indication to customers that the business is legitimate. General licences have been used extensively in the UK for value-added services and indoor wiring. They have worked well as a means of avoiding the need for each individual operator to obtain a customised document. However, they raise a question about what the licensing is achieving that a completely unlicensed regime could not achieve just as well, unless there is an effective mechanism for enforcing the licence provisions.

Where market entry is restricted to companies or other bodies holding licences to operate and the number is small, the licensed operators will usually obtain a degree of economic power. This, unless restrained, will permit them to make abnormal profits. Dominant telecommunications operators usually have price or profit controls in the terms of their licences, but cable operators have largely been free to set their own prices. US cable TV operators with a monopoly franchise have at times commanded a market value well in excess of the value of their assets, see Section 15.4.10.

The granting of a limited number of licences for cellular mobile communications, as was the case initially in the UK and is common practice in all but the most developed countries, has been accompanied by a mixed bag of benefits for users. There are large infrastructures to be constructed and roll out has been reasonably rapid. It remains true that two cellular operators will feel safer and less inclined to competitive price cuts if they are assured that no more entrants will be allowed into the market. The granting of four or more licenses for CT2 digital cordless telephony and mobile data services in the UK reached a limit in market entry. The process which was used resembled the parallel franchising of commercial television broadcasting. Potential PCN and CT2 entrants had to produce detailed business plans to sustain their claims for operating licences. Most of these plans proved impractical, despite being approved by the regulators. Some of them were not taken up and other licence holders merged their interests. Market entry in these areas was initially very limited.

Limiting the number of licences may not achieve more effective use of the radio spectrum, and it can lead to economically inefficient use. Other operators might make better use of the spectrum than those who have the licences, especially if it is allocated free of charge. Free provision of a limited number of licences can discourage innovation and lead to inefficient use of spectrum.

Telecommunications licences are usually granted for a long period, matching the long life of the assets involved. BT has a 25 year licence which is renegotiated at intervals, so that the unexpired period is never less than ten years.

Another form of licence is the franchise, commonly used commercially to allow the right to distribute goods such as motor vehicles of a particular make. Franchising is also used by government when granting the right to supply television and, in France, water supplies and other utilities. Franchisees do not necessarily own the assets which they operate. Water infrastructure may be owned by the government and managed through a franchised private company. Telecommunications services in less developed countries are sometimes run through a management contract with a foreign operator. Cable & Wireless provides management services through such contracts in various parts of the world, including the Caribbean, Africa and the Middle East (Section 10.2).

12.4.2 Fixing operator obligations by contract

Licensing permits any necessary social obligations to be placed on the operator as a condition of being allowed to provide service. It also provides a flexible means of specifying and enforcing tariff objectives. Within the specified framework the operator can be allowed to maximise profits.

This has led to a fruitful examination of regulatory incentives, which can lead to social objectives being willingly met by a profit-maximising operator with potentially more success than can be achieved by trying to impose them on a state-owned operator of uncertain motivation. The operator may be encouraged to meet the social and economic objectives of the regulator if they can be made profitable and hence natural objectives for a profit-maximising company. Where possible, the operator should have an incentive to comply with any desired public-service obligations. Quality of service and other public obligations should be clearly set out in regulations and meeting these obligations should be a condition of obtaining price increases under the guidelines.

12.4.3 Broadband networks

There are potential economies of scope in the provision of delivery systems for local telecommunications and television. The regulatory issues are concerned with how far these economies would be achieved and passed on to customers if both were provided by the same supplier, telecommunications or cable. Prices based on incremental cost would be efficient but may be anticompetitive.

A dominant telecommunications operator could cross-subsidise broadband services from the higher prices in the markets which it dominates, driving out the cable TV competitors and later raising prices all round (see Section 12.8.7). In principle, regulation could enforce a prohibition of such cross-subsidies by demanding separate accounts for the broadband business, refusing to allow higher profits on basic service and applying legal sanctions. The procedures required, in a network with many joint costs, would be complex and often arbitrary. The dominant operator might press to be allowed to base its broadband prices on incremental cost, reflecting the economies of scope, since the business decision to extend into TV transmission will properly have been based on incremental costs and revenues. The cable companies, by the same token, might base their telephony prices on incremental cost.

In the UK it was feared that, if BT was allowed to create a network for the combined purpose of telephony and TV distribution, it would extend its monopoly power in an unacceptable way and prevent the growth of an independent cable TV industry which competed for telephony revenues in the local loop. An asymmetry was therefore imposed, in which cable TV companies could transmit telephony over part or all of the network used for television transmission, but BT could only transit TV signals over a dedicated network, or lease out lines for others to do so. BT has argued that, if it could not exploit the scope economies made possible by transmitting telephony and television over the same cables, the investment in a full broadband network would not be worthwhile. Crossownership rules have prevented American telephone companies from owning cable operators within their own service area.

A monopoly of cable-based broadband capacity would not be immune from the pressures of a contestable market. If powerful economies of scope exist, and a monopoly were permitted to evolve, it does not follow that the telecommunications operator would be the winner. Economies of scope could be exploited by a new entrant to displace the incumbent and might do so unless there were strong barriers to entry. The most obvious barrier is the sunk cost of the incumbent. However, this may not be sufficient if entry costs are low or the new entrant is well funded. Nor is the incumbent safe from being outflanked by an alternative technology. There are at least seven services with some overlap in their markets and delivery systems:

- telephony;
- narrowband data;
- broadband data;
- analogue radio;
- digital radio;
- analogue television;
- digital television.

These have cable, terrestrial radio and satellite delivery systems. Most could be delivered by any one of the three. The costs which have already been sunk by existing operators may not be able to deliver all the services required and may therefore not be significant.

New developments may further increase the potential capacity of the existing copper infrastructure but not be sufficient to cope with rising bandwidth requirements. Even if they were, the cost might be excessive. These are matters to be settled by research and market experience rather than theory.

There are three ways of meeting rising broadband demand in the local loop. One is for the service providers to develop their own delivery systems, using set-top boxes and other terminal devices where necessary to control the reception of broadcast services, and leasing or constructing transmission media in a way which minimises their costs. The terminal apparatus then becomes a significant part of the fixed cost. To the extent that it can be recovered from customers and used somewhere else, the cost is not sunk and specific to particular links.

The second is for one supplier to provide a common infrastructure which can satisfy the needs of more than one service. Such infrastructures have already emerged for terrestrial-radio and satellite links. A broadband optical-fibre cable network could do the same for cable-based delivery systems. As mentioned in Chapter 7, there have been moves towards this in France and Germany, with mixed commercial success.

Provision of services over the local infrastructure could still be in the hands of competing companies (see Section 6.8). The leasing of capacity from an incumbent monopolist could be an unattractive alternative to self provision among companies in competition for services with the monopolist. Strict regulation of the local monopoly assets could control prices but it would be more difficult to solve inherent problems of quality, efficiency and access. There has been a history of bad feeling between US cable TV and telephone operators.

Greater competition in both markets may lead to a relaxation of restrictions and crossmedia mergers have already taken place, especially in the US. Concerns about cross-subsidy carry less weight if the local broadband telecommunications network operator can provide capacity to a cable TV operator without an internal conflict of interest. This might be the situation if the network operator only provides transmission services and does not compete with other companies for call revenues, but carries call traffic from other operators through a standard interconnect charge mechanism.

A third possibility is that a dominant telecommunications operator would provide network upgrades only to those customers who wanted one. This would require new electronic circuitry in the exchange, tests of the local loop quality and replacement of those local lines not capable of handling the broadband services wanted by the customer. Good quality copper exchange connections will carry television and telephone service on the same line by the use of DSL technology, avoiding the need to provide a separate (and costly) optical fibre link. This is the basis of the thrust into video-on-demand being made by BT and some US companies. Not all connections are good enough, but for most it will work. A piecemeal approach would avoid the cost of total replacement at the expense of a loss of ubiquity and possibly a higher cost per line. It would provide better value net of externalities but would miss larger welfare benefits.

Many of these arguments have not been tested. As in other regulatory questions, there is a choice between imperfect competition and imperfect regulation. The balance of consumer losses and gains is as yet unstruck. The US experience cited in Section 6.7 suggests that the dynamic benefits of competition may be greater than the technical benefits of monopoly provision.

12.4.4 Implementation through economic legislation and treaties

Policy implementation can be at several levels. Primary national legislation is usually used for the specification of principles for the regulation or liberalisation of markets. There may be a special telecommunications law, such as the Telecommunications Act 1984, which provided for the privatisation of British Telecommunications, set up an independent regulatory body and laid down the separate duties of the responsible minister and the regulator. Laws which apply to markets generally, such as the Fair Trading Act 1973 [4] in the UK, may have special provisions which exempt specified industries (which can include telecommunications) from some or all of the provisions. In the US, the 1934 Communications Act covers interstate commerce in telecommunications and other industries. It contains structural provisions and confers substantial rule-making powers on the Federal Communications Commission.

International treaties and agreements may also have provisions relating to telecommunications. The most important of these are the General Agreement on Trade and Tariffs (GATT) and various regional agreements such as the Treaties of Rome and Maastricht. The GATT agreement of 1994 covered telecommunications services for the first time. It established the World Trade Organisation (WTO), which was set up in 1995 and charged, among other things, to obtain a worldwide agreement on telecommunications liberalisation. This finally took shape in the Basic Telecommunications Service Agreement of 15 February 1997, which was effective from 1 January 1998. The agreement was signed by 69 countries, accounting for 95 per cent of the world market in telecommunications services. National ratification and implementation was to be through domestic legislative processes. The EU treaties contain clauses covering competition in telecommunications and are implemented by putting the provisions of negotiated directives into national laws. At a more operational level, regulation is carried out through the issue of licences and by orders, regulations or rules under regulatory powers granted by statute.

12.4.5 Regulatory bodies and associations

Industry regulators such as AUSTEL in Australia, OFTEL and the FCC are at the front line of regulatory implementation. They monitor the industry, take steps to ensure that the terms of licences are kept and keep the public informed through statistical and other reports. They are consulted on legislative changes and may suggest them. The more effective bodies carry out or sponsor research studies. Regulators may interact with bodies with a wider brief, such as the Monopolies and Mergers Commission in the UK, which was established under the Fair Trading Act. OFTEL may propose changes to the licences granted to UK operators. If a change is not accepted by the operator concerned, OFTEL may refer the question to the MMC for an opinion; opinions are not binding on the regulator or the government. The MMC, like antitrust bodies in other countries, usually does not tackle market dominance issues in telecommunications.

In Australia the Bureau of Transport and Communications Economics (BTCE) carries out economic research on behalf of the government and the regulator, and similar bodies exist in a few other countries. The Rand Corporation performs a similar function in the US. Extensive use is made of consultants in many countries. A large amount of statistical and economic data covering virtually every country in the world is published by the ITU and a huge volume of US data is available from the FCC.

12.5 Regulation of profits

12.5.1 Direct control of the profit rate

A direct control on profits through the imposition of a maximum rate of return on capital comes closest to the idea of holding the operator to making normal profits. It can work at the level of the sum of all the regulated parts of the business or there may be a different maximum rate of return in each part of the business. Cost and asset allocations are part of the process. The allocations must separate the regulated from the unregulated parts of the business and may split up the regulated business into services. They can be manipulated, overtly or covertly, to shift the balance of profitability between business divisions and services.

There can be a high degree of fine tuning, creating extensive work in four areas:

(i)   Cost and asset allocations, for the calculation of achieved rates of return on capital on the regulated services.

(ii) Establishing the cost of capital.

(iii) Estimating price elasticities so that the post-tariff increase in revenue (and hence profit) and the future rate of return can be accurately forecast.
(iv) Special regulatory financial accounts to show the rates of return actually achieved.

Regulators may also encourage changes in capital structure, to incorporate a higher proportion of debt with a view to lowering the cost of capital. Some of these efforts risk damage to the business being regulated, by diverting attention away from the imperfect markets in which it works towards paper concepts. Rate-of-return controls may run into difficulties under conditions of rapid inflation. A historic-cost accounting base would be unrealistic, but current-cost accounting would not be fully reliable.

12.5.2 Criticisms of rate-of-return regulation

Critics of rate-of-return regulation have focused on three main problems:

(i)   Management incentives are poor.
(ii)   Information requirements are high and costly.
(iii) Regulators get drawn too far into operational decisions.

The method presumes that the actual and correct rates of return for the operator can be determined. Its numerical value depends on the accounting conventions used. The value for one year conveys little information about the return actually achieved on any past tranche of investment (Section 7.7.1). Investment decisions will be based on returns expected in the future rather than past rates, but future rates require many speculative assumptions. The correct rate may not be obvious, given the range and variability of rates actually made by different companies. Calculations of the cost of capital are themselves estimates based on past experience, rather than future certainties.

Markets provide one test of the accuracy of all these calculations. If the allowed rate of return really does equal the cost of capital then, over a long period, the stock price of the regulated operator will follow the trend in the market. Its variance will be less and the level of the return may be above or below the market average, but it will retain its market position. The stock of a company with prices which are consistently held down to a profit level below that expected by the suppliers of capital will lose its relative position. Each year, the stock price will lose ground as buyers preserve their rate of return by bidding less for the stock. The underperformance, compared with the market, of the stocks of British Telecom and some other privatised utilities in the UK between 1985 and 1997 appears to be owing partly to unintended or, in the case of British Gas, deliberate downward pressure on the allowed rate of return. The overperformance of others was a reflection of unexpectedly high profits.

Rate-of-return regulation is basically a 'cost plus' system, because it adds the profit allowed by the regulator to any pre-existing structure of costs to arrive at prices. Direct pressure on the cost base is often absent, although regulators may be able to exert influence if they have enough information. Tight control of profit weakens incentive to cut costs. The operator may see little point in reducing costs to enhance profits if the regulator is going to take the extra profit away. Since the profit control is based on capital employed, known as the rate base, there is a perverse incentive to overinvest provided that the regulator can be persuaded to allow a rate of return in excess of the cost of capital. A more promising way of increasing profits would be therefore to persuade the regulator to agree to:

- the need for extra investment;
- a rate of return exceeding the cost of capital;
- excessively long accounting lives for assets;
- the overvaluation of assets.

The information requirements of rate-of-return regulation are large and complex, making it an expensive and bureaucratic system to operate. It also creates a risk that the regulator will be captured by the operator, the main supplier of the information needed.

Some of the work is a waste of effort since the basic premise, that the cost of capital can be accurately estimated, is probably misplaced. The methodology usually used, which is fully allocated historic cost, is of limited relevance as a market signal to potential new entrants. Limitation of the maximum rate of return may get the regulator too closely committed to individual investment decisions.

12.5.3 Direct control of prices

No controls should be necessary for products and services sold in fully competitive markets. Their inclusion in a price-control system which is applied as an average over the whole range of prices has positive disadvantages, since it opens the possibility for cross-subsidy from the monopolistic to competitive part of the market.

In less competitive markets, regulation may control the rate of profit directly by placing an upper limit on the permitted rate of return, or indirectly by placing upper limits on prices. The two are related but not the same. There are four merits of direct price controls:

(i)   They meet the direct concern of the user, who understands prices better than a rate of return.

(ii)   The fixing of price limits in real terms provides a stronger consumer guarantee on prices than can be obtained from rate-of-return regulation, where it is the operator's profit margin that is guaranteed.

(iii)   The operator has a strong incentive to cut costs to enhance profits.

(iv)   A relatively low level of regulatory effort is required. Rate-of-return regulation requires more information for its implementation.

Direct price control has some dubious antecedents, including rent controls, but the method has been successfully used from time to time by telecommunications and other regulators. One example is the pharmaceutical price regulation scheme in Britain. The State of Michigan was a pioneer in the use of telecommunications price controls. The first national application was the RPI–X rule, which was agreed for the privatised British Telecom in 1984. Since then, the method has been widely employed in other countries and industries. Western regulators are shifting from profit controls based on a maximum return on capital to a maximum increase in real tariffs.

The origin of RPI–X was a report by Littlechild [5]. In discussions with BT it had been agreed that the most direct customer concern is with price, not profit. RPI–X addresses this directly. The essence of the scheme adopted is that a limit is placed on the weighted average price change by reference to the consumer price index (CPI) or its local equivalent (in the UK this is the retail prices index (RPI)). An X factor is used to set the upper limit to price changes. Since prices are usually expected to fall in real terms, X is usually positive, i.e. the allowable price increase is less than the RPI increase. The formula is presented as CPI–X, or RPI–X. As an example, if inflation (as measured by the CPI) rises by five per cent over a year and X is set at two (a two per cent real fall), the maximum permitted price increase is three per cent. There are alternatives for the price reference period. Data requirements and the potential for regulatory disputes are minimised by using the last known inflation figure for the period prior to the one during which prices are controlled.

A more cautious approach is to base the price control on the rate of inflation during the price-control period. Since this will not be known in advance, it has to be forecast. Forecasting error may be ignored, or used to adjust the allowed prices in the following period. The requirement for forecasts has several disadvantages. It makes the system more costly to operate, introduces another area for negotiation and dispute between the operator and the regulator and gets the regulator more involved with management process.

12.5.4 Implementing a CPI-X pricing limit

Implementation of a CPI-based price control scheme leaves a number of choices to be made:

- frequency of price changes;
- which prices to control;
- control of an average price of a basket of services or individual service prices;
- the balance of prices;
- choice of X in a CPI-X per cent pricing limit;
- review period for X in the limit;
- information to be collected for monitoring;
- public or private process.

The system works well with relatively stable inflation, but it can cause systematic deviations during long periods of rising or falling prices. Rapid and unstable inflation can create unacceptable risks for the company, which may find its costs rising while prices are held at a low level. The twelve-month cycle will then be too long. Quarterly systems have been used in some developing and East European countries to cope with the problem. Scope for fine tuning is limited and direct indexation to the prior period CPI (i.e. if the CPI rises by 30 per cent, prices are allowed to do the same) may be a viable alternative for the first few years.

Detailed regulatory intervention in the setting of prices should be avoided where this can reasonably be left to the commercial judgement of the operator. The direct control of prices is best imposed as an average percentage change permitted for all prices, to minimise the possibility of political or other noneconomic considerations getting into the process by the back door.

Regulators may, in spite of this risk, want to go further. Some schemes use a set of average changes allowed within several groups of prices, or as a limit on each individual price. Subsidiary limits on particular prices, for example on residential rentals, may be applied within the overall limit. Prices may be encouraged to move towards cost over a period, although attempts to force the pace may arouse discontent among consumer groups and lose revenues from those more willing to pay, at least in the shorter term.

The median bill (Section 4.2.5) has been used by OFTEL as a subsidiary limit to regulate UK residential call bills. It was chosen in preference to the average for being closer to the experience of the typical customer, and being more stable than the mode which may (with UK consumption patterns) shift erratically from year to year. Median residential bills were not to increase by more than RPI. Since 1997, separate (but identical) price caps have been applied to each of deciles three to eight of the residential call distribution.

Calibration of a price-control limit need not be precise to achieve effective results. An X factor based on the historical trend in real unit-cost reductions would, theoretically, keep price in line with cost if past trends continued and leave scope for profit improvements. Once fixed, the formula needs to be left unchanged for a number of years if the full incentive benefits are to be achieved. Given the long time horizons of telecommunications investment, four or five years is about the shortest period likely to be effective. A period much longer than that will leave the operator open to substantial risks from unseen factors and, to some extent, the customer will also be vulnerable.

Regulators are likely, in practice, to make projections of future cash flows with allowance for changing market size and share and expected efficiency improvements. Operators will do the same. The final value of X may then emerge from a bargaining process. Tightening the factor allows some or all of the profit improvement to come through as lower customer prices. Carried too far, the operator will see the tightening as a direct control on the rate of return, and act accordingly. The independence of the regulatory body will be strengthened if it does its own research, minimising its reliance on data from the operator. Information requirements presented to the operator should be kept to the minimum needed for an effective system.

Regulatory processes have been more secretive in Europe than in North America, where public hearings of proposals for rate increases are the norm. Public hearings are expensive to operate, but are likely to become more common as privatisation and competition raise the commercial stakes in regulatory processes.

12.5.5 Price control experience in UK telecommunications

Until 1983 British Telecom and the Post Office telecommunications arm which preceded it, were subject to rate-of-return controls which, in their nature, allowed cost increases to be passed through to the customer to preserve the rate of return. The operator suffered no risk to profit if cost control was poor. In the RPI–X system subsequently used to control BT prices, the reference figure is the rate of inflation during the year to July for the year commencing August, e.g. the percentage retail price increase from July 1995 to July 1996 is the base for BT price changed during the period August 1996 to August 1997.

Prior to the introduction of RPI–X in 1984, customers were often faced with large and erratic price increases, as illustrated in Figure 12.1. The large spike in the mid 1970s resulted from the Government first holding tariffs down in an unsuccessful attempt to control inflation in the economy as a whole, and then allowing such a large tariff increase that none was needed again for a lengthy period. These manipulations were economically inefficient and they were no longer possible after the introduction of RPI–X. BT was provided with an incentive to cut costs and the cost-control risk was shifted from the

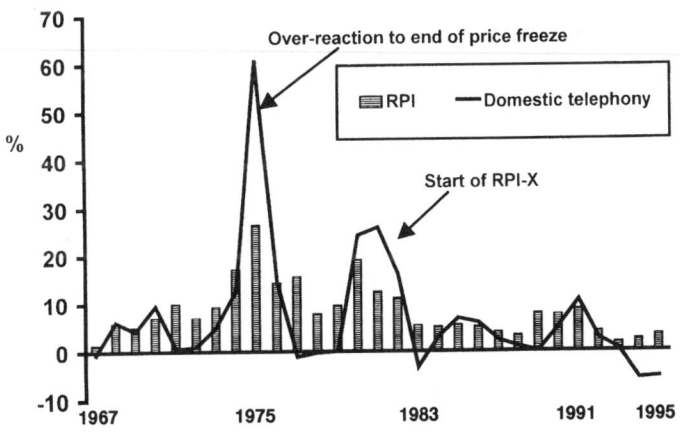

Figure 12.1 UK telephony and retail prices, 1967–95
Source: OFTEL [6], Post Office Annual Reports,
Government Statistical Service

Note: Post Office telecommunications tariff index to 1974, domestic telephony price index 1975–95

customer to the shareholder. Since then, the average customer has experienced smoother price changes and the downward trend has been more consistent in real terms. BT has managed to keep its overall average price changes well below RPI, although the domestic customer did badly in the early years.

12.5.6 Criticisms of price control regulation

The main criticism which has been made is that the rate of return is not explicitly controlled, allowing excessive profits to be made by companies clever or lucky enough to get an easy price limit from the regulator. An alternative view is that price control is rate-of-return control regulation only at a short remove, since regulators will probably make assumptions about the rate of return when setting price limits. The validity of this depends on how far the operator sees a chance of retaining abnormal profits. For example, a system that was allowed to run for five years without retrospective correction would encourage a medium-term view of investment. If the scheme is operated in such a way that it is viewed by the operator as a rate-of-return scheme in disguise, the implementation will have failed.

A scheme which clawed back all supposedly excess profit within a year would suffer from the poor incentive effects found with explicit rate-of-return regulation. A background of political uncertainty, in which a future government might attempt a clawback, would have less effect on incentives but would raise the cost of capital. The windfall tax laid on what were claimed to be unexpected profits of privatised companies in several industries by the British government in 1997 is the type of action which might have this effect.

As with rate-of-return control, the regulator does not have the knowledge to reproduce the price structure which would prevail in a perfect market. In this sense, economic regulation is bound to produce an imperfect result. This is to be balanced against the market imperfections prevailing in a market that is unregulated.

The RPI–X scheme has been widely adopted around the world on account of its simplicity and sound incentive properties. The North American operators and regulators were among the first to adopt it outside the UK. It was subjected to careful appraisal against the prevailing rate-of-return profit controls, but the operating companies decided that it gave them a balance of risk, cost and freedom which was appropriate to the fast-evolving home market.

The scheme has found favour among investors in developing countries because it offers protection against inflation uncertainties.

The operators in these countries like it because there is a more secure guarantee that tariffs will keep up with costs.

The inclusion of an element Y of profit sharing in the form of RPI-X-Y for the customer has been proposed, and sometimes included, as an element of the formula for political reasons.

12.6 Regulation to promote operational efficiency

Direct price controls provide an incentive to improve efficiency because the operator can retain the extra profits made, at least for a while. In Pareto terms, the operator should be encouraged to improve productive efficiency by minimising the resources needed to produce its output and to promote allocative efficiency by minimising the prices paid for the goods and services which it uses. Dynamic efficiency comes from such things as developing production methods which are better than the best existing practice. It is stimulated by competition and, less certainly, by some types of regulation.

Measurement of productive efficiency presents its own problems and the choice of the wrong measure as indicator can introduce distortions. Simple measures, such as the number of exchange lines per employee, are often used for initial comparisons, but more substantial methods are used for reliable work, as described in Chapter 15.

12.7 Regulation of service quality

A monopolist may pay less attention to quality and offer poorer value than a supplier under competitive pressure. Regulators, and the public, often feel that they must intervene to maintain quality standards especially in private industries. The quality performance of publicly-owned monopolies has been uneven and often downright poor, although without any great consistency.

Whichever method of profit control is chosen, parallel provisions concerning the quality and availability of the service provided and the nature of any public service obligations are also needed. These should not go beyond what would provide good value for the customer (see Section 14.4.2). They should be made part of the regulatory deal, so that the profit limit allows for the costs of providing reasonable, but not excessive, quality and meets agreed norms in other respects. To maximise economic efficiency the quality provisions should equate marginal utility to marginal cost as discussed in Section 15.7.1, but there is little sign of any such precision being attempted as part of regulatory practice.

The two charts below suggest that poor countries cannot afford high quality. Figure 12.2 shows that waiting lists are longest, in relation to system size, where penetration rates are low. Much of this is probably an income effect. Penetration rates rise with income levels, and some of the extra exchange capacity required to reduce waiting lists would add to the average cost of providing service, and hence its price. Telecommunications services are not, after all, unique in being provided to a higher standard in more prosperous countries. Longer waiting lists may also indicate a shortage of funds to develop the service. Such marginal funds that are available might yield a higher social return if spent on, for example, roads.

Figure 12.3 shows that waiting lists are correlated with revenue per line and here there is a stronger suggestion of consumer preference. Higher revenue per line is associated with higher income, reflecting tariff and usage differences. Tariffs tend to be higher in more prosperous countries [7]. As income increases, the extra marginal resources available to consumers are shared between those needed to get onto the system more quickly and those needed to use it more intensively. Annual revenues as low as $300 per line are not enough to pay for the extra capacity needed to shorten the waiting time.

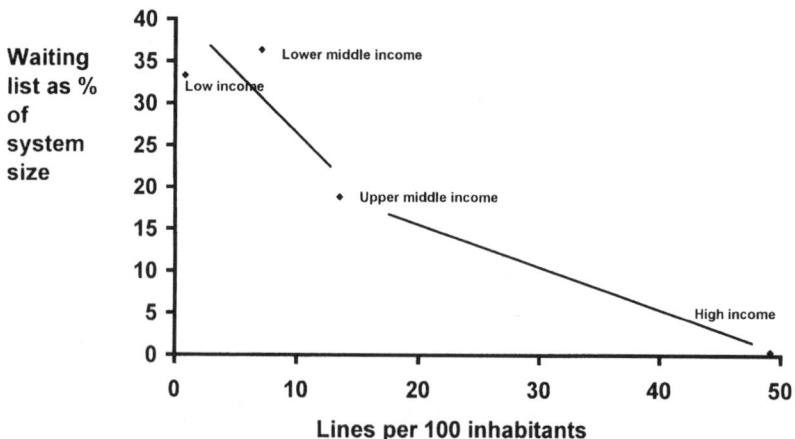

Figure 12.2 Waiting lists and penetration rates by world income groups, 1992
Source: ITU [7]

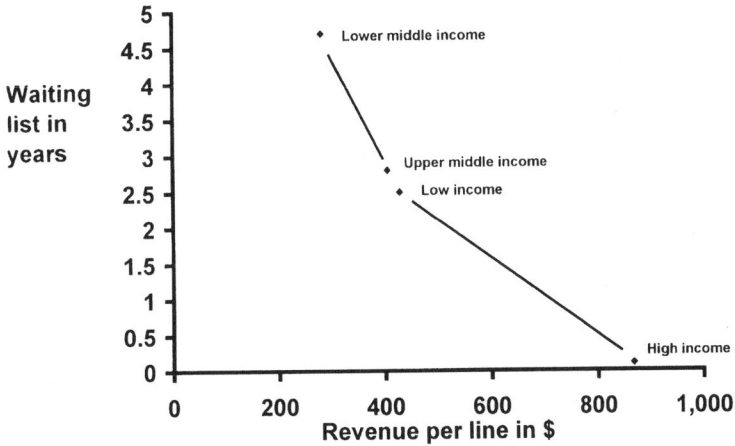

Figure 12.3 Time on waiting lists and revenue per line by world income groups, 1992

Regulators are pragmatic and vary in the way in which they deal with service quality issues and the trade-off with cost. One approach is to attempt the creation of the average price and quality levels that would exist in a competitive market. A better one is to encourage the provision of alternative levels of quality, pricing them in a way which reflects cost differences. If the operator feels that money can be made out of quality improvement within a framework of tariff flexibility, this improvement is more likely to be seen. As with other obligations, any quality requirements need to be clearly specified. For example, fears that British Telecom would cut costs by reducing the quality of service were met by placing minimum service requirements on the operator in its licence.

A public outcry about service quality, stimulated by users or the regulator, may inflict considerable damage on a privately-owned regulated company. Some examples are given in Chapter 10. OFTEL began collecting and publishing statistics of service quality for all the main operators to exert peer pressure on them, after the issue became sensitive in the UK.

Another example concerns the Latvian operator Lattelekom. When the government sold 49 per cent of the stock by a public tender process in the autumn of 1993, a detailed schedule of objectives for the improvement of service quality was specified in the tender documents. The successful tenderer was expected to achieve these and or suffer penalties for failures.

12.8 Improvement of market efficiency

12.8.1 Regulation of dominant firms

Economic regulation requires special rules to deal with the market power of dominant operators. For example, it may be sufficient to operate price or profit control on the dominant operator only and leave the competitors, which by and large are price takers not price makers, to charge what they like. The dominant operator may be given special public-service obligations not laid upon its competitors and recognised in the price or profit controls. Nondominant firms requiring a licence to enter the market may be subject to rules affecting price discrimination and other matters, in recognition of the favoured and partly protected position given to them in granting a licence.

Dominant operators are usually required to show transparency in pricing, working from published tariff schedules and only granting discounts to large users within an agreed nondiscriminatory framework. Regulation intended to prevent cross-subsidy and anti-competitive behaviour is likely to be applied to dominant firms in recognition of their market power and may not be applied to their competitors because of the weakness of their market position. Various regulatory provisions may be applied to dominant firms for monitoring and preventing anticompetitive practices. There are two main remedies, both aimed at preventing cross-subsidy:

(i)   Accounting separation — the operator keeps separate accounts for the monopoly and competitive businesses and is required to show a reasonable rate of return on each.

(ii) Structural separation — the operator has to sell off the competitive business, or run it as a fully separated subsidiary with its own board, accounts and capital market access.

There are pros and cons to both. Accounting separation enables economies of scope and other synergies to be preserved, but it may require complex and subjective allocations of costs and revenues between the business divisions. Ultimately, since the separate parts of the business are all answerable to the same board, the separated accounts may look more like different pockets in the same suit than anything of management significance. Internal accounting transfers between the divisions make no difference to the aggregate group profit. The transfer prices are not the product of market forces. Policing the process can be difficult.

Structural separation is a drastic remedy. The creation of an arms-length business relationship between each unit removes opportunities for synergy and empire building alike. Genuine economies of scope could probably be accommodated by appropriate market prices in fully competitive markets, but may be difficult to achieve in many areas of telecommunications.

A separation between wholesale and resale service operations has been proposed e.g. by Harper [8] for a structural model of this kind in the UK, see Section 6.8.

12.8.2 Preservation of common carrier principles

Dominant operators are common carriers. They have to provide service to all comers at reasonable, standard, rates and are carriers of last resort. The emergence of pluralistic networks, in which common carriers face competition from others without common-carrier obligations, requires regulation to ensure that, as far as possible, they work together as an integrated service, thus preserving the network externality. This involves attention to interconnect arrangements, quality of access, technical compatibility, quality, privacy and other matters where there may be a conflict of interest between the carriers, or where consumer interests may fall between carriers.

12.8.3 Supply of information to customers

The consumer needs good information about service and price options to make an informed choice. This requires:

(a) Bills with major elements itemised so that their cost can be seen and compared with services against which they compete.
(b) Bills that are accurate.
(c) Price changes to be published well before implementation. Information should be made readily available about the full range of services provided and the quality of the service which is obtained.
(d) Customers should be given advance individual notification of price changes which affect them.

Market efficiency can be further improved by the provision of good telephone directories and dial-up directory-enquiry services.

In practice, the arrangements may be rather less perfect. Dominant carriers are usually required to notify price changes to the regulator in advance, but the complexity of the schedules may leave small users unaware of the exact prices, as mentioned in Chapter 9. Nondominant carriers are sometimes not required to publish their tariffs and are often willing to offer individually-negotiated prices for the business of large customers.

12.8.4 Protection from unfair trading

Customers may require regulatory protection in a variety of ways apart from the pressure on tariffs and costs discussed earlier. Most of the pressures stem from market dominance, which broadly correlates with market power. The customer needs indirect protection by safeguards to prevent competitors with efficiency which is greater than that of the incumbent from being driven out of business by unfair trading practices.

Dominant operators may try to bundle services together or even force customers to agree not to buy from other suppliers (full-line forcing). Protection against this may be necessary. There may also be restrictions on use. These might, for example, prohibit the use of a fax machine except on a line with a higher charge, or the use of a leased line to carry traffic for third parties.

Antitrust legislation may restrain suppliers with excessive power but not eliminate it. Policy aims are normally confined to protection of the consumer against excessive prices or undue restriction of choice.

12.8.5 Price discrimination

The essence of price discrimination is that the same product is sold at different prices to different customers, because some are willing to pay more. There are three degrees of discrimination:

(i)   First-degree price discrimination exists where a monopolist charges each customer the highest price that the individual will pay, appropriating the whole of the consumer's surplus.
(ii)   Second-degree price discrimination exists where a monopolist charges by price bands. All those willing to pay £10 or more are charged £10, those willing to pay £9 to £9.99 are charged £9, etc.
(iii)   Third-degree price discrimination segments markets rather than individuals and is the one usually encountered in telecommunications.

Suppliers working in imperfectly competitive but unregulated markets often practice a measure of third-degree price discrimination to maximise sales and profits. With switching equipment this could result in the main domestic customer, particularly one unwilling to buy imported goods, paying a price which includes a large share of overheads. Sales to domestic operators which are happy to buy imported goods and to operators abroad are then made at prices closer to marginal cost. The latter is one of the international trade issues discussed in Chapter 16. Procurement strategies are discussed further in Chapter 17.

Lawyers distinguish between undue and other price discrimination. Regulators tend to speak of fair and unfair prices. Little British case law exists, but the general implication is that price differences should be justified by cost differences and are then not discriminatory. Some have difficulty in accepting that price discrimination may enhance welfare. There is an underlying feeling that one customer's discount must be another's increase. The bag of benefits is seen as fixed. This is not necessarily true and conditions do exist where it is possible to grant a selective discount without making any other customer or the shareholder worse off. In special circumstances, a selective price reduction can be Pareto optimal in benefiting some customers without being detrimental to others.

Court cases on the matter are rather more common in the US, particularly in the area of patent law. A case reported by Hausman and Mackie-Mason [9] concerned a patented synthetic material which was being sold at different prices in at least 17 markets, with restrictions on resale between markets. American patent law is to some extent in conflict with antitrust law on the rights that patent holders have in respect of price discrimination between users, patent law being more sympathetic. There are three conditions for third-degree price discrimination to be Pareto optimal:

(i)   No reduction in allocative efficiency.
(ii)   Better incentives to innovation.
(iii) No losers.

The necessary conditions for a welfare improvement which leaves nobody worse off are then:

- segmented markets, so that customers do not switch between markets, because they do not know of others or want to participate in them, or because resale between markets is prohibited;
- price elasticity <-1, so that a cut in price increases sales revenue in the market segment where prices are to be reduced.
- low marginal costs, so that the extra sales incur costs which are less than the increase in revenue;

If these conditions are met, it may be possible to cut the price in the market segment, generate extra revenue and pay for the extra production costs incurred by meeting that demand out of the extra revenue. This is the basis of the net-revenue test discussed in Chapter 6 as a test for cross-subsidy. The test will only be passed in sectors where all three conditions are met. Customers in a sector where price elasticity is low could not be offered the same deal without net revenue loss being incurred.

The advent of competition has led to an urgent search for a means of justifying rebates to profitable large users without giving them to unprofitable small ones. UK licence provisions qualify the abuse to undue or unfair price discrimination. They are loose enough to allow operators to look at the total profitability across the whole bundle of goods and services which a customer buys. Bulk discounts are not held to be unfair if:

- they are available on equal terms to all users with the same bundle of requirements;
- Other users also have options appropriate to their level of usage;
- They do not contain elements that are provided below marginal cost.

These schemes break the simple link of price with cost on a service-by-service basis, but they increase the economic efficiency of tariffs by bringing at least some prices closer to marginal cost. They enlarge the market by giving the customer a choice of tariff plan. Some will choose one and some another; overall, they will use the system more and still pay at least marginal cost for service.

12.8.6 Prohibition of unfair cross-subsidy

A dominant operator may be faced with competition in some areas, e.g. the sale of PBXs, while having a monopoly in others, such as the provision of basic network services. There will be a temptation to cut prices where there is competition and raise them elsewhere. A competitor selling PBXs might be forced out of business if the dominant operator cut its PBX prices to well below cost and supports the operation by monopoly profits made from its network operations. The cash flow from network services to PBX sales would be a cross-subsidy. It would be an unfair cross-subsidy because of its effect on competition.

The regulator should be able to call for appropriate accounting information where there are reasonable grounds to suspect the existence of an unfair cross-subsidy and there is a public complaint about it. OFTEL has investigated BT's apparatus-supply business, monitors its financial performance and publishes the results.

12.8.7 Predatory pricing

Prohibitions on unfair cross-subsidy are meant to guard competitors and customers against the effects of predatory pricing. The theory behind this is that the dominant operator will sell goods at a price below its own costs and those of competitors until the competitors are driven out of business. It will then raise prices by a large enough margin to recoup all the losses and make an excess profit. Whether or not the operator would want to act in this way, consumer protection can best be achieved by prohibiting those cross-subsidies which can be shown to be unfair.

Cross-subsidy from excess profits in monopolistic markets is an alternative to making a short-term loss. Regulators tend to be wary of attempts to introduce prices based on short run marginal cost because of the risk that they are unfairly discriminatory or predatory. Predatory pricing is hard to prove and the bald scenario set out looks unlikely to fool a regulator. In domestic trading it may be less of an issue than the literature suggests. The issue is an active one in international trade and this aspect is discussed in Chapter 17.

12.8.8 Terms for interconnection

Interconnection provides a final customer benefit because it enables:

- choice between suppliers for basic services;
- the ability to speak across different types of system, fixed and mobile.

In the US, Finland and some other countries where service has long been provided through numerous independent local carriers, payment systems were devised for traffic exchanged between networks. The interconnection of national networks was achieved in the early days of telephony through the ITU. Payment mechanisms and traffic protocols were set up and worked quite well in a monopoly environment.

A new wave of operators in Australia, France, Germany, the UK and other countries was quick to press for reasonable interconnect terms and similar pressures later arose in France. The new operators include:

- foreign operators such as AT&T, wishing to offer local and long-distance service outside the US;
- international resellers;
- cable TV operators offering telephony.

In the early days of interconnection for the exchange of traffic between competing networks, two approaches were tried:

(i)   Leave it to the parties to agree among themselves, following the example of international accounting.

(ii) Treat the call starts and terminations required by a trunk operator from a local loop operator as local calls. The charge for the originating local call is paid by the caller or by the trunk operator. The termination is paid for by the trunk operator.

Experience in the UK and the US shows that successful interconnect negotiation can be:

- critical for the viability of other operators;
- difficult to negotiate if the incumbent operator sees its markets threatened, for example by providing interconnect to an operator competing for the same traffic.

The idea that the commercial basis for interconnect could be agreed between the operators has attractions. If the bargaining is carried out in a fully competitive market, it is likely to produce a better result, in terms of resource allocation, than any regulatory ruling could achieve. Most bargaining takes place in imperfect markets but, even here, bargaining between the parties may produce the best result if the right regulatory incentives can be provided.

The new carriers compete for long-distance and international traffic, but most of them do not have local networks through which they can access their customers. New entrants need to be able to pass calls through the networks of other operators if the full benefits of competition are to be achieved. Only in this way can a caller reach all other addresses without duplication of facilities. There is a substantial external benefit in the ability to speak to any telephone from any other telephone. Interconnect is arguably the key competitive issue in market evolution. Interconnect terms can, as illustrated in Chapter 5, erect substantial barriers and legal provisions may be needed to reduce them. The vertical integration of trunk and local operations, which creates a conflict of interest within the carrier, was eventually ended in the US by the break-up of AT&T.

The workings of agreements relating to the exchange of international traffic and the early experience of aspiring new entrants in the UK and the US both demonstrate the difficulty of providing adequate incentives, even where there is a will to do so, and there have been frequent regulatory interventions. AT&T originally refused interconnection to independent companies, but was later obliged to do so by legislation in 26 states and finally by the Kingsbury Agreement with the US Justice Department in 1913 [10]. It was a long series of legal disputes which eased MCI into the US long-distance market. There have been extended disputes in Germany and the UK. Mercury needed OFTEL to help it to obtain an agreement with BT. There have been examples of voluntary interconnect agreements in UK mobile services.

Voluntary agreement remains the commonest basis for interconnect and regulatory incentives have been improved, but external rulings on rates are becoming more common. In the US, interconnect conditions have become part of a larger bargain in which the Bell Operating Companies would be allowed to enter the long-distance market provided that they fully opened their local markets to competition from other carriers. The US Telecommunications Act of 1996 set out a 'competitive checklist' of fourteen elements to test the quality of access and interconnection afforded to other telecommunications carriers and the FCC has been using this to adjudicate on applications. The Act also specified that the FCC should take the public interest into account when making decisions. Ameritech and SBC both had their initial applications turned down for failing to meet all 14 points and were given guidance on what they would have to do further to be successful.

The incumbent operator is likely to want an interconnect payment which includes a contribution towards supposed losses on its local loop. In the US, plans to recover surcharges from long-distance competitors such as MCI and Sprint proved difficult to achieve. They are discussed further in Chapter 14. A structured and open regime, with published interconnect rates, provides for greater market efficiency. There should be provision for interconnect with mobile and other operators. Interconnect requirements will depend on the extent of network competition expected.

Arbitration procedures are required. The regulator may have to lay down interconnect conditions if the parties cannot agree. An economic basis of adjudication is desirable and will raise issues such as the use of fully allocated or long-run marginal costs, historic or current-cost accounting and the treatment of cross-subsidies in the pricing of local service. A long-distance operator may want, for example, access to a local loop which is provided to users at a tariff well below cost.

12.9 Regulation to improve public welfare through externalities

12.9.1 Potential benefits of regulation

Networks can have their usefulness increased by features which the operator cannot implement alone, or has no incentive to implement. These features are of five main kinds:

(i)   Standards, which make it cheaper to develop equipment and add to its value for the user.
(ii)   Protocols for data exchange, enabling different computers and networks to exchange data.
(iii) Numbering plans and signalling systems which speed the routing of calls round the network.
(iv)   Service specifications. Interconnect between operators provides one kind of externality, the use to which other users may put facilities is another.
(v)   The nature of the basic service and its availability, discussed in Chapter 13.

12.9.2 Standards

The analysis of standards involves a consideration of the process by which innovation works its way out through the economy. This is a large subject with its own literature on diffusion processes and there is only space here for a few of the more important economic elements. Economies of scope and scale are important. A *de facto* industry standard can emerge without external intervention if scale economies lead to the dominance of one of the suppliers. The standard itself may not be the best, but if compliance is a small part of the total cost, it may come in on the back of a dominance driven by other factors. David [11] gives the QWERTY typewriter keyboard as an example. Developments in computing provide others, notably in the dominance of the IBM personal computer design and its Microsoft operating systems.

   The process may 'lock-in' a particular way of doing things and 'lock-out' others by its own momentum. Lock-in and lock-out factors come from large investments on R&D and production facilities by manufacturers and from the possession by customers of an installed base of equipment that they know how to use, wish to add to and need to keep in working order. A minor, but curious, example is the opposite layouts of the key pads on telephones and calculators. Users of equipment which has been locked out face costs such as premature obsolescence and expensive maintenance. In a situation where technology is changing rapidly, this is unavoidable. The costs can be offset by the development of adaptive devices to enable pieces of incompatible equipment to work together. Lock-out seems to be a less serious problem in telecommunications than in computing. Network operators have long been used to passing messages between network elements of different vintages and have usually managed to make this invisible to the customer. Computer users, on the other hand, have had to cope with physical problems such as floppy disks of different sizes and capacities, as well as obvious differences in operating systems.

If a standard is to be imposed by external forces, there may only be a limited time window in which to do so. This is when the direction of market evolution has been determined but the supply and demand sides of the industry are still willing to accept change. Unfortunately, the view through such a window may not be clear enough to improve on the market process by regulatory initiative alone, especially where technology is evolving rapidly. In the US, there has been a reluctance for regulators to enter too readily into the standards arena for this reason, although initiatives from leading private firms have often set the pace for international standards. Recent British practice has been similar, with the Department of Trade and Industry encouraging the emergence of industry standards such as a common air interface (CAI) to enable communications between different CT2 systems. There has been rather more intervention in other parts of Europe, for example with modems, where a European standard for modems resulted in an overpriced product, the use of which was widely undermined by supplies of adequate but cheaper pieces of equipment available from the US by mail order.

ISDN exemplifies the benefits and problems of forward-looking standards. It has little to offer the ordinary voice traffic for which it can immediately be used. Data traffic, which benefits from the faster transmission speed, requires the development of terminal equipment capable of using the system. Equipment manufacturers were initially slow to put this on the market in the absence of a demand, but users will not sign up for ISDN service without the equipment. In the meantime modems available for ordinary lines have become faster, reducing the ISDN advantage.

The approach towards ISDN implementation has varied between countries and is discussed in Section 8.5.3. The European Commission tried to force technology push by issuing targets for ISDN penetration, seeing it as a vehicle for wider social benefits, connected with the so-called *information society* and a means of improving the competitiveness of European industry. Critics of this approach would like to see more demand pull, and have questioned whether it was the right priority or whether facilities for which there was a more obvious demand, such as a Pan-European freephone service, might have been a better use of resources. Some of the national state monopolies went for a rapid spread with low and probably subsidised tariffs. Others, including more market-driven operators such as BT, proceeded more slowly, as did most of those in the North America. Current marketing efforts in the UK are directed towards small businesses and to those working from home.

Standards for terminal equipment have frequently been used to keep products out of the market. The reasons most commonly advanced are:

- protection of the user from injury;
- protection of the network from damage;
- prevention of interference with other users of the network.

All have a validity. The first two concern market and pricing imperfections without the involvement of third parties. There is an externality in the last reason, where the interests of other users of the network are damaged by equipment which causes interference.

Equipment standards are barriers to entry and may protect a dominant operator from competition. For example, as long as the operator only allows its own telephones to be attached to the network it has little incentive to improve their design or reduce their price. The plethora of choice which followed the liberalisation of terminal equipment markets in the US and the UK showed the result.

Standards may be enforced through an approvals process in which equipment is legitimised or banned. The standards-making and approvals processes need to be independent of the operator to avoid potentially anticompetitive pressure. Separate approvals bodies now exist in many countries, such as the British Approvals Board for Telecommunications (BABT) in the UK. The approvals process can be expensive. Its main cost elements to the industry are:

- approvals fees;
- fees of test laboratories;
- supply of equipment for testing (which may be destructive);
- administrative time in answering queries from approvals staff;
- delays in bringing the successful products to market.

Some countries have no approvals process. They buy equipment which is in use elsewhere, and so presumed acceptable.

International standards negotiations tend to be dominated by political considerations and are of great industrial significance. In recent years there has been a struggle between european, Japanese and US interests, both with analogue television then later with digital TV systems. Mobile communications have seen a contest for world dominance between the european GSM, the Japanese personal handyphone system (PHS) and the US personal communications system (PCS).

Standards making in telecommunications tends to follow rather than lead. In the past, the process has been dominated by operator interests, although in the UK the manufacturers, the Telephone Managers Association (TMA) and the Telephone Users Association (TUA) have representation in the British Standards Institution (BSI). More recently, consumer groups including the International Telephone Users Group (INTUG) have obtained representation on international bodies including the CCITT and the European Telecommunications Standards Institute (ETSI) which was established in the 1988. International approvals can be costly to obtain. In general, a fresh approval has to be obtained for each country, although progress has been made towards a single approval covering the whole of the EU.

12.9.3 Protocols for data exchange

The TCP/IP protocol and others such as X.25 for data exchange on other data networks are introduced in Chapter 2. Protocol creation, like the development of standards, has to cope with rapid change in equipment technology. TCP/IP was originally designed by the US Department of Defence for the system which is now the Internet, and it was quickly adopted by companies for the integration of their own networks and spread rapidly in the private sector. Like X.25, it has proved to be robust in coping with technological change.

The economic benefits created by these protocols are impossible to measure, but are extremely large. A lower bound must be the amount spent on sending messages which could not otherwise be sent. An upper bound includes the value of the messages to the recipients and to third parties which might benefit from them. Data transmission depends on standards and other features of networks, and the value of the benefits has to be shared with them.

12.9.4 Numbering plans

The allocation of telephone numbers is a process with economic and commercial as well as purely technical dimensions. Seven economic interests are:

(i)   The advantages for the user and the operator in having a single worldwide system of numbering, with a unique code for each country and, within a country, a unique number for each line. The evolution of a such a numbering plan is an externality reached by agreement rather than competition between systems.

(ii)   There can be tariff information in the number, for example in the country codes for international calls and abbreviated dialling for local service. This improves market efficiency. An OFTEL survey showed that users found it useful to associate prices with dialling codes on the fixed network. Calls from the fixed network to mobile phones are often more expensive, making it useful for mobile numbers to be identifiable as a group.

(iii) Changes in the numbering plan generate costs for users, possibly followed by benefits over a longer term.

(iv)   Some numbers have more value than others to users. The area or exchange code may indicate a desirable or undesirable address, such as the 212 code for Manhattan in New York City. Easily-remembered codes such 0800 500 300 are used extensively in freephone services. Alphanumeric dials and keys are still found on telephones in some countries but have fallen out of use in national numbering plans. They have not been used in the UK since the 1960s, when the letters were removed from telephone numbers to permit a world numbering scheme for international subscriber dialling. Prior to the change, the letters were used to form acronyms for the names of telephone exchanges. Alphanumeric keypads for personal numbering (e.g. JACKO) or for advertising (e.g. TAXI) are available from specialist suppliers. These uses of letters are for value-added purposes and not part of the numbering plan.

(v)   New types of service may need special number ranges to enable them to operate and be identified.

(vi)   Ownership and control of the numbering system. There may be competition issues such as entry barriers with some types of numbering system. This makes it necessary to move the ownership of the system from the dominant operator to the regulator, as has been done in the UK.

(vii) Long numbers may enable better services to be provided, but they take longer to dial and involve a greater probability of misdialling.

12.9.5 Number portability

Number portability was first introduced in 1993, when it was brought in for freephone services in the US. Fixed network applications began with Hong Kong and the UK in 1996; by 1997 there were plans to introduce it in Denmark, France, Germany and the Netherlands. The NPV of the net benefits of portability to the UK economy were estimated at £1.4 billion by the consultancy firm NERA in a 1993 report to OFTEL [12]. Portability was expected to confer substantial direct savings through such benefits as not having to change stationery and a reduction in dialling errors. The creation of greater competition gave an estimated gross benefit of £1.28 billion through better prices, innovation and quality. A later estimate of the NPV of the net benefits of number portability in UK mobile communications came out at £98 million [13]. It was thought that competition benefits might be less than for the UK PSTN because:

- mobile competition is more effective;
- there has never been a dominant firm;
- the main barrier to entry is the need to obtain a licence; no extra licences were expected in the next five years.

12.9.6 Open network provision

The European Community developed an open-network provision (ONP) concept to promote public benefits through improved network standards and access. The equivalent in the US is open network access (ONA). ONP was introduced in a 1987 Green Paper on telecommunications [14] and adopted in a Council Directive in 1990 [15]. The stated object of ONP is to '*harmonise access to and use of telecommunications networks and services throughout Europe, and to encourage the provision of new competitive telecommunications services, by ensuring a 'level playing field' for all market entrants*' [16]. The ONP framework involved the harmonisation of three elements:

(i)   Technical interfaces and service features. New services should use existing interfaces at network termination points wherever possible.

(ii)   Usage conditions. These should cover delivery time, minimum contractual period, quality of service, maintenance, fault reporting, conditions for resale, shared and third-party use, access to frequencies and conditions for interconnection with public and private networks.

(iii) Tariff principles. Tariffs should be based on objective criteria, nondiscriminatory and cost oriented. They must be transparent and published. Services should be unbundled and specific service features should be charged independently of transmission via the network.

ONP was tackled separately for leased lines, data, ISDN, broadband, voice telephony, telex, mobile services and local access. Practical implementation could not proceed quickly without the active support of member states and their main national operators. This was not readily given in all cases. In the event a movement of tariffs towards cost and the liberalisation of access has owed as much to pressure from resellers and other new operators, and to the World Trade Organisation talks taking place at about the same time, as to the ONP policy. Some at least of the external benefits which might have arisen from ONP have been internalised by the process of competition.

12.10 References

1   LORD ACTON: 'Historical essays and studies, Appendix' FIGGIS, J.N., and LAWRENCE, R.V., (Eds.), 1907
2   HAYEK, F. A.: 'Knowledge, evolution and society' (Adam Smith Institute, London, 1983)
3   SCHUMPETER, J. A.: 'Capitalism, socialism and democracy' (George Allen & Unwin, London, 1943)
4   'Fair Trading Act 1973': HMSO, London, 1973
5   LITTLECHILD, S.C.: 'Regulation of British Telecom profitability' (HMSO, London, 1983)
6   'Pricing of telecommunications services from 1997, Annexes to Consultative Document' (OFTEL, London, December 1995)
7   'World telecommunication development report 1994' (ITU, Geneva, 1994)
8   HARPER, J. M.: 'Monopoly and competition in British telecommunications' (Pinter, London, 1997)
9   HAUSMAN, J. A. and MACKIE-MASON, J. K.: 'Price discrimination and patent policy'. Paper presented at NERA seminar, 23 October 1986, London
10 VIETOR, R. H. K.: 'AT&T and the public good: Regulation and competition in telecommunications, 1910-1987' *in* BRADLEY, S. P. and HAUSMAN, J. A. (Eds.) 'Future competition in telecommunications' (Harvard Business School Press, Boston MA, 1989)

11 DAVID, P. A.: 'Some new standards for the economics of standardisation in the information age' *in* DASGUPTA, P. and STONEMAN, P. (Eds.) 'Economic policy and technology performance' (Cambridge University Press, Cambridge, 1987)

12 'Number portability: modifications to fixed operators' licences' (OFTEL, London, 1997)

13 'Economic evaluation of number portability in the UK mobile telephony market' (OFTEL, London, July 1997)

14 'Towards a dynamic European economy: green paper on the development of the common market for telecommunications services and equipment, COM(87) 290'. (European Commission, Brussels, 1987)

15 'Establishment of the internal market for telecommunications through the implementation of open network provision, OJ No L 192' (European Commission, Brussels, 1990)

16 'Proposal for a council directive on the application of open network provision (ONP) to voice telephony, DG XIII/F/1' (European Commission, Brussels, 1992)

Chapter 13
Practical problems with regulation

13.1 Economic behaviour of regulators

13.1.1 The problem areas

Regulation is itself not a perfect process. The regulators have an imperfect knowledge of the markets that they regulate and will make mistakes. *Why regulate?* is not a trivial question where imperfect competition exists and the deficiencies of regulation have become obvious or its cost excessive. Some American state regulators have opted for minimal interference or even none, leaving the efficient functioning of the market to be settled by such factors as the threat of entry, the behaviour of competitors, peer group comparisons and the wish of the dominant operator to avoid legal and regulatory action over market abuse.

Practical problems arise in two regulatory areas. The first concerns the objectives of the regulators, arising from their position as agents. The second is whether their actions have unintended consequences which are perverse.

13.1.2 Regulators as agents

Just as the owners of a large firm are not the managers, the staff of regulatory agencies are not the customers. Regulators act as agents for the community interests which they are appointed to represent. The theory of agents suggests that a proper study of regulation should take into account the factors which actually drive the behaviour of regulators. There are four main theories of regulatory behaviour; all have elements of truth and there is a degree of overlap between them. They are outlined below, following the well documented account in Reference 1 for the US material.

13.1.3 Public interest

The public interest theory of regulation is the oldest and most altruistic, holding that regulation is carried out to protect the consumer from abuses of market power and to follow broad public-interest objectives which might not otherwise be achieved because of market failure. The broader objectives to be promoted might include externalities adding to public welfare, or distributional considerations based on a right of universal access to service. Public interest is always a primary objective of regulatory legislation and it is frequently invoked by regulators to justify their decisions. Several examples of this have been quoted in earlier chapters. The question at issue is whether regulators follow the public interest in what they actually do, rather than invoking it to justify decisions taken for other reasons.

The cost of regulation has to be set against its benefits. Some writers have assumed the costs to be too low to be of any consequence. More recently there has been a good deal of attention to regulatory costs. In the 1980s, the State of Nebraska moved faster than most American States with a scheme which gave pricing freedom while retaining the monopoly. This saved the cost of public rate hearings and left the operator with the task of limiting its own profit rate. In the UK, 'light rein' regulation was pioneered with the establishment of OFTEL. Its first director, Sir Brian Carsberg, made public statements about the need to balance the benefits of regulation against its cost. OFTEL is funded almost entirely from the licence fees paid by operators. In the early licences of the main operators, annual fees were capped at 0.08 % of the network turnover of the previous year, sometimes with a minimum payment provision.

It is widely believed that the compliance costs incurred by the regulated industry greatly exceed those of the regulator, by a factor of ten in popular US telecommunications estimates, although not usually quantified. When the Securities and Investments Board (SIB) was established to regulate the UK financial services industry one estimate put the cost at £20 m per annum for the regulator and £100 m for the industry. Others thought the cost might be higher (both estimates in Reference 2). Subsequent problems experienced by investors at Lloyds and customers of the pensions industry suggest that neither expenditure was enough.

The cost of complying with specific laws and regulations is to be distinguished from the cost of negotiating and contesting them. Figures quoted for the financial services industry appear to include both kinds of cost. Those for US telecommunications relate only to negotiation and contest. They are high because the tariff-setting process is carried out through public hearings at which expert evidence can be given. Continual upheavals in the market structure arising from new technology add more cost by causing frequent legislative changes.

A rational regulator, like a rational company, would continue to increase regulatory expenditure until the marginal return exceeded the marginal cost. For the regulator the return is, in the public-interest case, some measure of consumer utility usually constrained by a minimum rate of return for the regulated company. The return for the company is the additional profit or profit opportunities allowed. In each case, rational decisions would be based on a benefit flow discounted at an appropriate rate and summed over time.

There are a number of reasons why regulators may spend less than companies:

(i) Effectiveness of action: regulators may be able to obtain bigger results with less effort.

(ii) Different discount rates: if corporate discount rates were lower than those used by regulators, the same cash flow would be worth less as valued by regulators than by those they regulate. There may be some truth in this, given that the cost of borrowing is usually higher for individuals than for telephone companies. To make the point stick it would be necessary to show that regulators use the personal borrowing rate and not something lower like the social time preference rate referred to in Chapter 8 when taking their decisions.

(iii) Wage differences: regulators of comparable negotiating skill may be paid less than the company employees with whom they deal.

(iv) Different pattern of marginal returns: diminishing returns may set in earlier for regulators than for regulated companies.

(v) Real money is worth more than notional money: regulators are negotiating for benefits on behalf a large number of small customers, but they do not directly experience them except as customers themselves. Profit benefits show up more directly in the balance sheet of a company and may have a more direct and measurable effect on the wellbeing of the company employees who are involved.

Several of these factors have a direct bearing on the extent to which public-interest theory holds. Regulatory bodies are not immune to the problems caused by the private agenda of agents. The dedication of public servants to the public interest may be instinctive, or because it is written into the terms of reference of the job. Conflicts of interest may arise unless their career prospects are advanced by following the public interest.

13.1.4 Regulatory capture

A regulatory body set up to protect the interests of consumers may later tend to serve the interests of the industry which it is regulating. This, if it happens, is known as regulatory capture. In some cases, regulation may have been part of a bargain accepted in exchange for allowing a cartel structure to operate in the industry from the outset. Company and consumer interests are not necessarily in conflict. An obligation laid on the director general of OFTEL was to ensure that '*any person by whom such services fall to be provided is able to finance the provision of those services*' (Telecommunications Act 1984 3 1 (b) [3]). This provision is usually taken as meaning that the rate of profit must be at least as high as the cost of capital. It was often referred to by OFTEL representatives in the early years of the office.

Factors tending to tilt the balance towards the interests of the regulated company and strengthen the risk of capture are:

(*a*)   Information: the regulator may be excessively dependent on the regulated industry for operational and financial information.

(*b*)   Career prospects: the staff of the regulatory body may look to the regulated industry for future jobs. If the regulated industry comes to be seen as a source of future employment or patronage the regulatory body will be compromised.

Regulatory independence can be improved by strengthening independent sources of information and ensuring adequate career prospects.

Empirical studies testing the capture hypothesis have been carried out in the US, where the widest and oldest range of regulated private industries exists. There have been few studies in Europe, where the equivalent industries are mainly state owned and, until the 1990s, self-regulating. Self-regulation is the ultimate in capture development. UK regulators have differed in their degree of exposure to this risk. It may have come closest in the water industry, where large information requirements have been generated by fine-tuning RPI–X to the requirements of large capital programmes year by year. Faced with modernisation and environmental costs forced on it by politicians, the regulator has chosen to control profits by a highly detailed price-control scheme which more strongly resembles a rate-of-return regime than the scheme based on the direct control of retail prices that it nominally is.

Capture theory has some explanatory power, but as originally formulated it does not explain why industries can be damaged by adverse regulatory developments. On a broader view, consumer pressures may prove stronger than those of the industry, leading to capture from another quarter.

13.1.5 The lifecycle theory

A more comprehensive view is provided by the lifecycle theory, which describes the development of a regulatory body as a progress through four stages: birth, youth, maturity and senility. In the initial stage, there is a prolonged debate about the objectives and methods of the proposed regulation and the terms of the statutes bringing it into being. The process has many participants and may produce vague, ambiguous or inconsistent objectives and may take decades to complete. In its youth, the regulatory body is likely to be aggressive and innovative. The first generation of regulators may be crusaders, possibly ambitious people anxious to clarify the new rules and break the new ground necessary to establish these.

As the regulatory body matures, the original political impulse behind its creation fades and its own regulatory agenda becomes established. The need for support leads it to become more dependent on the industry which it regulates and the capture process begins. In its final stage the rules, practices and objectives of the body lose relevance to what is happening in the regulated industry. Work slows down and the process of industry capture is completed.

The lifecycle theory has been used to explain developments in some US industries and contains elements of truth. In the terms of the theory, British telecommunications regulation was still in its youth in the 1980s, as was the regulation of gas, water, electricity, aviation and railways. The gestation period for all these industries was shorter than the decades expected for the American cases. The bones of the whole system were worked out between 1980 and 1984 in an intense period of negotiation between the government, the newly appointed regulators and the industries. There was considerable clarity in the objectives, set out in concisely-worded policy documents produced by government and the prospective regulators themselves. The designers of the new regulation were anxious to avoid what was seen as the failings of the American system, especially its litigious nature and the poor incentive qualities of its rate-of-return focus.

Some of the initial provisions were unduly optimistic about the speed with which competition would develop, especially in gas and telecommunications. Others, notably the branch systems general licence (BSGL) for telecommunications, turned out to have an almost impenetrable complexity born of the compromises made in negotiations. On the whole, the new regulatory bodies were established quickly and they were effective. The first generation of British regulators were mainly people of strong personality and established academic, business or public service reputations. Bryan Carsberg (telecommunications) and Stephen Littlechild (electricity) were professors. Ian Byatt (water) was a top treasury civil servant with a strong background in investment appraisal. They were given a good deal of personal discretion and were quick to find ways of obtaining agreement to price reviews and licence changes without the need for further legislation.

Initially, the regulated companies were mostly able to maintain unexpectedly good profitability under the price-capping rules. Competition was expected to develop quickly and the precision of the control was not, except in the water industry, seen as critical. Within a decade rate-of-return control was frequently being seen as a regulatory objective because competition had not developed sufficiently strongly and a rationale was needed for regulatory decisions. The regulators, over this period, proved increasingly resistant to capture. UK experience does not so far match the lifecycle theory very closely.

13.1.6 Interest-group theories

Interest-group theories see regulation as a product of industry and customer pressure groups. The industrialists want higher profits and the consumer groups lower prices or better service, both at the expense of unorganised consumer groups. The necessary regulation is made part of a political programme to gain votes. The Wenders analysis of pressures for cross-subsidy (Chapter 9) is an example of the process.

The theories have an element of plausibility deriving from the effectiveness of small, active groups in other aspects of politics and have empirical support in some aspects of American experience. They do not (with the exception of the Wenders analysis) have much relevance to the development of telecommunications regulation and liberalisation in the UK. This was largely driven by the strong views of politicians, backed by a shift in public opinion on a broad range of issues.

13.1.7 Evaluation of the theories

All these theories have been true for some regulatory bodies at some time in their history but they do not generalise easily and have poor predictive power. Regulatory bodies are public organisations and are not driven by the single profit focus of the privately-owned companies which they regulate, but by a mixture of motives dependent on personalities, local legislation, government pressure and social factors. In this, they are close to the PTT model, a state-owned enterprise capable of acting in many different ways according to the demands made on it.

Social goals are given prominence in theories which see regulation as a way of moving from economic efficiency towards considerations of equity and fairness. There is a long history of this in public policy. Uniform telecommunications tariffing was general in the UK during the period of public monopoly ownership from 1911 to 1984 and common elsewhere, but it began to break down with the market reforms of the 1980s.

13.2 Unintended consequences of regulation

13.2.1 The reason for perverse consequences

Regulation may provide an incentive for perverse and unanticipated behaviour on the part of those regulated, if the economic process being regulated is not fully understood. In extreme cases overvigorous regulation in favour of the consumer can destroy the industry, and hence the consumer interest which it was created to protect. For example, British rent-control legislation covering most of the twentieth century reduced the private rented accommodation market to very small proportion of the whole.

13.2.2 A telecommunications example: Averch–Johnson effects

Badly devised regulatory profit constraints may direct the firm towards resource allocation which is not fully efficient. This dilemma was first researched by Averch and Johnson in the US in the context of 'fair rate of return' regulation [4]. In the Averch–Johnson model, the monopolistic firm seeks to maximise profit subject to a constraint on its rate of return on capital. If the allowed rate of return is higher than the firm's cost of capital but lower than the return which could be achieved without any constraint, the firm has an incentive to substitute capital for other factors of production and operate at an output level where cost is not minimised. The incentive to enlarge the rate base with wasteful capital spending was mentioned in Section 12.5.2. This is a separate problem, although a related one. It is commonly referred to as 'gold plating'.

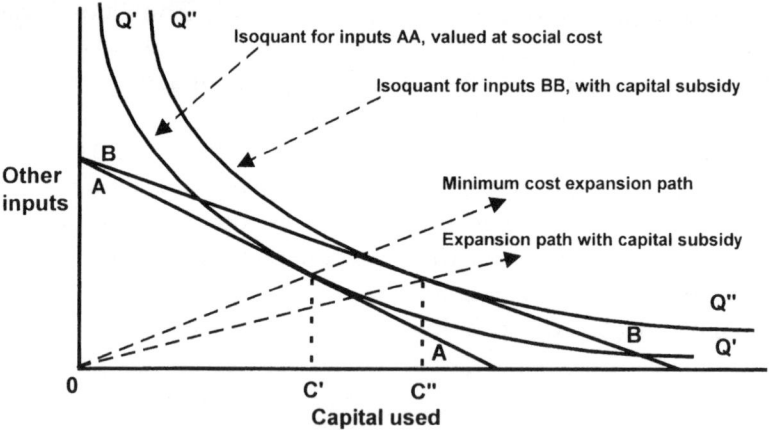

Figure 13.1 The Averch–Johnson effect

Following the discussion of efficiency frontiers in Section 5.3, isoquant Q'Q' in Figure 13.1 is a curve representing different combinations of inputs which produce the same volume of output, and isocost **AA** is a tangent representing the highest level of output that can be produced for the level of expenditure which it represents.

A regulator who allows the rate of profit to exceed the price which the firm pays for capital effectively reduces the cost of capital to the firm. **BB** is the corresponding isocost curve for the same overall cost to the firm as **AA**, with the reduced cost of capital, and Q"Q" is the isoquant to which it is a tangent.

The action of the regulator does not change the market value of capital, only its effective cost to the firm. The reduced cost of capital to the firm comes about through the regulator allowing higher sales prices and profit margins than would prevail under conditions of perfect competition. Q' and Q" are two of a set of isoquants covering all possible output levels. The expansion path which minimises the market cost of resources used for each production level passes through the point where **AA** is a tangent to **Q'Q'**. It need not be a straight line, the exact shape depends on the production function.

The expansion path preferred by the firm would follow the more capital-intensive path shown on the chart passing through the point where **BB** is a tangent to **Q"Q"** This path uses more resources for any given level of output. They are additional resources which would have been used elsewhere had they not been diverted to the firm by the action of the regulator in allowing excessive prices. These results sometimes known as A–J effects.

The extent to which the chosen output level is higher depends on the nature of the production function. Regulation that did not allow an excessive rate of return would keep the firm on the minimum-cost expansion path. This would increase allocative efficiency but not eliminate the output restriction resulting from the monopoly profit-maximising rule of equating marginal revenue with marginal costs.

Several factors may be involved in allowing an excessive rate of return including:

- overestimation of the cost of capital;
- overvaluation of property assets;
- conservative depreciation provisions.

The cost of capital is not easy to measure (Section 7.7). Asset valuation involves taking a view on replacement costs allowing for inflation and technological change, a difficult and often subjective procedure. Empirical evidence for the existence of the effect, which applies to privately-owned monopolies with rate-of-return profit controls, is mixed. Most of it has been gathered in the US. State-owned monopolies may get capital subsidies from a variety of sources or they may be denied access to the capital market and be short of cash for investment. They over or underinvest mostly for political reasons.

13.3 Assessing regulatory performance

13.3.1 Competition and peer pressures

Methods for measuring the performance of regulators are useful, given the power that regulatory bodies usually have. Most are monopolists of a kind, but competition can bring similar benefits to those which it brings to the regulated industries. There is an analogy with the capital market competition to which private companies are exposed even if they are monopolies. If the performance of regulatory bodies in different industries can be compared it should improve regulation by:

- creating peer pressure between the regulators;
- encouraging useful innovation and the exchange of ideas;
- enabling promotions to be made on comparative merit.

The large number of regulatory commissions in the US, spread across industries and mostly duplicated in each state, creates these pressures on its regulators. Many members of the American commissions are directly elected, giving a populist dimension which is lacking in the UK. It is possible to see peer competition between different regulators within the UK, although the regional dimension found in the US is lacking. The creation of regional regulatory bodies in Northern Ireland, Scotland and Wales, charged with local responsibilities could be a positive step in this regard. At a european level, separate national regulators may be stimulated by comparative pressure into greater efficiency than would be found in a single european regulatory body.

The regulations themselves are potentially subject to competitive pressure. In the US, where each state can decide its own rules for the establishment of companies, many companies are registered in the state of Delaware (the Delaware effect) because of its relatively liberal company law. States such as Liechtenstein and Luxembourg and UK dependencies, including the Cayman Islands, the Channel Islands and Gibraltar, have attracted the providers of financial service by similar means. The arrangements are seen as conferring some benefit on the larger centres by providing an alternative level of service.

13.3.2 Predictability of regulatory rules

All business people face uncertainties. In competitive markets they arise from the unpredictability of the economic environment, customer choice and the behaviour of other firms. Regulated industries are partly protected from market risks, especially where they have a legal monopoly. However, they will be exposed to risk arising from the regulatory process, some of which is inevitable. Regulation is essentially a political process trying to achieve economic objectives. In a democracy, political processes are liable to change.

Regulatory processes with discretionary or vague powers for regulators will deter investors. The powers of the regulator should be clearly set out, particularly in respect of the process for setting prices. Failure to do so will raise the cost of capital for the operator.

13.4 References

1 PHILLIPS, C. R.: 'The regulation of public utilities' (Public Utilities Inc, Arlington, 1985)
2 LOMAX, D.: 'London markets after the Financial Services Act' (Butterworths, London, 1987) *as quoted in* GOODHART, *et. al.*: 'Financial regulation – or over-regulation?' (IEA, London, 1988)
3 'Telecommunications Act 1984 (HMSO, London, 1984)
4 AVERCH, H. and JOHNSON, L. L.: 'Behaviour of the firm under regulatory constraint', *American Economic Review*, 1962, **52**, pp. 1052-1069

Chapter 14
Universal service

14.1 A brief history

Universal service is an umbrella term for many policies to enable poorer consumers, and those in rural and other areas which are expensive to serve, to join the telephone network. It has been a widespread regulatory objective in North America for many years and has less formal antecedents in the history of posts and telegraphs. The British magazine *Punch* wrote of a telegraph network that might reach every house, as early as 1858 [1].

The notion of reasonable prices goes back at least to the nineteenth-century origins of telephony and the use of regulation in US law is older than the telephone. In 1878 Alexander Graham Bell speculated on a time when telephony would be cheap enough to be used even by the poorest man and virtually all houses were connected to the network [2]. Although the early writers expected telecommunications to become cheap enough for even the poor to be able to afford the service, they expected the universality to come from falling costs and not from subsidy. Various formulae such as 'affordable service' and the 'satisfaction of all reasonable demands' have since appeared in legislation and the cross-subsidy of local service from long-distance revenues eventually became well established. The existence of huge, sparsely-populated areas of difficult terrain in North America and Australia added political pressure to any business reasons for cross-subsidy. The loneliness and isolation of life on farms in the prairies was well understood by the pioneers of telephony, making universal service a natural and desirable social objective to a degree not experienced in Europe.

In the US, the idea of universal service was given shape in the 1909 annual report of AT&T, which stated the business objective as '*One system, one policy, universal service*' [3]. The company was then buying up other operating companies. The Interstate Commerce Commission (ICC) was given the power to regulate telephone companies in 1910. Its initial work was mainly concerned with establishing a uniform system of accounts, based on its experience with regulating railways. The Justice Department began antitrust investigations in 1912. In the following year AT&T, under its president Theodore Vail, reached an agreement with the Attorney General whereby the company changed its strategy of buying out competing telephone companies to one of providing them with interconnect [4]. The policy bargain effectively made the Bell System a dominant operator regulated at state level. After a lull in acquisition activities these were resumed following permissive legislation in 1921. The number of residential subscribers fell by 25 per cent from 1929 to 1933 as a result of the Great Depression. Although not proportionately worse than in many other industries, the effect was magnified by the public perception of the telephone as a social necessity. This perception was confirmed by legislation in 1934, when the FCC was established and AT&T was subjected to regulation at federal level. The 1934 Communications Act, as amended in 1937 [5] provided for the regulation of interstate and foreign communications by wire and radio:

> '. . .so as to make available, so far as possible, to all the people of the United States a rapid, efficient, Nation-wide wire and radio communication service with adequate facilities at reasonable charges, for the purpose of national defence, for the purpose of securing a more effective execution of this policy by centralising authority . . .'

Common carriers providing telephone, telegraph and cable services were required to furnish adequate service 'upon reasonable request'. The FCC was given powers to order physical connections between carriers and to establish through routes and charges when this was found to be 'necessary or desirable in the public interest'. All tariffs, practices, classifications and regulations were to be 'just and reasonable'. Undue or unreasonable discrimination was made illegal. Tariffs were to be filed with the FCC and adhered to [5]. The FCC could only regulate interstate communications, but local regulators generated pressure to provide a minimum facility for rural and residential users.

The Rural Electrification Act of 1936 was amended in 1949, extending it to cover telephone services. Only 38 per cent of US farms had a telephone service at that time [6]. The Act promoted rural telephone services by making loans available to telephone companies at low interest rates through the Rural Electrification Administration (REA) and by guaranteeing loans made by other lenders. Rural co-operatives were given tax relief on part of their retained earnings. In 1976 there were 633 commercial companies and 243 co-operatives in receipt of REA assistance, none of them in the Bell system. Geographically, they were significant since the Bell system at this time controlled 83.3 % of US telephones but covered only 42% of the land area, or 35 per cent if Alaska and Hawaii were included. In Canada, as in the US, the regulators drew on their experience with regulating railways. Their emphasis was on affordability.

No specific service aims were formulated in the UK until the privatisation of British Telecom in 1984. Prior to that date, the extension of services was implemented more as an implicit obligation by its administrators and the politicians in charge of its investment. The 1984 BT licence placed the obligation to satisfy 'all reasonable demands' for basic service on the new company and the formula appears to have been working satisfactorily in marketing and political terms. From time to time BT has argued that its major competitors, especially Mercury, should also be given a similar obligation but this argument has been rejected.

14.2 The industrial and regional dimensions

14.2.1 *General principles*

In much of the writing about universal service there is an implicit assumption that market forces alone will not produce an economically optimal result, mainly owing to the presence of externalities, and that some form of subsidy is necessary. Translation of the rather general ideas outlined above into provisions that can add economic value in particular countries cannot be achieved with a simple universal prescription. There are a few common principles:

- there may be network externalities which would justify a subsidy or cross-subsidy;
- the compensation principle may be invoked to justify an income transfer;
- politicians may be able to persuade the public that there is a political or macro-economic benefit in providing an effective communications network in rural areas, and perhaps others;

- telecommunications can be a substitute for travel, cutting congestion and pollution costs.
- emergency and crime-prevention services are becoming more dependent on the telephone.

Some of these features also apply to the supply of other public utilities such as water and electricity. For these, in Britain at least, any universal-service obligation is less demanding, so the question has a rather broader context. British electricity distributors are required to meet all reasonable demands, but they are not required to cross-subsidise. The customer pays the full cost. There are several regional companies and they each have their own tariffs, as do the many independent telephone companies in the US. British gas customers pay the cost of connection to premises more than 25 yards from an existing gas main. OFTEL [7] acknowledges the uneven responsibilities, but says that telecommunications differs in being a two-way service, generating externalities not present in the other networks.

The self interest of the operator may make external intervention unnecessary. The network externality which accrues to the customer from being able to ring more numbers and to receive more calls from others also creates a profit opportunity. Any residual problem is in remote areas. In countries with a high rate of penetration there will, nevertheless, still be social groups where few households have a telephone. Milne [8] quotes examples in the UK where, within an overall residential penetration of 85 per cent, there were groups with a much lower penetration, including those in Table 14.1. These are likely to be groups on low incomes, although no incomes are given. Penetration by income group is well documented in the US but published UK data is rather sparse.

Table 14.1 Groups in the UK with low penetration rates by percentage

	%
Northumberland coalfield council estates	40
Unfurnished tenancies	66
Furnished tenancies	45
Single pensioners	72
One-parent families	62

(Source: Family expenditure surveys 1984–90 [9])

Market research carried out by OFTEL in 1992, when the national penetration was 89 per cent [10], identified groups with the lowest penetration as unskilled manual workers (73 per cent), those with a gross weekly income of £50 or less (60 per cent), adults living on their own (79 per cent) and households renting property from the local authority or housing associations (71 per cent). By 1997, penetration rates had increased further and the disparities appear to have been reduced. Against a national average of 93 per cent of households with a telephone, penetration was 88 per cent for single pensioners and 80 per cent for single-parent, one-child families. The least penetrated region was the North East, at 87 per cent [11].

OFTEL has published a series of papers concerning the basic level of service which should be available to all those in the UK who make a 'reasonable request' — a qualification which implies some attention to price and cost. OFTEL [7] proposes four elements:

(i)   A connection to the fixed network able to support voice telephony, low-speed data and fax transmission as standard service,
(ii)   The option of a more restricted service package at low cost with charged outgoing calls limited in number and paid for in advance to assist budgeting.
(iii) Reasonable geographic access to public call boxes across the UK at affordable prices.
(iv)   Special facilities for the disabled, to enable them to enjoy benefits comparable to those of the able-bodied.

The paper also proposes that customers should be able to access emergency services free of charge, receive itemised bills, be able to choose selective call barring, have access to operator and directory services and have the opportunity to repay debt while retaining a service for incoming calls only (outgoing calls being barred), as an alternative to disconnection. The OFTEL proposals provide an example of the adaptation of universal service to rising standards. Itemised billing can be offered at low cost with digital exchanges and would not have been practicable at reasonable cost without them.

There are pressures from the European Commission and elsewhere to widen the definition further to include elements of communications technology such as access to leased lines, ISDN and information services [12]. These pressures arise out of fears that the 'information society' may otherwise have an underclass of the unconnected information-poor citizens beneath it.

A feature which has distinguished virtually all of the debate has been the lack of conclusive, quantified, economic arguments for departing from tariffs which reflect cost.

14.2.2 Service in rural areas

Universal service does not have to be identical service. Nor does it have to bring identical tariffs. Rural areas tend to have less sophisticated services because of lack of demand. Parker *et al.* [13] note that, in the US, '*all indicators show that current investments are improving existing services and providing new services. The rate of upgrade, however, may not be able to keep up with even more rapid advances in urban areas*'. Among the features found were:

- a lower penetration level, primarily owing to poverty, but compounded by geographical remoteness;
- multiparty telephone service, still provided on 7.4 % of independent telephone company lines in 1987, but due to be phased out;
- less digital switching, because the companies could not afford it,
- poorer quality of service, with ten to 12 per cent of lines worse than REA guidelines;
- poorer mobile cellular coverage in rural areas.

Increased subsidies and a wider availability of 'lifeline', or low-user tariffs with reduced rentals, were recommended to facilitate penetration. It has elsewhere been remarked that the enthusiasm of companies for phasing multiparty service out may not always be welcomed by users if the result is higher bills.

A study [14] carried out in the UK for the Rural Development Commission and OFTEL found that penetration rates in rural parts of the UK were similar to those in the UK as a whole. This finding reflects the more homogeneous nature of the UK in respect of both geography and regional income distribution. The study also found no evidence that rural telephone exchanges were being modernised less quickly. It pointed out that some rural exchanges were among the first to be modernised. This finding can be accounted for by the higher maintenance costs in rural areas, giving a greater incentive to replace them. There were, nevertheless, several findings similar to those in Parker *et al.*:

- some advanced data services were not available in rural areas;
- relatively expensive dial-up access to packet-switched service;
- more limited coverage by mobile cellular and paging services, especially in upland districts.

14.2.3 Service in poor countries

So far, the discussion has been in terms of what is appropriate for developed countries. Milne [15] describes five stages in network development and summarises their typical universal service goals, extracted in Table 14.2.

Table 14.2 The evolution of universal service objectives

Stage	Penetration business	residential	Service objectives
1	0–30%	0–10%	long distance service linking major cities, public telephones where demand warrants
2	20–80%	5–30%	telephone service in all population centres, widespread business use
3	70–100%	20–85%	widespread residential take-up, meeting all reasonable demands
4	100%	75–100%	telephone affordable to all, services for special needs (e.g. disabled)
5	100%	100%	telephone affordable for all, public access to advanced services

(Source: adapted from [15])

A distinction has to be made between universal access, which means having a public or private telephone not too far away, and universal service, where all households have their own telephone. In developing countries, it will be a waste of resources to make public-service obligations too rigorous. There will almost certainly be other infrastructure developments that yield more benefits, internal and external, once telephone service provision much exceeds the income norm for the group. The report of the Maitland Commission [16] contains one of the broadest expressions of the universal-service ideal in the concluding part of its recommendations. It says that '*All mankind could be brought within easy reach of a telephone by the early part of the next century and our objective achieved*'. Maitland implies an access to communal facilities such as public call boxes rather than telephones in homes. Facilities can be provided even in remote areas by satellite.

The Maitland Commission's aim to provide access to telephone services even in the poorest communities was followed up by the ITU and led to the launch of WorldTel in 1995. WorldTel was established as an independent company with the objective of organising investment funds for countries with a telephone penetration of less than 1 per cent or an average time on the telephone waiting list of more than five years [17]. In the 85 or so least developed and low income countries (LDCs and LICs) of the world, the Maitland objective can be achieved through village access even where personal lines cannot be afforded. These countries are all at Milne's stage 1.

In 1993 the Indian state of Kerala had 1530 villages, all with at least one telephone, whereas only 17 per cent of the other villages in India had telephone access. Kerala was one of the poorer Indian states and the high penetration was achieved as part of a local policy to improve the infrastructure [18].

In developed countries, universal service is generally seen as a matter to be managed, if at all, by cross-subsidy and the adaptation of tariffs and service features to make a telephone service which is affordable for the poor and does not let them run up excessive debt. In developing countries, structural reform of the supply side is likely to be needed if the long waiting lists and overloaded networks which are commonly found are to be improved.

Service provision will inevitably be uneven during the first two stages of network development. Investment economics dictate that local networks are built region by region, making it impractical to try to meet 'all reasonable demands' for individual service in all regions until each has a network to provide it.

14.3 Options for tariffs

14.3.1 Uniform tariffs

The application of a single scale of charges anywhere in the country has been common practice and this often has universal-service connotations. It also prevents price discrimination and, in monopolies, has administrative convenience. Uniform tariffs in the UK have a history going back to 1840 when Rowland Hill introduced them to popularise the postal service. They live on in a uniform basic letter rate across the EU. The UK telecommunications service was originally provided by private local companies which had divergent tariffs. It was their absorption by the National Telephone Company, which was eventually taken over by the state in 1912, that led to uniformity. By then, the same rental and call charges were applied to all users no matter where they were located. The only significant departures from this were:

- higher rentals in London;
- different tariffs in the city of Hull, which has an independent municipal telephone business;
- higher connection charges for connections requiring more than 100 hours of work to install.

OFTEL has sometimes discussed regional de-averaging, but has not yet felt ready to sanction this. Inhibiting factors are:

- fears of predatory action against BT's competitors;
- potential price discrimination;
- lack of adequate accounting data.

The existence of a uniform BT tariff has not prevented the emergence of a substantial diversity in line and call tariffs after market liberalisation.

The UK and the US differ in their tariffing of rural areas. The US, with its numerous local telephone companies and a strong state tier of regulation, has never had a uniform national tariff. Since the companies are regulated individually there will, even with all the capriciousness said to plague local regulatory processes, be a tendency for service provision to be priced at cost. To a lesser extent this is true in Canada also. Thus, there is a huge variety of local-rate structures across North America. Many countries with more than one operator have not applied uniform tariffs, neither have some with only one operator. In 1986, Denmark had four operators which, between them, had three different line rentals. In the same year, Sweden had one local operator but two regional line rentals. In Norway the single operator charged a monthly rental which varied with the number of customers in the local call area, the largest areas being over twice as expensive as the smallest (Eurodata Foundation Voicebook, 1986).

14.3.2 Lifeline and related services

Higher rentals can be made more acceptable by the introduction of special tariffs for low users, allowing them a cheaper rental in exchange for higher call charges, or similar variants. The idea is that if a special scheme is introduced to cater for the really needy, it will provide headroom for rentals to be raised for the customers who can afford to pay. The various lifeline tariffs, developed first in the US, had this objective and were brought into being after the divestiture of AT&T made it more difficult to cross-subsidise local services out of long-distance revenues. Lifeline tariffs are not confined to telecommunications. They have been used for gas and electricity in parts of the US since the 1970s [5].

The US has the most developed range of schemes, produced by the interaction of local operating companies and regulators in the various states. In California, a rather open-ended scheme was originally introduced in the 1970s at $2.50 per month with a 30 unit free-call allowance and a subsequent charge of five cents per unit. In other states the operating companies have had more success in handling the main potential problem with this type of scheme, that of limiting access without getting involved in means testing. The schemes can be useful in providing a minimal basic service for people who cannot afford the standard service but need a telephone to receive calls or for emergencies. They involve the abandonment of the simple but inefficient uniform tariffs common in public monopolies.

British Telecom introduced a low user rental rebate scheme in 1983. It was replaced by Supportline in 1991, which in turn was replaced by the light-user scheme (LUS) in 1994. In 1997, LUS offered a maximum rebate of 61 per cent off line rental, and was available for the 21 per cent of customers with the lowest bills. Second lines and lines in second homes were not eligible. OFTEL (1997) criticised LUS for allowing the build up of debt, being complicated for the customer to control and not being targeted at households without a telephone. As in the US, the BT schemes defused some of the political resistance against rental rises and tariff rebalancing.

14.4 The cost of public-service obligations

14.4.1 Measuring the cost

In rich countries, where at least 90 per cent of households have telephones, there will be at least nine lines providing revenue for every unconnected household, making the cost of any residual service provision almost a marginal consideration. For countries with a lower penetration, such as the US in the 1920s, the immediate provision of service to all those without a telephone would be impracticable and a suboptimal use of resources. Even so, an obligation to provide service in more limited circumstances may still provide benefits.

Privatisation and market liberalisation have brought disputes about the cost of universal service, compared with the network that would be provided on purely commercial criteria. There are two approaches:

(i)   To measure the incremental cost of providing service to scattered areas, or groups thought to be unprofitable.

(ii)   To identify loss-making services on a fully allocated cost basis.

The second includes a share of overheads omitted from the first and produces a higher figure. The difference matters if the operator is to be compensated in some way for universal provision. The incremental approach represents the opportunity cost, but operating companies tend to stress fully allocated costs. In Australia, a study of costs by the Bureau of Transport and Communications Economics produced results sharply lower than the fully allocated costs used by Telecom Australia.

REA and FCC figures for 1988 in Table 14.3 show that the Bell Operating Companies had lower costs per line than rural borrowers and the larger of the independent operators in the US, but also lower revenues.

Table 14.3 US telephone operating cost and revenue per line in 1988

Operators	Cost per line ($)	Revenue per line ($)
Bell Operating Companies	463	646
Large independent companies	531	751
REA borrowers	522	682

(Source: calculated from FCC [19], REA [20])

These figures use a fully allocated cost methodology, where, as is commonly the case, part of the call revenue is used to cover the fixed cost of providing a line, customers who make few calls may not generate enough revenue to pay for the cost of providing service to them. Other customers may contribute less than the full cost of providing service because they live in places which are expensive to serve. The cost of providing service for remote rural areas in developed countries may not make up a high proportion of operating costs because very few people live there. OFTEL (1997) provides estimates of the marginal cost to BT of providing three types of uneconomic service, shown in Table 14.4. The figures relate to long-run avoidable costs, after the deduction of revenue in the year ended March 1996. Revenue includes customer bills and the value of incoming calls. Emergency services and obligations placed on all operators or funded by other means have been excluded. The figures are calculated at BT's actual level of efficiency and do not include any allowance for trading advantages arising from ubiquity.

Table 14.4 Cost of uneconomic services in the UK

Type of service	Cost (£m)	Number of lines
Uneconomic customers	10–15	less than 0.5 % of UK lines
Uneconomic areas	45–55	6–7 % of UK lines, nearly all in Scotland
Uneconomic payphones	10–15	about 20 % of payphones
Total	65–85	less than 1 % of network revenue

Bergendorff [21] puts the Swedish cross-subsidy from urban to rural areas at SEK 1bn (£95 m) per annum on total telephony revenues of SEK 14bn (£1330 m). In the US, the regional element of cross-subsidy runs from densely populated north east to Alaska and the more scattered south west [22].

None of these figures includes any offset for the benefits which a dominant operator may derive from being a universal-service provider. Dominant operators enjoy a variety of advantages from their size and market power, including:

- economies of scope and scale, giving a cost advantage against competitors;
- the ability to charge more than smaller competitors through price leadership;
- access charges, which can make competitors contribute to cost;
- interconnect agreements which base payment on the volume of traffic passing between networks and involve the disclosure of traffic flows, often to a dominant competitor.

These provide a margin of competitive advantage, from which an element of cross-subsidy can be afforded and extra profit from which universal-service obligations can be financed or imposed without market distortion.

OFTEL [21] proposes three others:

(i)   Life-cycle effects. Customers increase their calling rates as they get older and richer.
(ii) Ubiquity. New customers tend to go to the operator which they know best.
(iii) Brand enhancement and corporate reputation, which also attract new customers.

The benefits of ubiquity were quantified by OFTEL with a model on the following lines:

- about 11 per cent of UK households move location each year;
- about 7.8 million households had an alternative local-access supplier (the local cable operator) in the third quarter of 1996;
- about 858 000 (11 % × 7.8 m) of those moving house move to one which has an alternative supplier as well as BT;
- it is assumed from survey evidence that 66 per cent of these movers do not know that they will have a choice of supplier;
- it is assumed that 60 per cent of those ignorant of the second supplier would still have chosen BT, but 40 percent would have chosen the alternative, BT thus gains 26 per cent (40 % of 66 %) of the 858 000 lines required by the movers.

The survey was not large enough to be conclusive, but it illustrates the mechanism. The gain from ubiquity fades away each year as more people move and the alternative suppliers, which may market aggressively where they provide service, become better known. OFTEL's view was that the overall benefits, including those from ubiquity, were likely to be as big as the costs.

In 1997, the FCC noted that the annual subsidies for universal service in the US were variously estimated at from $4 bn to $20 bn and suggested around $12 bn (£7.7 bn) for an initial estimate, mainly for residential tariff assistance in areas with high operating costs [23]. This is approximately one third of local service revenues, a surprisingly high figure by comparison with the other estimates. It reflects, perhaps, the low population densities found in much of rural America, and differences in methodology.

14.4.2 Specifying and financing public service obligations

As indicated in Section 12.7, public service obligations need to be clearly set down. They should be satisfactory for the operator and for users. The regulator should see that they are complied with, as they are part of a regulatory contract between the company and the government.

A direct financial grant targeted at potential customers would, neglecting externalities, be more economically efficient than subsidising the service which they get, because it would offer the choice of using the grant on a preferred purchase. For example, suppose that a tariff reduction of £1 per week would be just enough to induce a marginal tenant without a telephone to take up service, but the £1 per week was given the cash instead. Would the tenant still take up telephone service, or would the £1 be spent on something else that was rated as being even more desirable? If the latter, a tariff below cost would be second best to a cash grant (Chapter 8). If, in spite of this, a subsidy is directed to the operator and not the customer, any net costs for the provision of universal service will require a financial contribution which may come from:

- government subsidy to the operator;
- other customers, by cross-subsidy;
- competitors, as a surcharge on interconnect payments;
- all operators, through a Universal Service Fund.

Direct government subsidy is rare. Telecommunications operators make profits in all but the direst circumstances and, with their market power, they can usually be pressured into cross-subsidy.

As noted by Vietor [4], cross-subsidy in the US had for many years been formalised in the *separations* process by which the allocation of costs was loaded towards long-distance service. This resulted in higher interconnect charges for the delivery of long-distance calls accepted by local companies, and lower charges for local calls and access. The system began to come apart under pressure for competition when companies outside the Bell system, such as MCI, were granted discounts on the interconnect rates. In 1982 the FCC introduced a new plan whereby:

- Customers paid a flat rate subscriber line charge (SLC).
- Carriers paid a per minute carrier line charge (CLC).

Subscriber line charges rose to about \$3.50 per month, a level which was calculated to eliminate most of the cross-subsidy, over a period of six years to 1991. Being paid at a flat rate, the SLC was effectively an increase in the rental which did not increase marginal call costs to the user and was allocatively efficient. The CLC was a per-minute fee paid to the local operator for delivering a long-distance call. It fell as the SLC rose and was eventually phased out.

In the UK, OFTEL introduced access charges. These, like the carrier line charges, were a surcharge on the interconnect payments paid by long-distance carriers to local operators to recover the supposed losses from the underpricing of access. In practice, most of the payments were to flow from the new carrier (Mercury) to the incumbent (BT). It was envisaged that the charges would be phased out as local-service charges rose to cover cost. The plan was contentious from the outset. There were numerous deferments and waivers as new companies entered the market and much of it fell into disuse.

14.4.3 Universal service funds

In the absence of government funding, the elimination of cross-subsidy leaves a funding gap for universal service. This gap could be filled by a fund which levies a tax on all the operating companies and provides subsidies to meet the cost of providing cheap local service. The FCC proposals of 1982 included a fund to replace the cross-subsidy revenues. This was eventually provided for in the 1996 Telecommunications Act, though implementation was to prove complex. OFTEL [7] proposed a fund of this kind for the UK, but it seems unlikely to be introduced.

The European Commission has considered a universal service fund and in November 1996 published guidelines for funding universal service. Funds were to go only to efficient operators, they were to be managed by national regulatory authorities and they should take operator benefits, such as ubiquity and brand recognition, into account. The reason for the Commission's interest was a fear that universal service might be damaged by the introduction of full network competition from January 1998. A study was commissioned to advise on how universal service obligations should be costed and financed. The resultant report, which was published in 1997 [24], was mainly concerned with methodology and it included a useful survey of the costing schemes currently in use. The familiar themes of uneconomic areas, uneconomic customers and uneconomic pay phones were covered, with a warning that funding schemes must be designed with care to avoid damaging economic efficiency. The track record of the EC in managing price support and subsidy schemes is not such as to encourage confidence that this can be fully achieved. There is also the problem of data collection. Information requirements for the assessments of claims on the fund would be extensive, and probably contentious.

As with cross-subsidy, the most efficient way of distributing the income of a universal fund might (neglecting any additional management costs) be to give it out as cash to the client groups rather than as subsidies to operators for the provision of cheaper service, or as vouchers to deserving customers. This seems unlikely to happen with any of the schemes so far proposed.

14.4.4 Universal service under competition

Economies of scale and elements of natural monopoly may give rise to a belief that competition leaves incumbent operators unable to afford the provision of universal service as their more profitable lines of business are picked off by competitors. Practical experience and the arguments advanced by OFTEL about the benefits of ubiquity suggest that operators in developed countries would usually prefer to offer service everywhere, improving their image as reliable, universal providers, provided that they could levy a charge on other users and carriers to assist with costs. The universal service fund is an example of such a levy. Four factors limit the effectiveness of the levy strategy:

(i)   The operator may be less efficient than the new competitors.
(ii)   There may be too much cross-subsidy in the tariff structure.
(iii)   Even with weak competition, competitors may not be willing to pay a levy.
(iv)   With strong competition prices will be driven towards cost.

The introduction of cost-based prices and, implicitly, higher rentals, was thought by critics to be prejudicial to the continued existence of universal service in the US, Canada and the UK. The fears seem to have been misplaced. Tariffs have been rebalanced, as Table 9.4 shows, but the proportion of American households with telephones has continued to rise. The number of households with a telephone rose from 91 per cent in 1983 to 94 per cent in 1994. As Figure 14.1 shows, the penetration growth rate was virtually unaffected by the initially large real percentage increases in local charges for access and calls or by the subsequent decreases. This may have been because overall residential telephone bills fell, in response to falling unit costs and the

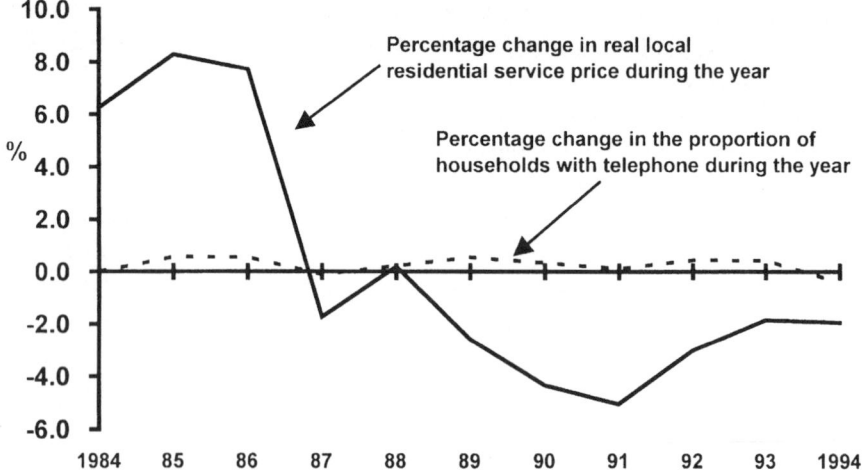

Figure 14.1 Penetration and residential telephone price, US 1984–94
(Source: FCC [25], calculations)

reduction in long-distance rates. Tariff rebalancing towards rentals in the US evidently occurred at a slow enough pace for adverse demand effects to have been avoided.

Other FCC statistics in Table 14.5, showing telephone penetration by household income, confirm that there was no substantial reduction of telephone penetration among the less well off. For some of the lower groups, it increased.

Table 14.5 Percentage of US families with a telephone

Income group ($ per annum)	1983 (%)	1992 (%)
Under 5000	71.7	72.0
5000–<7500	82.7	83.2
7500–<10 000	88.2	87.5
10 000–<12 500	89.7	90.5
12 500–<15 000	92.1	91.5
15 000–<20 000	95.2	93.3
20 000–<25 000	96.9	95.9
25 000–<30 000	98.0	97.1
30 000–<35 000	98.8	98.2
35 000–<40 000	99.0	98.6
40 000–<50 000	99.2	99.2
50 000–<75 000	99.4	99.5
75 000 +	99.4	99.4

Comparable figures for the UK in 1992 can be derived from OFTEL market research [10] and are shown in Table 14.6.

(Source: FCC statistics in Reference 26)

Table 14.6 Penetration by household income, the UK and the US, 1992

Gross household income		Households with telephone	
		GB	Equivalent US
£ per week	$ per annum	(%)	(%)
Up to 50	Up to 4,680	60	71.7
50.01–100	4680.01–9360	73	
100.01–150	9360.01–14 040	84	
150.01–200	14 040.01–18 720	88	95.2
200.01–250	18 720.01–23 400	90	
250.01–300	23 400.01–28 080	91	
300.01–350	28 080.01–32 760	94	98.0
350.01–400	32 760.01–37 440	95	
400.01–450	37 440.01–42 120	96	
450.01–500	42 120.01–46 800	95	99.2
Over 500	Over 46 800	99	99.4

Notes: $ per annum at 52 weeks, $1.8 per £, percentages from Table 14.4. Figures not fully comparable because of definitions.

These figures show the strong relationship which exists between income levels and telephone penetration at the bottom end in both countries.

14.5 References

1   POOL, I. de S.: 'Technologies of freedom' (Belknap, Cambridge MA, 1983)
2   'Universal service and rate restructuring in telecommunications' (OECD, Paris, 1991) p.27, quoted in **3**
3   COMPAINE, B. M. and WEINRAUBE, M. J.: 'Universal access to online services', *Telecommun. Policy*, 1997, **21**, (1), pp.15-33
4   VIETOR, R. H. K.: 'AT&T and the public good: regulation and competition in telecommunications, 1910-1987' *in* BRADLEY, S. P. and HAUSMAN, J. A. (Eds.) 'Future competition in telecommunications' (Harvard Business School Press, Boston MA, 1989)
5   PHILLIPS, C. F.: 'The regulation of public utilities' (Public Utilities Reports, Inc., Arlington, 1985)
6   MEYER, J. R., WILSON, R. W., BAUGHCUM, M. A. and CAOUETTE, L.: 'The economics of competition in the telecommunications industry' (Oelgeschlager, Gunn and Hain, Cambridge MA, 1980)
7   'Universal telecomunications services: consultative document on universal service in the UK from 1997' (OFTEL, London, 1995)
8   MILNE, C.: 'Universal telephone service in the UK, an agenda for policy research and action', *Telecommun. Policy*, 1990, **14**, (5), pp. 356–371
9   Central Statistical Office: 'Family expenditure survey' (HMSO, London, 1990)
10   'Households without a telephone' (OFTEL, London, 1994)
11   'Family Spending in the United Kingdom 1996-97, ONS (97)297'. Office of National Statistics, London, 1997.

12 'One step forward, two steps back', *Communicate*, February 1997, pp.9–11

13 PARKER, E. B., HUDSON, H. E., DILLMAN, D. A. and ROSCOE, A. D.: 'Rural America in the information age' (Aspen Institute and University Press of America, Lanham, 1989)

14 'Telecommunications in rural England, a report by Economic and Transport Planning Group for OFTEL and the Rural Development Commission'. Rural Development Commission, London, 1989

15 MILNE, C.: 'Universal service for users: recent research results – an international perspective'. Paper presented at the 25th annual Telecommunications policy research conference, September 1997

16 Maitland Commission: 'The missing link' (ITU, Geneva, 1984)

17 'On the threshold of Telecom 95'. ITU, Geneva, 1995

18 KRISHNASWAMY, G.: 'Where poor have access to telephones: telecommunications and development in the Indian state of Kerala'. Paper presented at ITS 11th Biennial Conference, June 1996, Seville

19 FCC: 'Statistics of the communications common carriers', (FCC, Washington D.C., 1988, 1988/89 edn.)

20 '1988 statistical report, rural telephone borrowers'. Rural Electrification Adminstration, Washington D.C., 1988

21 BERGENDORFF, H.: 'Swedish policies towards universal service provision', *SPRU/PICT* conference paper 4, October 1990

22 WENDERS, J. T.: 'The economics of telecommunications, theory and policy', (Ballinger, Cambridge, 1987) p.277

23 HUNT, R. E.: 'Statement on implementation of the Telecommunications Act of 1996 before the subcommittee on telecommunications and finance committee on commerce, U.S. House of Representatives July 18, 1996', (FCC, Washington D.C. 1997)

24 'Costing and financing universal service obligations in a competitive telecommunications environment in the European Union, final report'. WIK, Bad Honnef, October 1997.

25 'The statistics of the communications common carriers' (FCC, Washington D.C. 1995)

26 MITCHELL, B. M. and DONYO, T.: 'Utilisation of the US telephone network' (Rand, Santa Monica, 1994)

Chapter 15
Performance measures

15.1 Introduction

Quantitative measures of industrial performance are useful to shareholders, managers, planners, consumers, regulators and research economists. Shareholders may consider selling their holdings and buying assets better suited to their needs. Those who own or run a business want to know how well it is doing and where its performance can most easily be improved. Consumers, regulators and researchers have an external focus and are more interested in quality, prices and ways in which the business can be pressured into giving better value for money.

Researchers may want to analyse the dynamics of the industry with a view to enlarging understanding of how it works and where it is going. They will do this to assist in policy formulation if there are policy makers to advise. Financial institutions use analysts to appraise company performance. The emphasis may be on the current situation in a single company but more often it is comparative, of prices, productivity or service quality across time, industry sector or country. The principal dimensions of comparison are:

- one company over time;
- a cross-section of companies in one industry in a single year;
- cross-sectional international comparisons within the same industry;
- cross-sectional comparisons with companies in other industries.

15.2 Economic issues in performance measuresment

15.2.1 Economic objectives

Performance comparisons have an implied objective function with a measurable quantity to be maximised. At the level of society as a whole, the maximand would ideally be a constrained form of net social benefit, a quantity not readily translatable into indicators used at the level of the individual company or operator and not at all easy to measure. Constraints may arise from distributional considerations. Net social benefit may be increased through improved efficiency. Physical indicators are widely used in efficiency measurement.

15.2.2 Welfare considerations

As the private owners of companies working in fully competitive conditions are profit maximisers, the indicators of most value to them and to potential owners are those which have a bearing on the measurement and maximisation of present and future profit. The people who manage these operators on behalf of the owners will want indicators showing the effect of changes in the many different operating procedures which may affect profit. They may hope to identify procedural changes which act as the most effective levers in achieving profit improvement.

The regulatory bodies which intervene in telecommunications markets need performance measures which throw light on the economic efficiency of the operators, not just their profits. Benchmarking is an area of common ground between regulators and commercial operators. It has become common for companies to look for industry best practice as a means of increasing the efficiency of their own operations. One significant difference is the level at which the comparisons are made. At company level, benchmarking may be highly disagreggated, focusing on specific processes which might be improvable. The regulatory and analytical interest is at a greater level of generality, where broad conclusions about relative overall efficiency are clearer at the expense of detailed information about the processes which might be improved.

Performance indicators are not of much economic value unless the rankings which they provide can be calibrated or ranked in terms of economic welfare. Some indicators readily pass this test. The production of the same set of outputs from fewer resources will release resources for the creation of other kinds of wealth and must be an economic gain. There are other indicators which do not. The production of a different set of outputs from the same inputs may be an economic gain or loss. Most comparisons of output are beset by this problem. Other cases involve trade-offs of cost against the value of a better result, especially in the area of marginal changes in service quality.

Information availability can be a problem too (Sections 1.1.2 and 10.3), except where there is regulatory pressure for the publication of operational data. The many state-owned enterprises tend to publish more detailed statistics than their privately-owned counterparts and they are often less concerned about commercial sensitivity.

15.2.3 Economic efficiency

Operating efficiency is usually measured by physical indicators such as:

- labour productivity;
- productivity of all inputs used;
- unit costs;
- international comparisons of these.

All physical indicators are ratios of some combination of inputs and some combination of outputs. They are often in the form of index numbers. Calculation requires:

- a good understanding of index number theory, such as that given in Reference 1;
- good data.

15.3 The users of performance measures

The different perspectives and concerns of shareholders, managers, customers, regulators and analysts are reflected in the performance measures of most interest to them, summarised here in a brief overview.

The shareholder perspective

- dividend level and its compound annual growth rate (CAGR);
- quality of earnings in terms of variability;
- risk;
- share price.

Financial markets measure the performance of private companies on purely financial criteria. The share price and statistics such as the price/earnings ratio encapsulate views about the ability of the company to make profits over a long period. The rate of return on capital, a statistic commonly used by regulators, does not seem to be much used in financial markets.

Managerial concerns

- short and long-term profit;
- market share;
- unit costs;
- performance relative to competitors.

Customer interests

- price;
- quality;
- variety of choice;
- innovation.

Measures looked at by regulators

- market share;
- prices;
- return on capital;
- performance relative to other operators.

Companies which have their profits regulated may not be adequately motivated towards efficiency and a variety of physical indicators may be used to examine performance. The indicators are of two basic types:

(i)   indicators of physical productivity, such as the number of exchange lines per employee;
(ii) indicators of service quality, such as the fault rate on private circuits.

Some aspects of performance are difficult to measure directly. It is hard to know if output is being carried out in the most efficient way.

15.4 Price, quantity and value index numbers

15.4.1 Definitions

Any series of numbers can be converted to index form by choosing one of them as a base and expressing the others as a percentage of it. A typical telecommunications example would be the number of exchange lines at the end of each year, with the first year used as a base and set to 100.

Performance analysis in telecommunications requires the decomposition of cost and revenue series into their volume and price components. A simple example is given in Table 15.1, based on the annual revenue from exchange lines and using the definitions below.

years run 0, 1, 2, ... i, . . .
year 0 is the base year
base year indices have the value 100

Table 15.1 Examples of index number calculations

Raw data for a single product				
Variable	Notation for year i	Year 0	1	2
Number of lines	Q_i	15 000	16 000	15 800
Rental per line, £	P_i	110	110	150
Line revenue, £m	R_i	1.65	1.76	2.37
Index numbers (year 0 = 100)				
Lines	I_{qi}	100	106.7	105.3
Rental	I_{pi}	100	100.0	136.4
Revenue	I_{ri}	100	106.7	143.6
Check calculation				
Price index × volume index/100				
$= I_{pi} \times I_{qi}/100$ = revenue index	I_{ri}	100	106.7	143.6

Where, as always in telecommunications, there are multiple inputs and outputs, the calculations require more data and involve choices about the type of index to be produced.

The notation used in the price and quantity index-number formulas in Table 15.2 is as follows:

P_{ij} is the price of the jth commodity in year i, per unit of the jth commodity ($j = 1, 2, 3, \ldots n$)

Q_{ij} is the quantity of the jth commodity in year i, measured in units convenient for the commodity and which may be different for each commodity ($j = 1, 2, 3, \ldots n$)

I_{pi} is the value of the price index in year i

I_{qi} is the value of the quantity index in year i

Two main types of index are in common use. In the base-weighted, or Laspeyres, form a single set of base-year weights is applied to the various series for each successive year. The indices for the ith year are given by:

price in year i, year $0 = 100$:

$$I_{pi} = 100 \times \Sigma_j Q_{j0} P_{ji} / \Sigma_j Q_{j0} P_{j0} \tag{15.1}$$

Quantity in year i, year $0 = 100$:

$$I_{qi} = 100 \times \Sigma_j P_{j0} Q_{ji} / \Sigma_j P_{j0} Q_{j0} \tag{15.2}$$

In the current weighted, or Paasche, form the weights vary, being derived from the current year. The indices for the ith year are given by

price in year i, year $0 = 100$:

$$I_{pi} = 100 \times \Sigma_j Q_{ji} P_{ji} / \Sigma_j Q_{ji} P_{j0} \tag{15.3}$$

Quantity in year i, year $0 = 100$:

$$I_{qi} = 100 \times \Sigma_j P_{ji} Q_{ji} / \Sigma_j P_{ji} Q_{j0} \tag{15.4}$$

The product of a base-weighted price index and a current-weighted volume index, like the product of a base-weighted volume index and a current-weighted price index, is an index of value, e.g. of turnover, since

$$\Sigma_j P_{j0} Q_{ji} / \Sigma_j P_{j0} Q_{j0} \times \Sigma_j Q_{ji} P_{ji} / \Sigma_j Q_{ji} P_{j0} = \Sigma_j P_{ji} Q_{ji} / \Sigma_j P_{j0} Q_{j0} \tag{15.5}$$

An example of these indices is given in Table 15.2.

Base and current-weighted indexes have different economic properties. Expenditure patterns change over time in response to changes in relative prices.

A rise in a single price with all others unchanged leaves the consumer of the product worse off. It also changes the consumer's preferred pattern of expenditure because the relative price of the product has changed. A consumer given an amount of money equal to the price rise may choose to spend part of this on other goods because of the change in price relativities.

Table 15.2 Indices for multiproduct case – two products

	Year		
	0	1	2
Raw data			
Lines	15 000	16 000	15 800
Minutes per line	5 000	5 100	6 000
Minutes (millions)	75	81.6	94.8
Rental (£)	110	110	150
Price per minute (£)	0.1	0.09	0.07
Revenue (£m)	9.15	9.104	9.006
Index numbers (year 0 = 100)			
Number of lines	100	106.7	105.3
Call minutes	100	108.8	126.4
Rental	100	100.0	136.4
Call price	100	90.0	70.0
Revenue	100	99.5	98.4
Quantity index weights			
Base-weighted lines	110	110	110
Current-weighted lines	110	110	150
Base-weighted call minutes	0.1	0.1	0.1
Current-weighted call minutes	0.1	0.09	0.07
Price index weights			
Base-weighted lines	15 000	15 000	15 000
Current-weighted lines	15 000	16 000	15 800
Base weighted call minutes (m)	75	75	75
Current-weighted call minutes (m)	75	81.6	94.8
Weighted indices			
Base-weighted volume index	100	108.4	122.6
Base-weighted price index	100	91.8	82.0
Current-weighted volume index	100	108.4	120.1
Current-weighted price index	100	91.8	80.3
Check calculations			
Base-weighted price ×			
current-weighted volume	100	99.5	98.4
Base-weighted volume ×			
current-weighted price	100	99.5	98.4
Revenue index	100	99.5	98.4

The question of how much extra money is needed to obtain the same overall satisfaction from purchases as before is therefore not a simple one. Current-weighted price indexes rise less quickly than base-weighted price indexes because the weights reflect price changes during the year. If the price of a product increases the volume of sales will fall. This reduces the weight given to the price increase in the current-weighted index but leaves it unchanged in the base-weighted index.

A base-weighted price index tends to overestimate the amount of extra money needed and a current-weighted price index tends to underestimate it. Other forms of index have been proposed to overcome this difficulty. Absolute accuracy depends on knowing the shape of the 'indifference curves' which map the detailed consumption patterns of individuals, information almost certainly not available, but there are various ways of getting a result which is likely to be closer.

Simple average

A simple average of the base and current-weighted indices can be used. The method is quick and obvious but lacks some desirable mathematical properties.

Geometric average

The geometric average (square root of the product) of the base and current-weighted indexes is known as *Fisher's ideal index* and was proposed by Irving Fisher [2]. One of its more important properties is that the product of the ideal price and quantity indexes gives a value index which is also given by the product of base-weighted price and current-weighted quantity indexes. It is the best measure of changes in real income, under fairly broad assumptions [1].

Linked Laspeyres

A third method is to update the weights periodically. This is done every five years in the UK national income and expenditure statistics; the weights in the UK retail prices index are updated at more frequent intervals from consumer expenditure survey data. The linked Laspeyres index changes them annually, using the previous year's weights for each successive year. This index also usually falls between the base and current-weighted index values. A specimen calculation is shown in Table 15.3.

Table 15.3 Linked Laspeyres indexes calculated from the same data

	Year		
	0	1	2
Quantity index weights			
Lines	110	110	150
Call minutes (millions)	0.10	0.09	0.07
Price index weights			
Lines	15 000	16 000	15 800
Minutes	75.0	81.6	94.8
Linked Laspeyres indexes			
Quantity	100	108.4	122.3
Price	100	91.8	81.8

15.4.2 Real price indexes

The price indexes described so far are related to nominal prices, that is to the money prices actually paid. A real price index can be calculated by adjusting the nominal price index for inflation. This is most easily done by dividing the nominal price index by a retail or consumer price index each year and expressing the result as an index. The real price index then shows how the measured prices move in relation to the overall price level, as in Table 15.4.

Table 15.4 Calculation of a real telephony price index

	Year		
	0	1	2
Telephony price index from Table 15.3	100	91.8	81.8
General price index	100	101.0	103.0
Real telephony price index	100	90.9	79.4

A linked Laspeyres index of real UK telecommunications tariffs was published for many years by the British Post Office and is shown in Table 15.5.

The index fell irregularly at a compound average annual rate of 1.6 % over the twelve years covered.

The FCC [3] has published telephone tariff changes in the US since 1935. From 1935 to 1994 the consumer price index for all goods and

Table 15.5 UK real telecommunications tariff index

Year ended March	Index
1970	100.0
1971	102.1
1972	94.0
1973	88.6
1974	84.2
1975	80.4
1976	103.9
1977	103.5
1978	89.8
1979	82.6
1980	74.0
1981	76.8
1982	76.8
1983	82.5

(Source: Post Office annual reports)

services rose by an average annual rate of 4.1 % against 2.1 % for telephone services, a long-term downward real telephony price trend of two per cent per annum in real terms. Figures for 1982 to 1994 showed a somewhat larger real decline, averaging 2.4 % per annum.

15.4.3 Labour productivity

The most widely used indicator of labour productivity, especially in international comparisons, is the number of exchange lines per employee. This is a crude measure, using only one input and one output, and it is affected by the extent to which, as described in Chapter 5, capital is being used to replace labour in the production process. Geographical system density has been put forward as part of the reason for variations in labour productivity but it does not seem to be the only factor at work, as Tables 7.1 and 7.2 in Chapter 7 show.

Another factor enhancing labour productivity improvement has been the trend towards buying in services such as catering and maintenance instead of having cooks and mechanics on the pay roll, as illustrated in Sections 7.2.2 and 15.4.10. Capital productivity has occasionally been calculated from output and capital-input volumes. It has not risen as fast as labour productivity because most growth is provided for by capital.

Operators often use their own labour force for part of their capital works and charge the cost of this labour to the capital account rather than to operating cost. British Telecom assigned nearly 13 per cent of its staff costs to its capital account in the early 1990s. The number of employees charged to the capital account is not normally available from other operators. The number of exchange lines per employee is therefore usually calculated inclusive of capitalised labour, even though the capitalised output is excluded.

The measure remains popular for four reasons:

(i)   It is readily understood.
(ii) Movements in lines per employee often correlate well with more comprehensive productivity measures.
(iii) Only two pieces of data are needed, both published by virtually every operator.
(iv) International comparisons are free of currency and exchange-rate complications.

Table 15.6 shows the number of exchange lines per employee for European operators over a period of 11 years. The main features of the measure are well illustrated. Productivity varies considerably between operators, and it is rising over time in all of them. Rising productivity is coming from rising output served by a static or falling labour force.

Labour productivity is better measured by using total output volume as the numerator instead the number of lines alone, although the data may not always be available. Productivity measured in this form is also rising. This may look like a high marginal productivity of labour, but part of it represents the diminishing importance of labour in the production process noted above for British Telecom.

15.4.4 Quantity indexes of resources used

Indicators of movements in cost at constant prices are essential for the analysis of physical productivity. Since the costs of different inputs may change at different rates, an overall quantity measure of costs will be some form of weighted index. The base-weighted, current-weighted and linked forms are available, among others. In Table 15.7 the costs streams of labour, capital (represented by depreciation) and other (bought-in) inputs are given. Labour volume is taken as the number of employees. Economic depreciation at constant prices is calculated from company asset registers and other inputs are revalued at base-year prices.

Table 15.6 Exchange lines per employee for European operators

		1993 values	1980-93 average growth (%)	Notes
Austria	lines (000s)	3579.2	3.8	
	employees	18 000	0.4	
	lines per employee	199	3.4	
Belgium	lines (000s)	4395.7	4.6	
	employees	25 900	−0.8	
	lines per employee	170	5.4	
Denmark	lines (000s)	3059.8	2.5	1980–93
	employees	16 700	−1.9	1987–93
	lines per employee	183	3.5	1987–93
Finland	lines (000s)	2760.7	3.6	
	employees	15 200	−2.2	
	lines per employee	182	6.0	
France	lines (000s)	30 900.0	5.2	
	employees	154 900	−0.3	
	lines per employee	199	5.6	
Ireland	lines (000s)	13 100.0	28.9	
	employees	14 269	−2.4	
	lines per employee	918	32.1	
Italy	lines (000s)	24 176.0	4.9	
	employees	115 000	0.8	
	lines per employee	210	4.1	
Netherlands	lines (000s)	7630.0	3.5	
	employees	32 000	1.1	
	lines per employee	238	2.4	
Norway	lines (000s)	2 334.8	5.3	
	employees	14 000	−2.0	
	lines per employee	167	7.5	
Portugal	lines (000s)	3260.3	9.6	
	employees	21 100	−0.6	
	lines per employee	155	10.3	
Spain	lines (000s)	14 235.6	5.4	
	employees	74 300	0.4	
	lines per employee	192	4.9	
Sweden	lines (000s)	5903.0	1.6	
	employees	26 200	−3.2	
	lines per employee	225	4.9	
Switzerland	lines (000s)	4265.8	3.2	
	employees	19 200	1.3	
	lines per employee	222	1.9	
Germany	lines (000s)	28 078.6	1.9	East & West 80–93
	employees	231 000	0.8	West 1980–90
	lines per employee	122	3.0	West 1980–90
UK (BT)	lines (000s)	26 061.0	3.0	
	employees	170 700	−2.6	
	lines per employee	153	5.8	
Total	lines (000s)	169 231.8	4.5	
	employees	891 874	−0.6	
	lines per employee	190	5.1	

Notes: UK: year to 31 March   Others: year to 31 December
Germany in total is West Germany 1990,
total excludes Denmark and Portugal

(Sources: ITU, BT annual reports)

Table 15.7 Calculation of an input cost index

	Year		
	0	1	2
Raw data			
Number of employees	120	119	119
Average paybill costs per employee (£)	25 000	25 100	26 000
Economic depreciation volume (£)	3 000 000	3 100 000	3 150 000
Supplementary depreciation (£)	0	30 000	90 000
Other input costs, year 0 prices (£)	2 800 000	2 900 000	3 000 000
Price index for other inputs (index)	100	101.5	104
Total operating costs (£ nominal)	8 800 000	9 060 400	9 454 000
Profit/Loss (£)	2 000 000	2 619 600	2 016 800
Calculated indexes			
Total costs at year 0 prices (£)	8 800 000	8 975 000	9 125 000
Index of nominal costs (index)	100	103.0	107.4
Index of total costs at year 0 prices (index)	100	102.0	103.7
Index of number of employees (index)	100	99.2	99.2

15.4.5 Total output per employee

Real output per employee can be calculated from the output and
employment data derived above. Table 15.8 provides an example of
the calculations. This is a labour-productivity measure, usually
presented as the annual percentage change.

Table 15.8 Calculation of an index of total real output per employee

	Year		
	0	1	2
Base-weighted output volume (index)	100	108.4	122.6
Index of number of employees (index)	100	99.2	99.2
Real output per employee (index)	100	109.3	123.6

15.4.6 Unit cost

It is difficult to calculate the cost of a unit of telecommunications
output, except at a very desegregated level, because of the varied
nature of the industry services. The calculations require detailed and
usually unpublished cost information and a robust methodology for

dealing with joint and common costs. Telephony, at least in terms of all-in cost per call, is cheaper in the US than in Europe because of higher calling rates, as well as cost differences.

Trends in the rate of change of unit cost can be more easily calculated at an aggregate level. Total costs in nominal terms should be calculated as an index number with the base year as 100. The index series for total output volume should be set at 100 in the same base year and calculated as indicated above. An index number series for nominal unit cost with a value of 100 in the base year can then be calculated by dividing the nominal-cost index by the output-volume index and multiplying by 100.

A related quantity of greater analytical interest is cost volume per unit of output. This shows the physical volume of inputs per unit of output and requires that each input series is measured at its own constant price, i.e. fixed wage rates for labour, a fixed price for paperclips etc. Derivation of these volume quantities may be difficult for all but the labour inputs and their estimation is the biggest single task in this form of performance measurement. The largest problem usually concerns capital inputs, represented in the accounts of the enterprise as depreciation, most often in historic terms. The search for a realistic measure of capital input volume may be easier if some form of economic depreciation has been used. In Table 15.9 the measurement problems are assumed to have been solved in a consistent way, permitting nominal expenditure changes to be split into volume and value effects. Remembering that base and current-weighted indexes usually compute to different figures, this involves using the same method for all the input series and if possible trying the calculation with both types of index to evaluate differences.

Table 15.9 Calculation of a unit cost index

	Year		
	0	1	2
Raw data			
Base-weighted volume index	100	108.4	122.6
Index of nominal costs	100	103.0	107.4
Index of total costs at year 0 prices	100	102.0	103.7
Calculated indexes			
Nominal unit costs	100	95.0	87.6
Cost volume per unit of output	100	94.1	84.6

15.4.7 Real unit cost

A real unit cost index avoids many of the difficulties encountered in revaluing individual inputs to base-year prices by taking the input cost total and adjusting it to a base-year value by a general index of inflation (retail prices index, consumer price index etc), see Table 15.10. This will usually give a different answer to that given by the index of nominal costs in terms of cost quantity change, because the individual input prices may have moved collectively at a faster or slower rate than general inflation, as it has in the example. The calculation is conceptually different in that it measures the cash cost of inputs per unit of output at a constant value of money, not the physical volume of inputs per unit of output. The difference between the two is the rate of change of real input prices. If the inputs tend to rise in price more quickly than prices in general (as wage costs, for example, usually will), the real-unit-cost index will rise more quickly, or fall less quickly, than cost volume per unit of output.

Table 15.10 Calculation of a real-unit-cost index

| | Year | | |
	0	1	2
Raw data			
Base-weighted output volume index	100.0	108.4	122.6
Total operating costs (£)	8 800 000	9 060 400	9 454 000
General price index	100.0	101.0	103.0
Calculation			
Total operating costs	100.0	103.0	107.4
Real operating costs	100.0	101.9	104.3
Real-unit costs	100.0	94.0	85.1

Real-unit-cost is the critical performance indicator. It determines the company's ability to sustain and survive long-term competition. It determines whether the expectations of the capital markets can be met.

British Telecom had a real-unit-cost reduction target which was set by the government for a number of years prior to its privatisation in 1984. The target figure was five per cent per annum and the achievement was about three per cent. Performance has since been rather better, under the combined pressures of private ownership and increasing competition. NERA [4] contains calculations showing an average annual real-unit-cost reduction of 3.7 per cent in the five years to March 1994. Output volume is measured by a quantity index in which the number of exchange lines, public payphones, local calls and long-distance minutes are weighted together in proportions derived from an analysis of BT incremental costs. NERA shows that in the US, where market restructuring and the breakup of AT&T have also increased competitive pressure, real-unit-cost reductions averaged 3.4 per cent per annum over the same period. This rate is also higher than the long-term average.

15.4.8 Total factor productivity

A measure of physical productivity growth which has found favour among analysts is total factor productivity (TFP), which is the difference between the annual percentage growth in total outputs and total inputs at constant prices. Data used in the preceding examples can be used to show the essence of the calculation. This is done in Table 15.11.

Table 15.11 Calculation of total factor productivity

| | Year | | |
	0	1	2
Raw data			
Base-weighted output volume index	100.0	108.4	122.6
Total costs at year 0 prices index	100.0	102.0	103.7
Calculation			
Percentage change in real output		8.4	13.1
Percentage change in real inputs		2.0	1.7
Difference = total factor productivity (%)		6.4	11.4

It is usually easier to construct the output index than the cost index. The example uses three input factors: labour, capital and bought-in goods and services. TFP is sometimes calculated in a two-factor form, from labour and capital alone. There may be numerous types of equipment and output volume series in the calculations. TFP uses input and output volumes and is closely related to cost volume per unit of output, using a similar method of calculation for input and output volumes. TFP differs from real unit costs in that it is not affected by changes in factor prices.

There are many theoretical studies and empirical results available for TFP. Canadian operators, regulators and academics have pub-lished a lot of information about the Canadian operating companies.

Some PTTs, such as Norway, have included TFP figures in their annual reports. The general drift of the published work has been to show annual rates of improvement in the range three to eight per cent. A representative summary is in Table 15.12.

The spread of competition and privatisation has shifted the focus away from TFP because of the omission of factor input prices. Liberalisation has often led to management exerting stronger pressure on both labour force and equipment suppliers and shown that this is an important area for management action.

Table 15.12 Total factor productivity of operators

Operator/country	Period	Annual average TFP (%)	Source (Reference)
Alberta Government Telephones	1968–83	3.42	5
Bell Canada	1953–79	3.65	6
Deutsche Bundespost	1984–87	1.40	7
France Telecom	1985–87	5.10	7
NTA, Norway	1977–85	3.40	8
NTT, Japan	1984–87	5.40	7
RT&T Belgium	1985–87	4.00	7
SIP, Italy	1985–87	4.30	7
Telecom Australia	1984–87	4.50	7
Telstra, Australia	1987–94	8.10	9
Telefonica, Spain	1984–87	4.40	7
US	1934–44	3.81	10
US	1945–54	2.20	10
US	1955–64	3.84	10
US	1965–74	4.73	10
US	1975–83	5.35	10
US	1984–87	3.58	10

15.4.9 Törnquist indexes

The index numbers so far discussed have limitations when used for productivity measurement in industries with changing mixes of inputs and outputs. More precise measures of change are given by the Törnquist index [11], which has been used for TFP studies and international productivity comparisons and is a discrete form of the Divisia index [12]. Following Reference 6, the Törnquist formula for an output volume index in period t is:

$$Q_{Tt} = \prod_i [Q_{it}/Q_{i,t-1}]^{1/2(r_{it} + r_{i,t-1})} \tag{15.7}$$

where Q_{it} is the volume of the ith output in period t, $t = 1, \ldots, n$ and r_{it} its revenue share of output, $\Sigma_i r_{it} = 1$.

The input volume index in period t is then

$$X_{Tt} = \prod_i [X_{it}/X_{i,t-1}]^{1/2(s_{it} + s_{i,t-1})} \tag{15.8}$$

where X_{it} is the volume of the ith input in period t, $t = 1, \ldots, n$ and s_{it} its revenue share of input

$$\Sigma_i s_{it} = 1 \tag{15.9}$$

The Törnquist productivity index is derived from eqns. 15.8 and 15.9 as

$$P_{Tt} = Q_{Tt}/X_{Tt} \tag{15.10}$$

The productivity gain is then

$$\log(P_{Tt}/P_{Tt}) \tag{15.11}$$

which can be scaled to a percentage.

Essentially, the Törnquist index weights change between periods in a neutral way, rather than using base or current weights. It has useful theoretical properties for the present purpose, including compatibility with forms of the translog production function. Kiss uses a translog function to separate technology changes and scale economies as two separate elements within the productivity change. The main disadvantage of the Törnquist index is the amount of data which it requires. This includes values and volumes for all the input and output series. Substantial data requirements are to some extent a disadvantage common to all volume-based productivity studies.

The Törnquist productivity index can be used to calculate the relative productivity of two operators instead of the change between a pair of years for one operator, by taking subscripts t and $t-1$ in eqns. 15.7 and 15.8 as referring to the two operators rather than two years [13]. In this case, the weights r and s are averaged between the operators rather than between the years.

15.4.10 Capital/output ratio

The ratio of capital employed to output has been used for a three main purposes:

(i)   As a measure of capital intensiveness.
(ii)   As a comparative measure of the productivity of capital among enterprises in the same industry.

(iii)   Macroeconomists have used the performance of the ratio over time to analyse productive potential and the dynamics of industrial growth.

Within telecommunications the ratio is most easily calculated as gross output (total sales) divided by capital employed. An alternative would be to use net output (Section 3.1 (i)). The two do not measure quite the same thing. As net output excludes all outsourced supplies, including those which have been used as alternatives to direct labour costs in the profit and loss account, it is a less stable basis for comparisons, as this example for a typical operator shows:

	Before more outsourcing	After more outsourcing
paybill	4 000	3 700
+ depreciation	2 300	2 300
+ gross trading profit	4 000	4 050
= net output	10 300	10 050
+ bought in goods and services	4 300	4 550
= gross output (turnover)	14 600	14 600

Here the operator saves 50 units by replacing 300 on the paybill with 250 for bought-in services, increasing the profit by 50 but reducing net output by 250 units. Gross output is unchanged, as is output volume. Labour productivity rises, illustrating its reflection of outsourcing changes. The macroeconomic equivalent of net output is the gross domestic product, in which intermediate goods and net imports are excluded. Gross and net outputs are dependent on the prices charged for outputs. They can be reduced to volume terms as shown above for volume indexes.

Capital employed as shown in company accounts includes the accumulated historic cost of past investments less the provisions which have been made for capital consumption through depreciation provisions. This may differ significantly both from the valuation of the enterprise on the stock market and the value of the assets employed if revalued at replacement cost. Stock market valuations are too volatile to be useful in capital/output ratio comparisons, although they have other performance measurement uses. Capital/output ratios for PTOs, calculated from turnover and balance-sheet values for capital employed, are typically in the range 0.8 to 3.0. Examples:

British Telecom 1997	1.1
Canadian Telecommunications 1995 (part of BCE)	2.5
NTT 1995	2.0
Telecom Australia 1990	0.9
Telecom Denmark 1992	1.2
Telefonica 1995	2.8

Those for manufacturers are similar (source: annual reports):

Alcatel Alsthom 1994	1.6
Nokia 1995	0.9
Nortel 1995 (part of BCE)	0.9

A related statistic from Tobin [14] is the market-to-book ratio, valuation ratio or Tobin's q, which is the ratio of stock market valuation to the net value of the assets. The assets should be valued at replacement cost, although this information is not usually available and book value is used in its place. A persistent value below unity may be the result of tax reliefs on investment, reducing the replacement cost of capital. A value in excess of unity may indicate an abnormally high profit. It can arise from the value of assets such as patents, licences, management skills and goodwill, which are not in the calculated asset base, and from market power. Cruder versions of the ratio, using capital employed as the denominator, have shown values in excess of unity in the US cable TV industry, which have been taken to indicate that the local monopolies commonly enjoyed scarcity rents at the time. Fournier and Campbell [15] found evidence for them during a period of intense cable regulation from 1969 to 1976 and again after the removal of ownership restrictions in 1984.

Hazlett [16] quotes market valuations of $1800 to $2500 per subscriber for existing monopolists in the late 1980s, compared with a new build cost of between $500 and $1000 and creating a q value of about three. Valuations of cellular mobile radio operators working in markets where entry is restricted have sometimes shown high valuation ratios. Macroeconomic studies in Reference 17 show that values from 0.5 to over three can persist in an economy as a whole for long periods of time, so the cable TV findings could, conceivably, not indicate market power.

The statistic was originally developed to indicate opportunities for profitable new investment. This implicitly assumes that the marginal value of Tobin's q is similar to the average value. In this, it has shortcomings because the difference might arise from technological change, where relative price changes are not fully reflected in the stock of capital actually employed for some years. The incremental capital/output ratio (ICOR) is a related marginal measure. Like the capital/output ratio itself, the ICOR can be calculated from increments of gross or net outputs. If calculated from net output it reflects any shift from in-house to bought-in labour resources. It is also sensitive to price changes, although it might be possible to net these off if adequate data can be obtained. ICORs are used in macroeconomic growth theory. The evidence from telecommunications suggests that they need careful interpretation owing to the effect of outsourcing and asset valuation methods on the numbers.

15.5 International comparisons

15.5.1 Problems to be solved

International comparisons are complicated by a variety of factors, as we have seen in the discussion of cross-sectional estimation methods in Chapter 8. The most important problems in the field of performance and tariff comparisons are:

(a) Currency exchange rates may be unrepresentative of local domestic purchasing power.
(b) Local procurement prices may be out of line with world prices.
(c) Varying patterns of demand. The composition of a typical basket of services varies from country to country, reflecting local tariff structures and other demand conditions.

Labour productivity is commonly used in international productivity comparisons.

15.5.2 Comparing GDP per head

International comparisons of monetary variables, such as the GDP per head of population or the price of a three-minute local telephone call, may be difficult to interpret where, as is usually the case, different currencies are involved. Figures for GDP per head are usually wanted as indicators of comparative wellbeing or disposable income. They are useful in cross-section studies of the demand for consumer goods, including telecommunications services.

The simplest and most widely used method is to reduce all currencies to US dollars using commercial exchange rates. This is relatively easy to do but it has two shortcomings:

(i)   Volatility. The rate will vary from day to day. Large changes may appear over a relatively short period of time.

(ii)   What they measure. Floating exchange rates are determined by market factors. They reflect the ebb and flow of currency used in crossborder transactions and swings of sentiment generating other capital flows. They are the exchange relatives experienced by traders. Domestic prices of goods and services which are not much traded across frontiers may be little affected by the factors determining relative currency values.

A third problem, which can be significant with comparisons involving Third World countries, is the exclusion of nontraded activities, such as subsistence agriculture, from measured GDP. Even in developed countries there may be a 'black economy' in which some trading activities elude all three measures of GDP. Others may be in one measure, such as expenditure, but not another, such as income. This is one of the reasons for discrepancies between the measures.

A related problem which arises with economies such as those of the old USSR is the difference in what is included in the local definition of GDP. Keynsian and Marxist methodologies produced different answers for the same country, developed or otherwise. The main differences involve the treatment of services. Accurate GDP figures may be difficult to collect for a variety of other reasons. However, in spite of their limitations, they remain a valuable basis for cross-section studies. GNP may be used as an alternative measure.

15.5.3 Purchasing-power parity

An alternative basis of comparison is through exchange rates based on the domestic purchasing power of each currency. A standard list of goods and services is priced in each country, and there are several ways of arriving at the volumes used for each item in the list. They have different economic properties and usually give different answers. Three ways that have been used are:

(i)   Home-country weights. The expenditure pattern is that of one country in the group.

(ii)   Average bilateral weights. Comparisons are carried out in country pairs. For each pair the comparison is carried out twice, using the local expenditure patterns of one country then the other, and the two results are averaged. This method gives a result which is intermediate with respect to a country pair.

(iii) Average basket weights. A set of volume weights representative of the average expenditure in all the countries is used for the comparison. Depending on how the average is calculated between countries of different size (e.g. whether it is weighted and if so, how) this method is not biased towards the prices in any particular country, but it suffers from the disadvantage of not exactly corresponding with the experience of any national consumer group. The method has been used by the Organisation for Economic Co-operation and Development (OECD) for an international comparison of telecommunications service tariffs.

There are two problems, one practical and one conceptual:

(i) Collecting the price data. Even in a single city there will be a good deal of variation in the prices of many goods, depending on shop location, consumer price awareness and product quality. The list cannot cover the full range of quality without being indefinitely long. There are differences between cities and regions as well as between countries. The prices are typically collected by an international organisation such as the OECD on a limited budget.

(ii) The conceptual problem lies in the nature of the list itself. A home-country list tends to make the purchases in that country look cheaper because local expenditure patterns will adjust to local prices, being lower for relatively expensive items. The use of a list which does not represent local expenditure patterns has an objective truth in saying what it would cost to buy a particular basket of goods in each of a number of countries but it does not properly measure local wellbeing. Local consumers will derive higher utility by using the same cash to buy products of their own preferences.

The list problem is particularly acute when comparing countries with different cultures. Comparisons between North American and European countries, for example, are dogged by this effect. The use of a US list in a European country will exaggerate transatlantic differences in real incomes. However determined, the relative costs of the total purchase are used as the basis of purchasing power parity (PPP) exchange rates. PPP rates are typically more stable than those derived from financial markets and their derivation from domestic purchases makes them a better indicator basis for comparing incomes.

Comparisons of the results of different bilateral pairs are difficult to interpret. Averaging methods may lack any exact meaning for individual countries, but they bring some stability to the comparisons and make them less dependent on the expenditure pattern for a single country. They are a useful means for deriving comparative data relating to real incomes.

15.5.4 Comparing telecommunications service prices

The use of PPP exchange rates will not help with the data problems which surround the collection of prices for comparable services in different countries. To take one example, which applies within as well as between countries, the area covered for the price of a local call varies greatly from region to region, both as regards its size and the number of exchange lines it contains. Figures in Table 15.13 are those of Cracknell [18]. They show a wide variation across Europe.

Table 15.13 European comparison of local call areas

	Average area (km²)	Local calls as % of total
Finland	3916	85
UK	2186	73
Sweden	1968	75
Germany	1260	56
France	1170	65
Belgium	764	30
Greece	530	90
Italy	156	65
Switzerland	100	45
Portugal	88	14
Netherlands	31	45

(Source: Reference 18)

Although it may be desirable to construct a basket of calls of varying lengths and locations to price against the local tariffs, this may not be possible at reasonable expense. An alternative, used by commercial agencies specialising in international tariff comparisons, is to select a few representative types of call, with the offer of more complex comparisons to those willing to pay for them. Eurodata Foundation figures for the cost in ecus, in each European country, of a local four-minute call, a long-distance four-minute call and monthly rental, and four-minute calls each way between each of the European countries, are published in the trade press. Simplified though they may be, they show striking disparities, indicating the likelihood of a high degree of suboptimality in pricing. For example, figures published in May 1998 show that a four-minute call from Ireland to Austria cost Ecu 2.405, a call of the same length in the opposite direction cost Ecu 3.774 [19].

15.5.5 Comparing productivity

Two methods of international productivity comparison have so far being discussed, lines per employee and total factor productivity. These both avoid the use of monetary variables. NERA [4] contains the results of a unit cost study in which BT's unit costs were compared with those of Telstra (Australia), Telia (Sweden), Telecom Finland and a group of US operators. Output categories were switched-access lines, public payphones, leased lines, local calls, local-call minutes, long-distance calls and long-distance call minutes. The costs of each operator were scaled to BT's output, by recalculating and allowing for economies of scale and other factors taken as outside the control of the operator. Input price differences were allowed for by adjustments based on national hourly wage rates and PPP currency-exchange rates. Adjustments were made for different methods of valuing assets. Pairwise comparisons were then carried out in which the efficiency $E_{i/0}$ of operator i relative to operator 0 (BT) was calculated as a Laspeyres cost index

$$E_{i/0} = \Sigma_k(u_{ki}q_{k0})/\Sigma_k(u_{k0}q_{k0}) \qquad (15.6)$$

where

u_{ki} is the unit cost of activity i in firm k

q_{ki} is the output associated with activity i in firm k

As noted in the NERA report, the use of BT quantity volume weights tends to produce results favourable to BT. The findings of the study, presented as provisional, were that BT had annual costs per line of £362 which were lower than those of Telstra (£461), Telia (£396), Telecom Finland (£444) and local exchange companies in the US (£407) other than the RBOC's. The RBOC's had lower costs (£359). A different picture emerged for costs per call. BT's calling rate was less than half that of the RBOC's and the cost per call minute was over twice as high, as shown by annual data in Table 15.14. In these figures, total network costs are divided by the number of lines in the first row and the number of call minutes in the second. Telstra, Telia and Telecom Finland also had lower costs per call minute.

Table 15.14 Comparison of BT and RBOC unit costs in 1994

	BT	RBOCs
Cost per line	£362	£359
Cost per call minute	£9.46	£4.04
Local calls per line	1015	2882
Long-distance call minutes per line	1180	2259

(Source: NERA)

15.6 Efficiency frontier methods

15.6.1 Data envelope analysis

The methods that have been discussed so far are mainly useful for tracking the performance of a single operator over time. Although some, such as labour productivity, show levels at different points in time and are used for intercompany comparisons, they omit too many variables for them to be very accurate when used in this way. Data envelope analysis (DEA), which has become widely used for intercompany comparisons, uses linear programming techniques to identify relative company efficiency. The theoretical framework derives from Farell [20], who desegregated technical and price

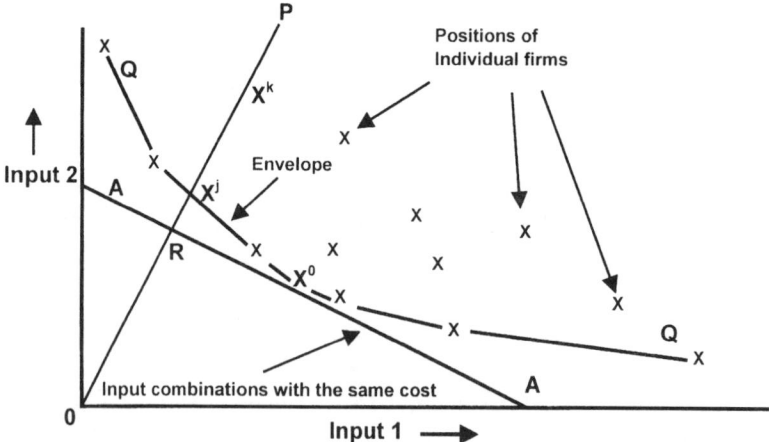

Figure 15.1 Data envelope anaylsis

efficiency using a framework illustrated in Figure 15.1, which relates to a firm with two inputs and one output.

The starting point is the efficiency frontier discussed in Chapters 5 and 11. In the earlier discussion, the frontier was used to identify the combinations of input volumes which can maximise output within a given total cost. The axes on the chart are scaled in terms of values of each of the two inputs. **AA** is a line marking the alternative combinations of inputs which could be bought within a fixed budget. It has a slope equal to the ratio of relative prices.

Figure 15.1 plots the scatter of input combinations used by different firms to produce a unit of output. **QQ** is now an estimate of the true isoquant for technically-efficient firms and represents industry best practice. It is a set of straight lines linked in such a way that:

- no firm is to the left;
- the approximated curve slopes downwards and to the right (implying a single optimum).

As we saw earlier, the point at which the true isoquant touches AA represents the maximum output that can be produced within the budget. All firms to the right of **QQ** are technically inefficient. Consider firms *J* and *K*, at $\mathbf{X}^j$ and $\mathbf{X}^k$, both on the ray **OP**. Firm *J* is one of the efficient firms on **QQ**, firm *K* is not. They use the same mix of inputs but firm *K* uses more of them to produce the same output. The technical efficiency of firm *K* is then defined as the distance ratio $\mathbf{OX}^k/\mathbf{OX}^j$. Neither firm is on the isocost **AA**, crossed by **OP** at **R**. The price efficiency of both firms is then defined as $\mathbf{OR}/\mathbf{OX}^j$, which is less than unity because neither firm is using the optimal mix of outputs at $\mathbf{X}^0$.

We are now dealing with a finite number of firms and it may not be possible to find any two of them with the same combination of inputs. There is a choice to be made. Efficiency can be measured against the most technically efficient of the other firms in a group with similar input combinations, or measuring against an efficiency frontier with a position which is estimated from the data but which may not be occupied by any firm (and may not be technically feasible). In the latter case, a hypothetical reference firm can be computed as a weighted average of firms which are on **QQ**, subject to the condition that none of them uses more of any input than the firm being examined, while producing the same output.

Since, again, no two firms are likely to be producing the same total output, the inputs are scaled to inputs per unit of output for the comparison. The use of this kind of scaling to produce the simple ratio

$\mathbf{OX}^b/\mathbf{OX}^a$ as a measure of technical efficiency is accurate for firms of different output volumes provided that they are all working under conditions of constant returns to scale. Modifications are needed if scale economies or diseconomies are present, because part of the difference in unit cost will result from the firms operating at different output volumes. Farell and Fieldhouse [21] suggest ways around the problem.

The term 'data envelope analysis' derives from Reference 22, a paper which proposes the use of linear programming to measure technical efficiency in the multiple input/output case. Since this publication, there have been many published case studies. Some of these show the results of hybrid studies in which economic and technical-efficiency measures are combined. Hawdon and Hodson [23] give a useful survey and some results of this type for purchasing expenditure in aerospace companies. The efficiency of a particular company was measured against a weighted average of other companies in which purchasing expenditure, number of employees and IT systems were, at most, those of the company being examined and in which output was at least as high. In a follow-up study, the object was to identify firms which could make an equiproportionate reduction in labour and IT systems (thus retaining the mix), while maintaining output and cost levels.

15.6.2 Stochastic frontier analysis

Pollitt [24] gives results for the electricity industry and compares DEA with some related techniques for measuring performance, of which the most important is stochastic frontier analysis (SFA). NERA [4] provides results for a comparison of British Telecom with a group of European and North American operators and provides numerical examples of adjustments for scale economies.

SFA attempts a refinement of the DEA method by allowing that observation errors may bias the estimate of the position of the efficiency frontier. Some firms which are in fact technically efficient may not be on the efficiency frontier in the data and others which are on the frontier may have got there by chance. The range of observation error is estimated from the data and used to adjust the estimated position of the boundary. Figure 15.2, based on an illustration in Reference 4, shows how this is done. Data requirements and assumptions are more extensive than for DEA.

NERA [4] used DEA and SFA to test BT's technical inefficiency against best practice. Both methods suggested that BT was 87–89 per cent efficient in a base case and at most 93 per cent efficient on

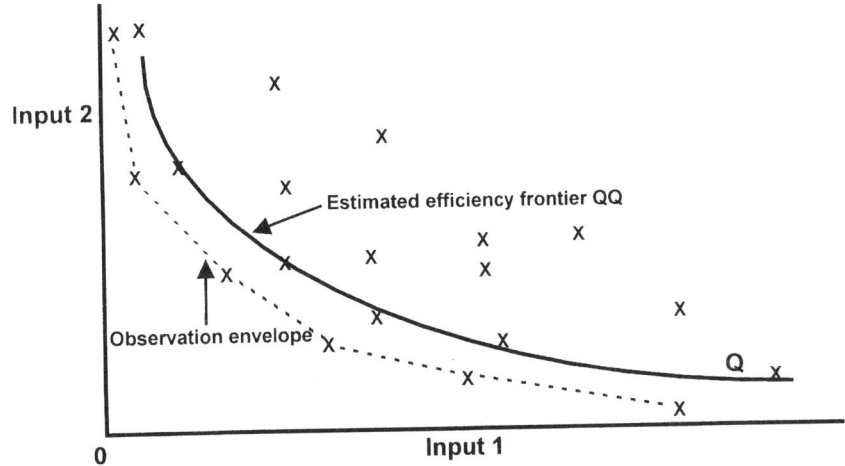

Figure 15.2 Stochastic frontier analysis

alternative assumptions. The calculations were not wholly accepted by BT and its reservations are discussed in Reference 25. International comparisons in this area are fraught with many problems of exact interpretation, as has been discussed above. Whatever its shortcomings, the study is useful as a benchmarking exercise to identify companies which may be explored further as models for greater efficiency.

15.7 Quality of service and network performance

15.7.1 Quality and grade of service

Quality issues have two main dimensions:

(i)   Network performance, equipment failure rates etc.
(ii)   People-related aspects, e.g. how soon did they come to fix the phone, were the staff rude or courteous?

Congestion can result in calls failing to get through. The relationship between the calling rate and the probability of failure owing to equipment not being free (the grade of service) is, as outlined in Chapter 2, a basic and well understood aspect of telecommunications engineering. Quality improvement comes at a price, since congestion can only be eased by providing more capacity. Marginal costs and benefits of improving service by the provision of additional plant can be calculated if there are adequate assumptions about the value of the time taken to make abortive calls and the value of the benefit forgone by their failure. Since a reduction in congestion benefits all callers, congestion costs can reasonably be taken as being embraced within the consumers' demand function rather than being an externality.

15.7.2 Waiting lists and spare capacity

Telephone-network operators need spare capacity in local exchanges and the access network to provide for expected growth and unexpected fluctuations in demand if they are to connect new customers to the network without undue delay. They also need staff resources to make the necessary connections. The number of consumers who have applied to join the network but not yet obtained service (the waiting list) can be reduced by installing a larger margin of spare plant. The length of waiting lists depends partly on how the spare plant is distributed from exchange to exchange. Additions to exchange and line plant have to be made in increments which may be quite large in relation to total demand and have to be planned and sometimes financed in advance. Operators can control costs by deferring capital expenditure at the expense of a growth in the number of people waiting to be connected. An optimum can be sought on lines originally suggested by David Bywaters in an unpublished BT research paper. Figure 15.3 shows the relationship between the size of the waiting list and the amount of spare local-plant capacity over a 17 year period in the UK.

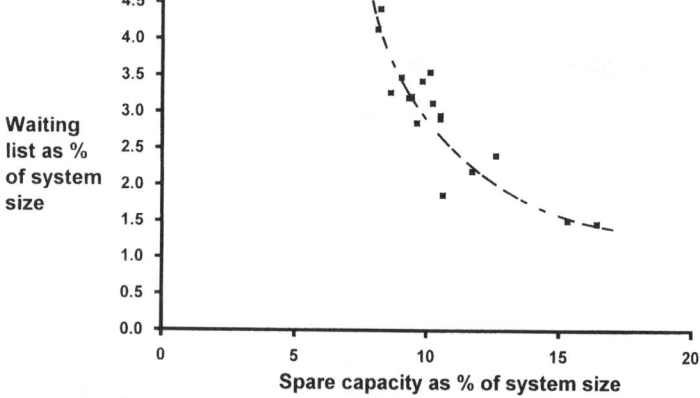

Figure 15.3 Waiting list and spare capacity, UK 1965–81

Source: calculated from Post Office telecommunications statistics 1974, British Telecom statistics 1983, Post Office annual reports.

Margins of spare plant and labour vary widely from country to country and from exchange to exchange, according to local service conditions, but an average margin of 25 per cent above the number of lines in service is not uncommon in developed countries. Margins for plant are often higher in markets where competition exists. Shortage of finance in poorer countries has led to the growth of waiting lists for basic service which may be several times the size of the entire system in extreme cases.

Waiting lists represent potential revenue, and an optimum could in principle be calculated. There would be three steps:

(i)   To establish the relationship between spare capacity and waiting-list length, as in Figure 15.3.
(ii)   To compute the marginal cost of reducing the waiting list, from the spare-capacity relationship and the cost of spare capacity.
(iii) To estimate the value to the customer of shortening the waiting list. A minimum figure for this value is the extra amount that would have been spent by a customer connected more quickly to the network.

Some of the lost calls will have been made by other means, such as the use of a public payphone. It may be possible to make an allowance for consumer's surplus on those which are completely lost. In this case, third-party interests are not, to the first order, involved, so it is reasonable to assume that such losses are also embraced by the demand function. There will be an optimum where the marginal cost of reducing the waiting-list matches the marginal value to the customers. This optimum has to take into account the incremental nature of plant provision, so that the costs and benefits are measured for a feasible increment and not a single line. Figure 15.3 suggests that the marginal cost of waiting list reduction rises steeply as the waiting list falls to one per cent of system size or less. At the other extreme, long waiting lists in developing countries can fairly easily be shown to be suboptimal and their reduction is often part of the case for development finance.

Waiting-list data are often not fully comparable from country to country or from operator to operator. For example, if waiting lists are very long they may include people no longer interested in taking service, but exclude those not bothering to go on the list because of its length. They may not include customers in parts of the country where no service is yet provided (Section 8.5.4).

15.7.3 Managing service quality

Modern management methods, in particular total quality management (TQM), have sometimes claimed that costs can be reduced at the same time as quality is raised. This may be true of interfaces involving people, but it is more difficult to achieve with machines. Two themes which run through management theories are the need to get the closest possible matches:

- within the firm, between corporate objectives, shareholder interest and the objectives of employees, thus addressing the principal/agent problem within large organisations;
- between the firm and its customers on quality supplied, thus reducing market failure and knowledge deficiencies.

As examples of the ways in which results are realised, the ordering and supply of stock may be performed more efficiently and in a more predictable way, reducing the amount of stock which needs to be held (reducing the capital/output ratio). Management activities that cannot be shown to have an adequate return against corporate objectives (improving labour productivity) can be ceased. Work that can be done more cheaply by others can be outsourced (reducing real-unit costs). Surveys of customer opinion, organised by suppliers, customers and regulators assist market efficiency.

15.8 References

1   MARRIS, R.: 'Economic arithmetic' (Macmillan, London, 1958)
2   FISHER, I.: Observations unpublished at the time, see Reference 1 p. 260
3   'Statistics of communications common carriers' (FCC, Washington D.C., 1995)
4   National Economic Research Associates: 'BT comparative efficiency study' (OFTEL, London, 1995)
5   McGOWAN, F. S.: 'Allocative efficiency and total factor productivity'. Paper presented at *The sixth international symposium on forecasting*, Paris, June 1986
6   KISS, F.: 'Productivity gains in Bell Canada' *in* COUVILLE, L., DE FONTENAY, A., and DOBELL, R. (Eds.): 'Economic analysis of Telecommunications' (North Holland, Amsterdam, 1983)
7   STARANCZAK, G. A., SEPULVEDA, E. R., DILWORTH, P. A., and SHAIKH, S. A.: 'Industry structure, productivity and international competitiveness: the case of telecommunications', *Info. Econ. Policy*, 1994, **6**, (2), pp. 121–142
8   FOREMAN-PECK, J., and MANNING, D.: 'Telecommunications in Norway' *in* FOREMAN-PECK, J., and MÜLLER, J. (Eds.): 'European telecommunication organisations' (Nomos Verlagsgesellschaft, Baden-Baden, 1988)
9   XAVIER, P., and GRAHAM, B.: Figures from Bureau of Transport and Communications Economics *in* 'Assessing the performance of Australia's telecommunications services industry, consultancy report prepared for AUSTEL' (University of Hong Kong Business School, Hong Kong, 1995)

10 NADIRI, M. I., and NANDI, B.: 'Productivity growth in the US telecommunication industry'. Paper presented at ITS *11th biennial conference,* June 1996, Seville

11 TÖRNQUIST, L.: 'The Bank of Finland's consumption price index', *Bank of Finland Monthly Bulletin,* 1936, **10,** pp.1–8

12 DIVISIA, F.: 'L'indice monétaire et la théorie de la monnaie' (Société Anonyme de Recueil Sirey, Paris, 1926)

13 FOREMAN-PECK, J., and MANNING, D.: 'How well is BT performing? an international comparison of telecommunications total factor productivity', *Fiscal Studies,* August 1998, **9,** (3), pp.54–67

14 TOBIN, J.: 'A general equilibrium approach to monetary theory', *J. of Money, Credit and Banking,* 1969, **1,** pp.15–29

15 FOURNIER, G. M., and CAMPBELL, E.S.: 'Shifts in broadcast policy and the value of television licences' *Info. Econ. Policy,* 1993, **5,** pp.87–104

16 HAZLETT, T.: 'Cabling America: economic forces in a political world' *in* VELJANOVSKI, C. (Ed.): 'Freedom in broadcasting' (The Institute of Economic Affairs, London, 1989)

17 'Economic studies no. 7', (OECD, Paris, 1986)

18 CRACKNELL, D. R.: 'Growing the telecommunications market — lessons from the past'. Paper given at ITS regional meeting, Crete, September 1994

19 Eurodata Foundation: 'European telecomms tariffs', *Public Network Europe,* May 1988, **8,** (5), p.56

20 FARELL, M. J.: 'The measurement of productive efficiency', *J. Royal Statistical Society, Series A,* 1957, **120,** pp.253–281

21 FARELL, M. J., and FIELDHOUSE, M.: 'Estimating efficient production functions under increasing returns to scale', *J. Royal Statistical Society, Series A,* 1962, **125,** pp. 252–267

22 CHARNES, A., COOPER, W.W., and RHODES, E.: 'Measuring the efficiency of decision-making units', *Eur. J. Oper. Res.,* 1978, **2,** (6), pp.429–444

23 HAWDON, D., and HODSON, M.: 'The use of data envelope analysis in benchmarking', *The Business Economist,* 1996, **27,** (3) pp.23–39

24 POLLITT, M. G.: 'Ownership and performance in electric utilities' (Oxford University Press, Oxford, 1995)

25 NERA: 'Assessment of BT's response to the NERA efficiency study', (OFTEL, London, 1996)

Chapter 16

Econometric modelling

16.1 Models and forecasts

16.1.1 The context of modelling

The pronounced economies of scale shown by telephone networks and the reasonably predictable peaks of traffic produced by their users lend themselves to the mathematical modelling of costs and business levels. Forecasting has become a major occupation of economists working in the marketing, corporate planning and finance departments of the larger companies in many industries. It plays a large part in the work of economists in operating companies because of the need to build now for future service requirements.

The emphasis on forecasting draws economists into areas of sophisticated statistical analysis and modelling. It also calls for judgmental skills since forecasting, like economics itself, is far from an exact science. Most good econometric forecasts owe at least as much to the assumptions under which the models are run as they do to the structure of the models themselves. The assumptions are termed exogenous variables because they are not determined by the model.

Some assumptions may themselves be the output of other econometric models from agencies specialising in macroeconomic (GDP etc) forecasting. They may be postulates of growth and inflation rates, set up to explore different outcomes for the future. For the long term, use may be made of projections of habits, lifestyles and attitudes, based on social research and speculation.

Good modelling is not possible without good data. Billing systems contain economic information covering the usage patterns of individual customers and the way in which they react to price changes. The type of information that is useful for this purpose includes time-of-day patterns of usage and average usage per customer at different tariff rates.

16.1.2 Modelling call durations

Call duration statistics are often wanted for revenue and call analysis and can be estimated by sampling. Where samples are not available, recourse is made to broader estimates current in the industry. The following discussion of distribution theory follows Cracknell [1]. Call durations are often assumed to be distributed exponentially, or to have a related distribution. The frequency function of the exponential distribution is

$$f(x) = (1/a)e^{-(x/a)} \tag{16.1}$$

where *a* is the average duration.

This is a continuous probability distribution, the area under the curve sums to unity and the variance is a². It can be adapted for use with a sample of N calls in an approximate way by multiplying by f(x) by N. The data would need to be grouped into class intervals. A more exact method is to obtain the proportion of calls which are, for example, between i–1 and i minutes in duration, by calculating f(x) for x = i–1 and x = i and subtracting the first from the second.

The distribution (also known as the negative exponential) is a useful approximation but it can be improved by the use of a gamma distribution, of which the exponential is a special case. The distribution can be adjusted for the typically greater variance encountered, arising from the mixing of traffic with different characteristics. The frequency is then given by

$$f(x) = (1/(a^b Gamma[b]))x^{b-1}e^{-(x/a)} \tag{16.2}$$

where *Gamma* [b] is the gamma function, *Gamma* [1] = 1

This has mean *ab* and variance $a^2 b$. Tables of the gamma functions are published in, for example, Reference 2. Figure 16.1 shows the two distributions fitted to a small sample of British long-distance and international calls made on the same line in 1992. The gamma distribution provides the better fit.

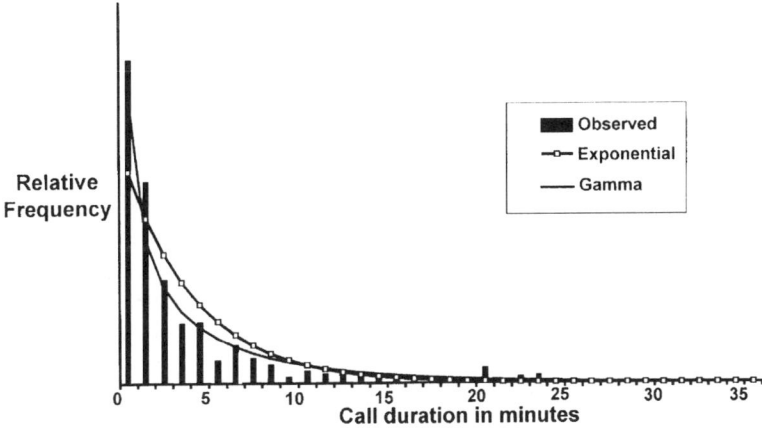

Figure 16.1 Distribution of a sample of call durations

16.2 Engineering models

The changing technology of telecommunications creates its own problems in forecasting optimal provisioning plans, adding the extra dimension of future options and their cost. On the uncertain ground of this forecast, world economists and engineers find that they have much in common in their approach. Engineering departments usually have network plans running many years into the future in which technology projections are combined with forecasts of future markets, prices and operating environment to develop plans for future investment. The relationship between outputs and inputs are examined in terms of volumes. The plans are in the form of computer models and can be rerun under different assumptions to optimise plant provision and system modernisation. The models are used to find the expansion path which minimises cost for a given pattern of expected demand growth. Marketing departments have similar plans for the short to medium term.

Optimisation of the mix of inputs used to produce a given output has to take account of the price of each input. This is possible within the framework of an engineering model. Costs may be analysed in operational terms by examining the resources used in individual parts of the production process. This kind of analysis may be very detailed, being carried out at a level finer than that of the individual product and often (as with billing) cutting across product boundaries.

16.3 Theoretical production functions

Economic analysts working on business planning and academic studies may find an engineering model prohibitively expensive to construct and run, or require data which is unavailable. General mathematical models or production functions which simulate the engineering within required limits of accuracy have been used in telecommunications and other industries for many years. All models making projections into the future need to take account of cost trends for the main resources used in the production process. These are usually analysed in real terms, i.e. shown as price changes compared with the general rate of inflation.

The Cobb-Douglas model dates back to 1928 and is for a single-product firm with labour and capital inputs, formulated as

$$Y = AL^\alpha K^\beta U \qquad (16.3)$$

where

  Y = output
  L = labour
  K = capital
  A, α and β are constants
  U is a random error.

Taking logarithms:

$$\ln Y = \ln A + \alpha \ln L + \beta \ln K + \ln U \qquad (16.4)$$

It is assumed that the mean value of log $U = 0$. The constants can then be estimated by regression analysis. The main interest lies in coefficients α and β. If both are positive and they add to unity, then the model exhibits constant returns to scale, a feature which is often approximated in empirical results. With this feature there are no fixed costs, and average and marginal costs are equal for all outputs. The model then shows linear homogeneity, whereby an equal proportionate change to the inputs makes an equal proportionate change to the output, e.g. doubling K and L doubles Y. With linear homogeneity and a further assumption that the input mix remains the same, the constants can be estimated by linear regression from time-series data for the firm.

There is a great deal of literature about Cobb–Douglas models. The form is too simple for some telecommunications applications and the assumptions of linear regression may not hold. If constants are estimated from time-series data, technology trends become confused with volume effects. McGowan [3] uses extra variables for materials and technology inputs in a productivity study of Alberta Government Telephones. For further reading see McGowan and standard texts such as References 4–6.

The translog production function from Brown, Caves and Christensen [7] is a more developed model which is better suited to telecommunications analysis. The function permits more types of input and is in terms of the aggregate cost of production. A time trend can be included to reflect technological change. It is usually presented in logarithmic form, as in the form used by Cronin *et al.* [8]:

$$\ln C = \ln A + \Sigma_i b_i \ln w_i + \tfrac{1}{2}\Sigma_i\Sigma_j g_{ij} \ln w_i \ln w_j + b_y \ln y$$

$$+\tfrac{1}{2}g_{yy}(\ln y)^2 + \Sigma_i g_{iy} \ln w_i y + \tfrac{1}{2}g_{tt}t^2 + \Sigma_i g_{it} \ln w_i t \qquad (16.6)$$

where

C = total cost
$w_1 = w_k$ = price of capital inputs
$w_2 = w_i$ = price of labour inputs
$w_3 = w_m$ = price of materials inputs
$w_4 = w_n$ = price of telecommunications inputs
y = output
t is a time variable
A, the b_i and the g_{ij} are constants, $g_{ij} = g_{ji}$

Linear homogeneity and the specification that costs should rise with increases in output and factor prices require that

$$\Sigma_i b_i = 1, \ \Sigma_i g_{ij} = \Sigma_i g_{ii} = \Sigma_i g_{iy} = 0 \qquad (16.7)$$

The constants can be estimated by simple regression analysis as before, although greater accuracy may be obtained by more sophisticated methods. Cronin used this formulation to study the impact of factor prices, production levels and technology on the telecommunications input to fifteen sectors of the US economy. One finding was that telecommunications services were being substituted for capital as their price fell. In a later study [9] Cronin *et al.* used a translog function to estimate cost savings realised by the use of telecommunications. It was found that in physical terms, the telephone was substituting for other factors and saving money in the process, cumulatively some $77 bn in 1991 dollars over the 1963–91 period. There are applications in the study of marginal costs, for example Reference 10, and productivity measurement.

16.4 Econometric models

16.4.1 The approach to modelling

Economists working on market structure, forecasting and policy analysis model telephone operations, although without the detail of engineering models and with other objectives including the derivation of marginal costs. The links between the two types of modelling are often close. Engineers want the minimum cost for given levels of traffic but the economists may want to know the change in cost for a change in traffic level. The engineers may ask economists for help in drawing up network and traffic forecasts. Economists, especially in North America, have therefore long been interested in engineering models of the network. There is a parallel with the electricity industry, where French economists pioneered the use of engineering economics in the theory of electricity pricing.

Econometric models relate changes in variables, such as the number of local telephone calls and their prices, to movements of output and price in the economy as a whole. They may model the world market, a regional market or the demand for the products of a single firm. Models for corporate planning include costs and derive measures of corporate profitability. Regulators are interested in models which can include measures of efficiency in the analysis. The equations use relationships between price, macroeconomic variables such as GDP and the level of demand, cost functions and other factors. They usually involve time series. Typically, they require assumptions about future movements in tariffs and the macroeconomic variables and produce forecasts conditional upon them. Coefficients are estimated from past data.

All models are at best a rather imperfect representation of the world, but they can be helpful none the less. Peter Whittle [11] put it well when he wrote that '. . *the model must still be 'local' in character, i.e. valid only over short runs, but at least one can hope to gather enough data that adequate estimates can be made . . .*'.

It is easy to overlook the degree to which subjective factors are involved. Behavioural equations stand in particular risk. The model may make precise and sometimes startling predictions about the world but there are, on paper, many possible worlds. Economists need to show that the world which they write about is one that we live in.

16.4.2 Time-series methods

Time series analysis has been extensively used to estimate coefficients. A series is usually assumed to be formed by a mixture of deterministic and serially correlated random disturbances. The two components are sometimes called the signal and the noise. Deterministic components are elements such as a trend in the volume of GDP, or prices charged, which do not follow the laws of random processes. Random components are seen as generated by a series of shocks, each of which has an expected value of zero but which may be correlated with earlier shocks. The deterministic and random components are not correlated with each other. In a simple case suppose the time series model is

$$y_t = a + bx_t + e_t \qquad (16.8)$$

where

y is an observed value of the dependent variable
a is an unknown constant
b is an unknown coefficient
t is the time period, $t = 0, 1, 2, 3, \ldots$
x is a linear function of t, say ct where c is an unknown constant
e is a random element of expected value zero

The random component may be serially correlated, so that

$$e_t = z_t + re_{t-1} \qquad (16.9)$$

where z is a random shock of expected value zero and r is an unknown constant. The shocks are assumed to follow a normal distribution, which is a continuous bell-shaped distribution, peaked over the mean and found as a limit for large samples in sampling theory. This is usually a reasonable approximation for a long series. See a textbook on statistics, for example Reference 12, for a further discussion.

There may be a lagged response to changes in the independent variables. The full effect of a price change, for example, may take several time periods to come through as customers adapt their habits and equipment to a change in relative prices. There are many ways of modelling this. One is to include the price series lagged by one, two, etc. periods among the independent variables. Longer lags can be allowed for with a Koyck lag structure, in which past changes show a geometric decay. Koyck lags were used by Breslaw and Pizante [13] to identify lags of several years in Bell Canada toll data. Cracknell [14] used them to show long-term elasticities coming through in up to six years for local calls in the UK.

The estimation of parameter values in econometric models requires repeated observations on the dependent and independent variables. The individual observations need to be independent of one another if they are to yield additional information. If one observation can be derived from another by mathematical manipulation it is not independent. Provided that the number of independent observations is at least equal to the number of variables to be estimated (two in this case), an ordinary least-squares (OLS) regression will give unbiased estimates of a and b regardless of the serial correlation structure in the random element. Serial correlation will, however, affect the accuracy of the estimate and its likely significance can be examined with the Durbin-Watson test of serial correlation in the modelled residuals. Another problem frequently encountered is the existence of strong correlation between the independent deterministic variables, known as multicollinearity. If a third variable had been added to eqn. 16.8 it might have revealed a strong time trend. Although this would not introduce bias into the OLS estimate, it would make for considerable instability. The reduction in accuracy which arises from the existence of random errors depends on the error process itself, the method of estimation and the number of independent observations.

The number of available observations in a time series is often restricted. One of the problems encountered is therefore to find ways of increasing the precision of the estimate, by the use of more sophisticated estimation methods or the inclusion of more observations. One approach is to take the residual errors from the OLS estimates and to use them to model the error structure of the e_t series and introduce this into the estimating procedure by iteration.

A linear trend, of whatever structure and coefficients, can be removed by taking first differences. Suppose, in eqn. 16.8, that $x_t = c_t$, a linear trend in which $t = 1, 2, 3, \ldots$ and c is an unknown constant. Then taking first differences

$$y_t - y_{t-1} = b(ct - c(t-1)) + e_t - e_{t-1} \qquad (16.9)$$

$$= bc + e_t - e_{t-1}$$

Differencing may also help in reducing the degree of an equation. Terms in t^2, for example, will reduce terms linear in t on taking first differences. Another use for differencing is to reduce the e_t series to stationarity, i.e. to a series which is trendless, which will improve the precision of the parameter estimates. Differencing can be used to remove seasonal effects, for example, in quarterly series, by taking differences between quarters a year apart.

16.4.3 The identification problem

When modelling economic relationships it may not be obvious which is the dependent variable, or there may not be one. For example, the constant-elasticity demand curve can be written as $Q = P^{-a}$, making demand a function of price, or as $P = Q^{-1/a}$ in which price is a function of the quantity purchased. A time series in which both prices and quantities vary from year to year may arise from:

(a)   A series of changes in the position of the demand curve, the supply curve remaining on the same place and data showing supply as a function of market price.

(b)   A series of changes in the supply curve (for example costs being reduced), the demand curve remaining in the same place and the data showing demand as a function of market price.

(c)   Changes in the position of both curves, reflecting both effects.

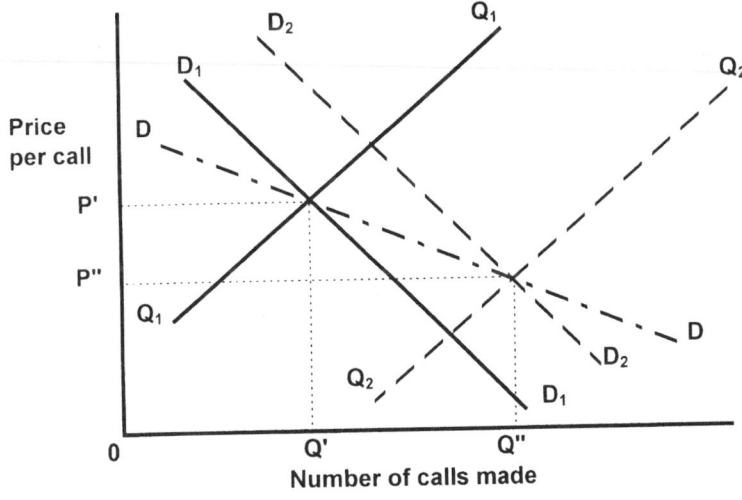

Figure 16.2 The identification problem

This is the identification problem. Figure 16.2 shows market clearing price/quantity observations, $P'Q'$ and $P''Q''$, relating to two consecutive years. The demand curve slopes downwards to the right and might at first be taken as observations on a single demand curve DD with a relatively high price elasticity. However, if the demand and supply curves had remained in the same place for both years the two observations would have been the same. They are not, so one or both curves must have moved. In case (c), the demand curve moves from D_1D_1 to D_2D_2 and the supply curve moves from Q_1Q_1 to Q_2Q_2 The true demand curve is steeper than DD, so the price elasticity is lower.

Movements in the demand curve may occur through changes in income levels. These can be picked up through macro-economic variables such as personal disposable income (PDI), which can be introduced as additional independent variables in the demand equation. It may be considered that, in the short term at least, the supply curve and the price elasticity are rather more stable, so that case (b) is unlikely to be observed, leaving the position of the supply curve to be estimated by other means.

In telecommunications, the most likely problem is that capacity constraints prevent changes in demand from being fully reflected in changes in supply. The effect of capacity contraints may have to be examined. Variables such as waiting lists or orders outstanding may help to resolve the problem when looking at connections. The percentage of succesful attempts may be helpful with call demand. A more extended treatment of identification problems and ways of dealing with them can be found in text books such as Reference 15.

16.4.4 Econometric forecasting

For reasons set out in Chapter 12, telecommunications econometric work is strongly focused on demand and revenue projections. The estimation of price elasticity is particularly important. With the advent of competition, total market elasticity is giving way to own-price elasticity as the main interest.

There are six main areas of work:

(i)   The demand for new access lines, usually modelled separately for business and residential users.
(ii) Call forecasting, split between local, long-distance and international and often split between business and residential.
(iii) Demand for new services,
(iv) Demand for mobile communications,

(v)   Market share under competition,
(vi) Large integrated models for business planning and profit fore-
      casting.

Economic modelling of demand for revenue forecasting uses advanced
mathematical and economic techniques to estimate parameters.
Whatever the exact form of a demand model, it is likely to assume that
the demand for a service depends on some combination of:

- what the national or regional economy is doing, in terms of
  gross national product, consumer expenditure and other macro-
  economic variables,
- nominal telecommunications service price,
- technical changes such as the effect of introducing STD and
  IDD (still not fully implemented in the developing world),
- inflation,
- factors personal to the consumer, such a recent move to other
  premises,
- new applications, such as the Internet
- advertising and marketing effort
- general trends.

Economic activity and telecommunications service prices are usually
treated in real terms, after inflation. It will usually be necessary to
make or obtain forecasts of external variables, such as GDP growth
and the rate of inflation, before the models can be run. Customer
perceptions of price, which may be poor, or differ markedly from
actual price (see Section 9.1.2), will affect price-elasticity estimates
which are derived from actual prices.

16.4.5 Model types

Cross-section analysis

Cross-section analysis is a useful alternative or compliment to time
series analysis. It uses observations from a number separate situations
for which the same model is thought to be valid. For example, in the
US the existence of numerous local or regional companies enables a
substantial set of independent figures for demand, price and income
to be collected during a single year. This would not be possible in a
country where the same tariff prevailed everywhere. Cross-section
analysis is used in Figure 8.3, where the penetration rate is plotted
against the level of income per head for countries across the world. A
regression analysis on this data, using penetration as the dependent

variable, indicates that the penetration rate rises by about one per hundred inhabitants for each additional $360 of income per head. Cross-section analysis tends to be better at picking out long-term effects, and benefits from having a greater variation among the observations.

Cross-sectional data can be increased by pooling, if parallel time series are available for different states, regions or routes. The Appelbe *et al.* estimates in Table 4.8 were based on pooled time-series and cross-sectional data. Pooling works if the price variables and incomes in each series are largely independent of each other, as might happen if they relate to different local operating companies in a country with heterogeneous regional characteristics. These conditions are mostly confined to Canada and the US.

Choice analysis

The decision process followed by individuals or subgroups of consumers is examined under this type of model. Thus, a decision to take up a telephone service may be influenced by such factors as a recent house move, family circumstances or education as well as on income and price. Choice analysis can be used as part of a more general regression model.

Spreadsheets

Widely used for business planning, where parameters estimated by other means are used in comprehensive projections of business volume, revenue and cost. They come into their own for trying out the effect of alternative assumptions and parameters and are equipped with good graphical facilities for illustrating the results.

Modelling by analogy

Used in forecasting the market for new products. Lifecycles and penetration rates for existing products believed to have similarities can be taken as a baseline, supplemented by choice analysis and other data believed to be relevant. The method relies heavily on judgement and its forecasting errors are difficult to estimate statistically.

Simulation methods

Used in large, complex models of systems with numerous supply and demand variables, where there may be nonlinear relationships and possible identification problems. They may be used in demand models where cross-elasticities exist between the demand for services. For example, the demand for lines may depend on the total bill size, covering line rental and call charges, as well as on the line rental itself. Trial values of the variables are used to set the model off, perhaps estimated by regression, and these are varied in subsequent iterations to minimise the residual errors.

Many models have been described. The papers appearing at international and North American academic and forecasting conferences are a particularly rich source. It is impossible to do full justice to these in the space available but the examples that follow give an indication of what is available.

16.5 Examples of models

16.5.1 Residential telephone connections

One of the earliest econometric studies of business connections growth was that of Simpson [16]. Independent variables were rental price and consumers' expenditure, both used in the form of a weighted average of the last 20 quarters. Optimum weights were found by iteration. First differences were found to be the most satisfactory form for estimation. More recently, modelling has moved towards service-sector output or detailed studies based on input–output analysis. New lines for office machinery, such as fax machines, have become a major element in business demand.

Binary-choice models of the demand for residential connections were pioneered by Perl [17]. Bodnar, Dilworth and Iacono [18] describe a binary-choice logic model in which data from 34 168 households were used to find the probability of having a telephone in the house. They give a summary of the binary choice literature. Their model is estimated from

$$\ln(p_i/(1-p_i)) = X_iB + e \tag{16.10}$$

where

p_i is the probability of having a telephone

X_i is a vector of telephone prices, household income and other household characteristics

B is a vector of parameters to be estimated

e is a random error term

No macro-economic variables are used. Price elasticities found were in the range −0.0002 to −0.032. They were most negative in rural areas, among young householders and among low-income groups.

16.5.2 International call minutes

Bhatia [19] describes models of traffic fitted to ten international routes from Australia. Independent variables included traffic lagged by one year, price, GDP of the country of origin, bilateral trade, migration and the introduction of international direct dialling. Price elasticities varied from −0.12 (Australia to Italy) to −2.03 (Hong Kong to Australia). GDP was the most consistent income variable with elasticities ranging widely, up to 4.1 (Hong Kong to Australia before IDD) but usually below unity. The use of lagged traffic enabled longer term effects to be picked up. Rabiee and Westlund [20] give results for traffic from Sweden to Denmark, Norway and Finland from linear, double-log and more general functional forms of model. Independent variables were price, industrial production indices for each of the four countries, imports plus exports and the previous value of the dependent variable. Longer term price-elasticity estimates ranged from −0.69 (Sweden to Denmark) to −1.14 (Sweden to Norway). Short-term elasticities were about half of these, in absolute value. Lagged traffic was included as an independent variable to obtain the long-term estimates.

16.5.3 Long-distance call minutes

Gatto *et al.* [21] describe an AT&T model of US interstate switched minutes of traffic. Traffic data consisted of pooled cross-sectional monthly time series. A polynomial distributed lag structure was used to pick up long-term effects. Independent variables were the lagged traffic for each of the previous nine months, real tariffs, personal disposable income and population, all state by state. The weights for the lag structure were symmetrically distributed over the ten months, with months 0 and ten being equal lowest and months five and six equal highest. A multi-stage regression procedure was used for the estimates. Results included the long-run price elasticity in Table 4.8 and an income elasticity of 0.827.

16.5.4 Forecasting the demand for new services

Business planners need to assess the commercial potential of new services. This involves taking into account the likely demand, production cost, competing products, the likely product life and any regulatory aspects. Market demand can sometimes be estimated by comparison with the growth in demand for comparably innovative products. These need not be similar in function but they should be similar in market price. The market prices of competing products will have to be estimated. As an example, video-on-demand supplied by the PSTN has alternatives, such as borrowing from video shop, which may place an upper limit on the price.

Market research is normally carried out. International experience is worth considering, especially for products which have already been tried in overseas markets. The penetration of cable TV in the US has been used as a model for the UK, where development took place later. Mobile communications have also been studied in this way. Care needs to be exercised, since market conditions may differ. Population density, income per head, geographical terrain and the quality and availability of competing products are examples of factors which may cause systematic differences. In the case of mobile services, demand in the poorer countries is often affected by the state of the public fixed network. Mobile services can be an effective, if expensive, way of bypassing it and this boosts demand for mobile services.

Marketing effort has begun to appear in econometric equations. Cracknell and White [22] identify the differences in the British and American call markets with market survey results for five different types of call, as shown in Table 16.1.

Table 16.1 Call patterns in the UK and the US

Call type	Ratio US/UK
Social calls to friends	3.4
Calls to relatives	1.2
Personal business	29.5
Household emergencies	0.8
Other	2.7
Total	2.5

This rather brief introduction to econometric modelling shows how a variety of telecommunications problems can be analysed mathematically. The application of most estimation techniques requires a grounding in statistical method which is beyond the scope of the book. Standard texts and papers in journals such as *Econometrica* and *Information Economics and Policy* should be consulted for further information.

16.6 References

1   CRACKNELL, D. R.: 'The analysis of revenue growth in telecommunications'. Paper for conference on *Forecasting and analysis for business planning in the information age*, Tokyo, 1986 (British Telecom, London, 1986)
2   PEARSON, E. S. and HARTLEY, H. O.: 'Biometrika tables for statisticians' (Cambridge University Press, Cambridge, 1970)
3   McGOWAN, F. S.: 'Allocative efficiency and total factor productivity'. Paper presented at the *Sixth international symposium on Forecasting*, June 1986, Paris
4   ALLEN, R. G. D.: 'Mathematical analysis for economists', (Macmillan, London, 1938, 1967 reprint)
5   BANNOCK, G., BAXTER, R. E. and DAVIS, E.: 'The Penguin dictionary of economics' (Penguin Books, London, 1992, 5th edn.)
6   WALTERS, A. A.: 'An introduction to econometrics', (Macmillan, London, 1970, 2nd edn.)
7   BROWN, R. S., CAVES, D. W., and CHRISTENSEN, L. R.: 'Modelling the structure of production with a joint cost function'. Social Systems Research Institute working paper 7521' (University of Wisconsin, Madison, 1975)
8   CRONIN, F. J., GOLD, M. A., HEBERT, P. L., and LEWITZKY, S.: 'Factor prices, factor substitution and the relative demand for telecommunications across US industries', *Inf. Econ. Policy*, 1993, **5**, (1) pp.73–85
9   CRONIN, F. J., GOLD, M. A., MACE, B. B., and SIGALOS, J. L.: 'Telecommunications and cost savings in educational services', *Inf. Econ. Policy*, 1994, **6**, (1), pp. 53–75
10   GREENE, W. H., and SMILEY, R. H.: 'The effectiveness of utility regulation in a period of changing economic conditions', *in* MARCHAND, M., PESTIEAU, P., and TULKENS, H. (Eds.): 'The performance of public enterprises' (North Holland, Amsterdam, 1984)
11   WHITTLE, P.: 'Prediction and regulation', (Blackwell, Oxford, 1984)
12   KENDALL, M. G. and STEWART, A.: 'The advanced theory of statistics, volume 2', (Griffin, London, 1967)
13   BRESLAW, J. and PIZANTE, G.: 'Lag structure in telecommunications demand analysis', *Inf. Econ. Policy*, 1989/90, **4**, (4), pp.325–345
14   CRACKNELL, D. R.: 'Growing the telecommunications market – lessons from the past'. Paper presented at ITS regional meeting September 1994, Crete
15   FOX, K. A.: 'Intermediate economic statistics' (Wiley, London, 1868)
16   SIMPSON, W. F.: 'The growth of business connections', Statistics and Business Research Department Report 13, General Post Office, London, 1967
17   PERL, L. J.: 'Economic and demographic determinants of telephone availability', *in* AT&T filing before the FCC, docket no. 20003, Exhibit 21, 1975
18   BODNAR, J., DILWORTH, P., and IACONO, S.: 'Cross-sectional analysis of residential telephone subscription in Canada', *Inf. Econ. Policy*, 1998, **3**, (4), pp.359–378
19   BHATIA, M.: 'Econometric models for ten international telephone streams', paper presented to the *Australian Meeting of the Econometric Society*, Armidale, NSW, 1989
20   RABIEE, M., WESTLUND, A.: 'International telecommunication demand', paper presented at ITS *Regional Conference*, Stenungsbaden, June 1993, Sweden
21   GATTO, J. P., LANGIN-HOOPER, J., ROBINSON, P. B., and TYAN, H.: 'Interstate switched access demand analysis', *Inf. Econ. Policy*, 1988, **3**, (4), pp.333–358
22   CRACKNELL, D. R., and WHITE, D.: 'The use of panel data in market analysis — British Telecom's experience', paper presented at ITS *Regional Conference* April 1989, Leuven

Chapter 17
International telecommunications

17.1 The benefits of trade

International trade strengthens competition in imperfect domestic telecommunications markets by opening them to goods and services from foreign operators and equipment suppliers. Extra competition reduces the market power of domestic producers, so that they are less able to make abnormal profits. Trade encourages countries to specialise in the production of goods and services where they have a comparative advantage, resulting in a more efficient use of resources and shared gains among the trading countries. Countries, companies and individuals have finite resources. The decision to use part of them in one way precludes their use in another. In the theory of international trade it can be shown that where markets are fully competitive, national wealth is maximised by specialising in products where the nation has the greatest lead, or comparative advantage, over its competitors.

A nation which has low costs over a range of exports will gain most by using its scarce resources to produce those exports where its competitive advantage, in terms of relative production, is greatest, even if this involves buying some goods from abroad which it could otherwise produce more cheaply at home. Proof was first given in 1817 by Ricardo [1] and it was developed by later writers. The theory can be illustrated by a numerical example. Suppose there are two countries, each of which has the capability of producing hardware and software, and that country A has lower costs than country B for both, as in the Table 17.1.

Table 17.1 Comparative advantage: an illustrative example

Product	Cost in hours of labour	
	country A	country B
1 unit of software	10.0	30
1 unit of hardware	20.0	40
Ratio of hardware to software costs	2.0	1.33

Country A has a competitive advantage for both products. It can produce three times as much software or twice as much hardware for a given labour cost. Its comparative advantage lies with software, for which its competitive advantage is greatest. Country B has a comparative advantage in the production of hardware, for which its competitive disadvantage is least.

If it is assumed that both countries have finite resources, taken here as 6000 hours of labour, and that all other goods and services can be ignored, the countries have production possibilities limited by production frontiers which are the dashed lines in Figure 17.1.

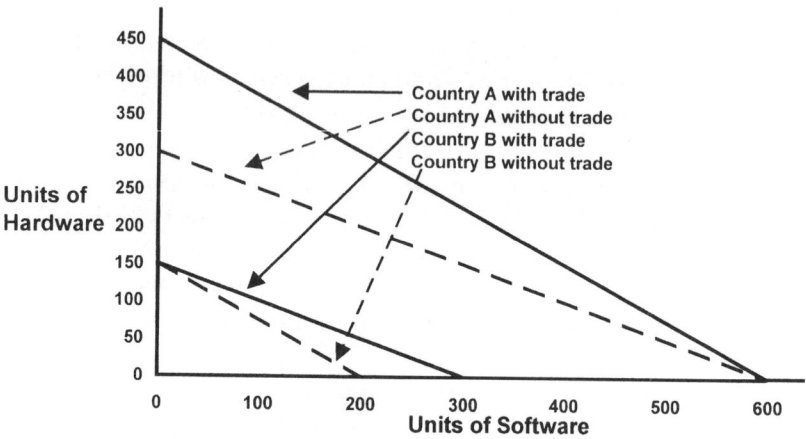

Figure 17.1 Production and trade possiblilities

Each country could sell to the other. Country A, for example, could trade part of its output of software to country B for hardware. Country B should be willing to trade at a price at least as low as its own cost of producing software. Any hardware/software exchange rate of 40/30 or more would get it software in exchange for hardware more cheaply than its own software production cost. For country A, a rate of 40/30 would get it hardware in exchange for software at a price less than its own software production cost, as would any price below 20/10.

The exact price would be settled by market factors. Suppose that the equilibrium price was 40/25. Then, by trading, the production frontier possibilities (including net external purchases) are enhanced for both countries. The largest gain is obtained when country A concentrates on software, where it has the comparative advantage, and country B produces only hardware. For any given level of hardware output and purchase, the chart shows that each firm can buy or make more software than without trading. A more extended analysis can be found in textbooks such as Reference 2.

17.2 Barriers to trade

17.2.1 Types of barrier

Tariffs are a common barrier to international trade. Other impediments, known generically as nontariff barriers, are common in the service industries. The more important nontariff barriers in telecommunications are:

- rights of entry into foreign equipment markets;
- restrictions on foreign ownership of domestic operators;
- terms for interconnecting networks;
- unreasonable local equipment standards;
- unfair trading practices.

17.2.2 Rights of entry into foreign equipment markets

Some countries keep foreign telecommunications equipment out by legal means or by only purchasing from domestic manufacturers, where imports would be cheaper and equally compliant with local standards. Since imported network equipment may require a considerable amount of customisation to suit the home market, discrimination in this area is difficult to prove, however real it may be. In some countries, local suppliers have been kept in business by the imposition of restrictive procurement policies on state-owned network operators on the grounds that a 'national champion' is needed to protect the national interest. This is a variant of the 'infant-industry' argument sometimes used to justify tariffs which protect emerging indigenous producers in other sectors.

Telecommunications service provision is always licensed, at least for basic services. A potential market entrant needs some form of government clearance to operate. An unqualified right of establishment is unlikely to be granted to foreign operators, although some countries are liberal, especially with data, value-added and mobile services.

17.2.3 Ownership and control

There are normally restrictions on the foreign control of operating companies, exercised directly or indirectly. Examples are the former golden share in the UK and the 1934 Communications Act in the US, mentioned in Section 10.2.

17.2.4 Terms for interconnect

A foreign competitor will want to connect its network to other networks in the host country. This commonly affects mobile services which are licensed to a foreign owner or consortium and need interconnection arrangements with the fixed network. The negotiation of reasonable terms for interconnect can be time consuming and unsatisfactory. It may be impossible.

17.2.5 Unreasonable local equipment standards

Where there are no legal restrictions on imports, local standards for equipment may differ significantly from those elsewhere without being any better, and favour indigenous operators by imposing heavy customisation costs on imported equipment.

17.2.6 Unfair trading practices

Various forms of price discrimination and unfair trading, financed out of the profits of a dominant operator, may impair the business prospects of new entrants

17.3 World trade negotiations

17.3.1 The global framework

Telecommunications operators work through the International Telecommunication Union on matters essential to the exchange of international telecommunications traffic, such as signalling protocols, the payment arrangements for crossborder traffic and the ways in which international leased lines can be used. The ITU was established as the International Telegraph Union by representatives from 20 European nations in 1865 and is one of the world's oldest international organisations. Codding and Rutkowski [3] give an account of its history and work. Negotiations on the terms of world trade in general have taken place at intervals since 1947. The original talks were mainly concerned with the reduction of tariffs on industrial products. The eighth set of negotiations, the Uruguay Round of 1986–94, broke new ground in covering services as well as goods. This put telecommunications onto the agenda for the first time and gave prominence to nontariff trade barriers.

17.3.2 Bilateral, multilateral and regional agreements

World trade negotiators attempt to achieve a multilateral agreement where possible, under which all participants agree to liberalise trade between themselves to the same degree. Broad agreements of this kind bring the largest benefits but they can be difficult to achieve because of the many special interests which have to be placated.

Regional agreements are sometimes easier to achieve, especially when built on a perceived cultural identity or relationship. A customs union was used in the 19th century to help bind the German states together. The British Empire had a system of Imperial Preference. The European Union has created a free-trade zone within its boundaries.

Bilateral agreements between pairs of countries with common interests may be made when broader agreements are too difficult to achieve. International trade economists tend to view them with disfavour because they may weaken the will to complete more broadly-based negotiations. An agreement between the US and Canada was concluded as part of a move towards a North American trading zone. It contained measures affecting telecommunications and set precedents for the Uruguay Round negotiations concluded later.

17.3.3 Reciprocity

In the bargaining that leads to trade agreements, individual governments are reluctant to make unilateral concessions even if these might be of some benefit to their citizens. They try to get corresponding concessions from others, which would enlarge the economic gains. This is a powerful incentive for the formation of bilateral agreements, usually in terms of reciprocity.

Reciprocity conditions are more restrictive than those in bilateral agreements. They usually match concessions of the same kind in the same industry. The United Kingdom's Department of Industry has only permitted simple resale from the UK to countries which allow it back along the route. In the US, the Federal Communications Commission has developed an equivalent competitive opportunities (ECO) test for access to the US satellite market. More recently, it has applied similar tests in crossborder network merger proposals such as that attempted by BT and MCI, which went to the FCC for adjudication in 1997.

17.3.4 Most favoured nation treatment

International trade treaties sometimes include a 'most favoured nation' (MFN) clause. This is a promise by the signatories to extend to each other any favourable trading terms offered in subsequent agreements with third parties. MFN clauses help to extend trade liberalisation.

17.3.5 Dumping and quotas

Predatory pricing of goods is an active international trading issue, where it is known as dumping and is countered by national anti-dumping duties. The most common complaints are against Asian and Far East suppliers of electronic goods, including some used in telecommunications. The offenders are claimed to sell at lower prices abroad than they do at home, to dispose of surplus stocks or to destroy foreign markets such as those for photocopiers and computers in Europe. The European Commission has procedures to test the cost of supply. Where dumping is regarded as proven, and a sufficiently broad backing can be obtained from member states, antidumping duties can then be imposed. The object of these is to raise the price of imports to a level where they are not a predatory threat to local manufacturers. Antidumping duties are found in many countries.

Quotas are another device used to protect suppliers from competition. Although usually used to safeguard suppliers at home, they have been used to protect regional suppliers, e.g. the protection of ex-colonial suppliers by the EC banana quotas. Duties and quotas can all too easily be used as a device to protect uncompetitive local suppliers and will then be against the interests of consumers. For this reason, they are looked upon with suspicion by the World Trade Organisation. Some of the worst economic effects of quotas can be mitigated by allowing them to be traded, so that they get into the hands of those able to make best use of them.

17.3.6 Trade in services

The GATT negotiations which were concluded in 1994 left many issues unresolved. The liberalisation of international telecommunications and the opening of domestic telecommunications markets to foreign competition made only limited progress. There was no substantive agreement on intellectual property rights (IPRs), although subsequent discussions have brought changes in patent and copyright law, at least on a regional basis.

IPRs remain an area of dispute between developed and developing countries. There are disputes between developed countries on cultural issues. Many products and services used in the information technology, entertainment and telecommunications industries are caught up in these. Economic questions which involve externalities include the value of protecting domestic culture from foreign (usually American) influences and free-rider problems such as the enforcement of patents to protect investment in pharmaceuticals and computer software. They may be distributional, with developing countries arguing that they have a right to exploit knowledge developed in the western world to create wealth of their own even if there are losses elsewhere.

17.3.7 Multilateral, bilateral and tied aid

International financial institutions make loans available to developing countries. This is multilateral aid and it is usually for specific projects, without conditions about where equipment is to be bought. Telecommunications loans accounted for about six per cent of lending by the six largest lending agencies in 1992. The biggest lender was the European Investment Bank (EIB) with almost $3 bn. Individual government agencies such as the Swedish International Development Authority, offer bilateral aid in the form of loans [4]. Governments may make the granting of financial assistance to developing countries conditional on the purchase of equipment from suppliers in the country providing the money. This is tied aid. Such conditions limit the usefulness of the resources to the recipient country and international aid agencies oppose the attachment of procurement conditions.

Developing countries may be offered attractive loan or other conditions making telecommunications equipment available more cheaply provided that it is bought from the donor country. Here, if prices are below marginal cost the loss is borne by the donor country, which may be acting for long-term strategic objectives or to prop up a weak domestic industry. Other losers, if any, are the firms from other countries who might otherwise have been able to provide the equipment and, in the longer term, possibly the recipient country. Overall benefits are difficult to evaluate. There might, for example, be a stimulus to the developing country which would not otherwise have arisen. Lower prices may represent the costs at the margin of products which would not otherwise have been sold. France and Germany are the countries most commonly mentioned as assisting domestic suppliers in this way. Suppliers in countries with less supportive governments, such as the UK and the US, have seen themselves as losers.

17.4 International services

International telecommunications services are of seven main types:

(i)   Ordinary telephone calls over the public switched   network, conveying voice and certain types of data traffic.
(ii)   International data networks, linking computers.
(iii)   Provision of international private leased circuits (IPLCs) to companies wishing to build private networks and to other operators as a part of their public networks.
(iv)   Network management services, involving the provision and management of private networks to companies operating internationally.
(v)   Global satellite services, such as those provided by INMARSAT to ships, and land mobile services provided by other operators.
(vi)   Mobile telecommunications based on the GSM or other systems allowing handsets to be used in more than one country.
(vii)   Value-added services such as international freephone, card-based services allowing users to have calls made in foreign countries billed to their home account and the provision of information from databases.

17.5 The international infrastructure

17.5.1 Economies of scale in the international network

The equipment that carries these services from one country to another shows the same general economies of scale economy as equipment used domestically. In terms of the cost per circuit, high-capacity undersea cables are cheaper to make and lay than smaller ones. Once laid, the capacity utilisation can be higher for larger numbers of circuits. For a given chance of blocking, a single set of 100 circuits can carry more traffic than two separate sets of 50. The manufacture and launch cost of satellites per unit of capacity also depends partly on how large they are.

17.5.2 Ownership of cables

Undersea international telecommunications cables were at one time usually owned by PTT consortia. A group of national operators would jointly plan the route and capacity, arrange for the construction and enter into agreements covering maintenance and the interfaces with other networks. Finance would be provided by the members in exchange for the right to use a block of capacity for the life on the cable, usually 25 years. Capacity not taken up would be leased to non-members with indefeasible rights of use (IRUs). IRU revenues would be shared among consortium members, which were free to lease additional IRUs from their own share of the capacity to other consortium members or to third parties if they wished [5].

The PTT consortium model was developed under conditions of international monopoly. Growing competition and the changing nature of international traffic have put it under pressure in two ways:

(i)   Members of a consortium may be competing against each other for the same traffic, creating an ambiguity of interest in the leasing of IRUs between members.

(ii) Voice traffic can be compressed to allow multiple circuits within a single channel without significantly worsening quality. The marketing of services which bundle voice and data in special packages for large users makes this more difficult to achieve, because data applications are typically less fault tolerant and cannot be compressed as much as voice. They may not be compressible at all.

For reasons such as this, the forecasting of capacity requirements is becoming more difficult, with a higher degree of risk. New forms of ownership have emerged. A few cables, such as C&W's PTAT between the UK and the US, are owned by a single operator. This removes part of the problem caused by competition among owners, but it leaves the forecasting risk.

A different pattern is represented by the FLAG (fibreoptic link around the globe) project, which is a 28 000 km cable system running from Britain to Japan, passing through the Mediterranean and the Suez Canal to India and the Far East and was put into service in November 1997. The system cost $1.2 bn to build. FLAG is a private venture owned by a consortium of eight business investors, including the US-based regional operator Bell Atlantic and trading companies in the Middle East and Asia. Carriers wishing to use the cable can lease capacity without previously having provided investment finance. The separation of ownership from usage rights is a transfer of risk from cable user to the cable owner, which will enable risk-averse users to enjoy more predictable costs in exchange for a usage fee which reflects the cost to the owner of carrying the risk.

17.5.3 Spare capacity and cost

Rapid technological change has led to the duplication of facilities. On routes that are heavily used, such as those linking major centres across the Atlantic, the balance of cost and service quality initially swung from copper cable to satellite and is now moving to optical-fibre cable. In a world with perfect foresight and provisioning flexibility, the obsolete equipment would no longer be used. In practice, new systems may be launched before the usefulness of the old ones has expired. The new systems may need an initial running-in period to establish their superiority. The older equipment is eventually withdrawn from service but until this happens it provides an extra margin of capacity.

The first seven transatlantic links, TAT–1 to TAT–7, used coaxial cables. The first transatlantic optical fibre cable, TAT–8, was laid in 1988 and by 1996 the last of the coaxial cables had been taken out of service (Section 7.4.2). TAT–12 is to be supported by TAT–13, running from the US to France and back to the UK, with facilities for switching calls between the two automatically if there is a cable failure. The cost of the completed system is reported to be about $1100 per channel. The Cable & Wireless Gemini system, which is also twinned, has a similar cost.

The fall in costs has been remarkable. The first transatlantic voice cable, opened in 1956, cost over $500 000 per channel in money of the time. By 1970 the cost had come down to $50 000 per channel. By 1990 the cost was down to about $5000 owing to a combination of volume growth, technology change and multiplexing (Figure 17.2).

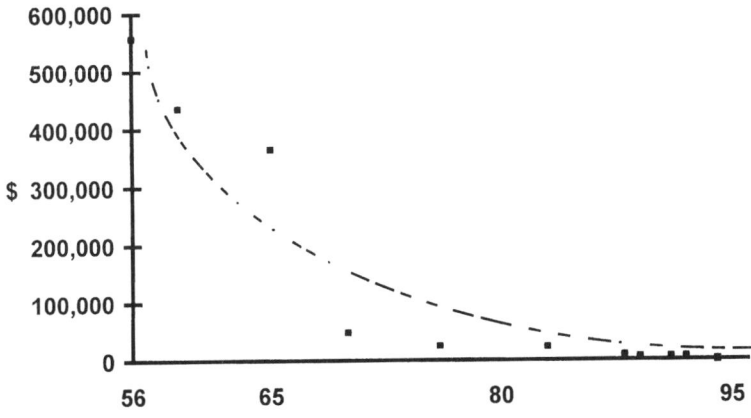

Figure 17.2 New transatlantic cable cost per circuit, 1956–96

17.6 Circuits on the main international routes

17.6.1 System capacities

The number of voice paths on the main intercontinental routes is shown in the Table 17.2.

Table 17.2 International circuits in 1996

Route	Number of voice paths	
	cable	satellite
Transatlantic	1 310 800	710 800
Transpacific	864 600	234 000

(Source: Telegeography [6]. Assumes 5 voice paths per 64 kbit/s circuit)

The potential for adding to this capacity is substantial. Gemini, a new twin-cable system, will add over 700 000 voice paths to the transatlantic route by 1999. FLAG will provide approximately 600 000 voice channels from Europe to the Far East. At $2 000 per voice channel, this is less than half the cost of the shorter transatlantic systems of the late 1980s. The system has a planned life of 25 years. Further out, wave-division multiplexing may increase the capacity on new and existing optical-fibre cables.

17.6.2 Market structure

If the market was mature and fully competitive, with a stable technology, the international links might be put up as an example of a natural monopoly and the duplication of facilities claimed to be a waste of resources. In practice, this argument has rarely been given a chance to run. The market is dominated by statutory monopolies and access is still highly regulated. Most cables have different owners at each end or are owned by consortia. An operator wanting to lay and own a cable between two countries needs permission from both and may not get it.

Legal restrictions on market entry have enabled operators to use monopoly power to keep international service prices well above cost. High profit margins have been used by state-owned operators as a source of cross-subsidy to cheapen domestic services. When international traffic passes between two operators with statutory national monopolies, there is a situation of bilateral monopoly. These factors have created substantial incentives to achieve market entry by operators with actual costs which may be no lower (and are often higher) than those of the incumbents, but which are willing to charge lower prices.

Technical factors have added a diversity of technology to the diversity of operator structure and removed natural monopoly characteristics. The digitalisation of satellite's analogue transmission systems assists this by providing extra capacity at a competitive price (Section 2.2.4).

Most international services involve several different operators. Arrangements vary from route to route, but at least five parties may be involved:

(i)   The local carrier in the country of origin.
(ii)   An international carrier in the country or origin.
(iii) The owners of the international cable or satellite.
(iv)   An international carrier in the country of destination.
(v)   The local carrier in the country of destination

There have to be prices for handling traffic passed between carriers and procedures for invoicing and payment.

Experience has shown that, for reasons of self interest or policy, the carriers, whether owned privately or by the state, are reluctant to cut prices themselves, and that national regulators are usually reluctant to make them do so. Even where competition is permitted, barriers to entry can leave the exmonopolists in a dominant position, still able to exercise market power. International service provision shows a diversity of organisational form some way removed from the natural monopoly model.

17.7 International settlement for calls

17.7.1 Basic accounting arrangements

At one time virtually all traffic-exchange arrangements were between national monopolies, effectively operating as a cartel. The picture is changing with the pluralistic nature of modern telecommunications service supply. An international telephone call on the public switched network incurs costs in the originating and destination countries. There are international agreements governing the rates and payment procedures. The originating country usually keeps the whole of the price paid by the caller and pays the destination country for the delivery of the call.

Three terms are in common use to describe the payments:
(i)   The collection rate is the charge made to the customer.
(ii)   The accounting rate is the per-minute rate used as a basis for international payments on a route. Accounting rates vary from route to route and may depend on the type of traffic.
(iii) The settlement rate is the per-minute rate paid for the delivery of an outgoing call. It is usually half the accounting rate and the same in both directions along a particular route.

The international cable and satellite infrastructure links most major countries directly, but some calls have to be routed via a third country to reach their final destination. Transit traffic, as such calls are known, passes through transit centres in the intermediate country where it is switched for onward transmission. If the traffic is routed via a third country the transit and delivering countries each take a third of the accounting rate.

17.7.2 The currency of accounting and settlement rates

Accounting and settlement rates are normally denominated in US dollars, special drawing rights (SDRs) or gold francs. The gold franc is effectively a function of the US dollar, being defined in terms of a quantity of gold of a fixed degree of fineness valued at a fixed (and artificial) price in dollars per ounce.

Most countries and operators keep their accounting rates secret. The main exceptions are the US, where the FCC has published them for many years, and the UK, where OFTEL began publishing rates between the UK and OECD countries in the mid 1990s. Secrecy about the rates reduces the efficiency of any market there might be in international message transmission and has become more important with the development of competition between carriers for transit traffic. In this, it is similar to the lack of transparency with other elements of cost.

Settlements are normally made net, that is the only payment made is by the country with more outward than inward traffic. Payment is then made on the net traffic balance. If the balance of traffic is exact, no payment is made. This is sometimes known as net accounting. An alternative would be for both countries to pay for what they sent abroad. This would be gross accounting.

Occasionally, a pair of countries agree on a 'sender keep all' arrangement whereby each country keeps the whole revenue from the call and no international settlements are made. Sender keep all invites economic inefficiency, since the delivering country has no direct incentive to provide an adequate service for incoming calls. The arrangement cannot be sustained under significant competition.

17.7.3 Parallel accounting and proportional returns

The introduction of competition on international routes can put heavy pressure on accounting arrangements and is seen by some to raise problems affecting the national interest. Suppose country A allows several operators to compete for international traffic to and from country B, where all traffic is handled by a monopolist. Country B could play the operators in country A off against each other and obtain cost-related rates for the delivery of its traffic in country A. However, country B's monopoly would enable its own charges for the delivery of incoming traffic to be kept well above costs, at a level which maximised its profit. There would thus be asymmetry in charges, with country B using its monopoly powers to get funds from country A. The situation is known as whipsawing in North America.

Two devices are commonly used to prevent whipsawing. These are:

(i)   Parallel accounting, whereby countries agree to maintain the same accounting rates for all carriers in both directions on a route

(ii)   Proportional returns, whereby country B returns to each international operator in country A the same proportion of its outward traffic as it receives from that operator. Thus, an operator responsible for ten per cent of country A's outward traffic to country B could expect to get ten per cent of country B's traffic to country A.

Proportional returns are important to operators when accounting rates are above cost, because they represent a source of profit. They may be difficult to negotiate and enforce.

17.7.4 The relationship of rates with cost

In circumstances of perfect competition, the payment made to the destination country would tend towards long run marginal cost. There would probably be asymmetry on a route, since the delivery costs in each country are likely to vary with local geographical and other circumstances. Thus, country A might get more for accepting and delivering a call from country B than it pays country B for delivering one of its own calls

Since most international calls are exchanged between monopolists or dominant operators, the charges for delivering incoming international calls may differ widely from any definition of cost. There have been various attempts to relate accounting rates to cost by international arrangement, but cost reductions arising from technological change have undermined them. In the absence of competition operators have little incentive to move rates, but market liberalisation changes that.

Parallel accounting and proportional returns are constraints of competition and their justification is based, like some other constraints on trade, on the need to prevent citizens of country B from benefiting at the expense of those of country A, even if that means less aggregate utility for the two countries combined. The nation is the political boundary of welfare maximisation. As in other matters of international trade, the hope is that reciprocal arrangements that bring greater benefit to both parties can be negotiated, in this case more cost-related collection and accounting rates.

17.7.5 International cross-subsidies

One factor inhibiting change has been the existence of large international cross-subsidies within the accounting-rate structure.

High international accounting rates are an important source of investment finance to Third World and developing countries, which typically receive more traffic than they send out and generate a large inward flow of settlements by consequence. When the rates are above cost, this acts as a subsidy from the rich countries to the poor and often from the heavily-used routes to the thinner ones. Not surprisingly, less well-off countries tend to resist reform.

A reaction built up in the late 1980s, driven mainly by US attempts to reduce a mounting balance of payments deficit on goods and services as a whole. Figures published by the FCC show the net outward telecommunications payment rising from $1.1 bn in 1985 to $5.4 bn in 1996 [7]. The Americans argued that the deficit arose from:

- lower, more cost-related international tariffs in the US which stimulate outgoing US traffic and depress outward traffic from its correspondents;
- accounting rates well above cost.

Insofar as the telecommunications deficit was part of a more general rising trade deficit, macroeconomic factors must also have played a part.

The political pressure continued to increase during the 1990s and by 1997 the US administration was threatening unilateral action to limit the amount that US carriers could pay to foreign partners [8]. The FCC estimated that at least 70 per cent of the annual settlement payments made by US carriers constituted an above-cost subsidy paid by US consumers to foreign carriers. An order made by the FCC established benchmark rates based on the publicly-available foreign tariffs and World Bank economic development data. US operators were asked to negotiate rates no higher than these and, if unable to do so by specified dates, would have them imposed. The benchmark settlement rates and dates are shown in Table 17.3. They were said to be still above cost. The threat of a unilateral imposition of maximum rates on bilateral agreements will not be effective unless it is backed by adequate market and political power.

Johnson [9, 10] drew attention to routes and times where the US carriers paid more to PTTs for completing the calls than they collected from their customers, for example losing money on every economy-rate call to some countries. Three possible explanations were offered:

(i)   Outgoing traffic stimulates incoming traffic on which a charge is collected, leaving the US carrier ahead.

Table 17.3 FCC benchmark settlement rates

Income level	Cents per min	Latest date for implementation
Upper	15	1.1.1999
Upper middle	9	1.1.2000
Lower middle	19	1.1.2001
Lower	23	1.1.2002 (1.1.2003 if fewer than 1 phone per 100 inhabitants)

(Source: FCC)

(ii)   Rate-of-return profit controls required losses to be made on some services to offset excess profits elsewhere. Profits were maximised by the chosen price schedule.

(iii) The losses were the outcome of bargaining positions in which the US carriers obtained greater benefits elsewhere.

The first two of these arguments depend for their validity on the level of international price elasticities and the degree to which they vary between routes. If the price elasticity for outgoing traffic is more than unity, a reduction in the accounting rate that is passed back to the customer in lower prices will increase the revenue on outward routes. Australia had a settlement deficit in the early 1990's. Ergas and Patterson [11] showed that, with price elasticities of –1.5 for collection rates below $US2.50 per minute and –2.5 for higher rates, almost all countries would gain from a reduction in accounting rates, the biggest benefits going to carriers with high accounting rates and/or collection rates and those in less developed countries. Australia would gain substantially. The argument has been slow to influence other supposed beneficiaries. BTCE [12] contains an analysis of Australian international telecommunications, with estimates of inward and outward price elasticities to ten countries including the UK, the USA and several neighbouring states derived from Bhatia 1989 (*op. cit*). Many of these were in the range –0.12 to –1, although elasticities of –1.3, –1.59, –1.64 and –2.03 were obtained for Japan, US, Canada and Hong Kong, respectively.

17.8 International private leased circuits

17.8.1 Pricing

By convention, international cable capacity between two countries is retailed in half-circuits; to get an international link you need to lease a half-circuit from the operator at each end of the line. Physically, of course, the two halves have to be part of the same cable if they are to be of any use. There may be restrictions on what traffic a circuit can be used for. Some operators control both ends of the same circuit.

Half-circuit tariffs are in national currencies and may be different for the two halves of a single connection. In the early 1990's, for example, AT&T charged less for the US end of transatlantic circuits than BT did for the UK end. In 1997 the half-circuit price for a 64 kbit/s line from Greece to Norway was more than twice the price of one from Norway to Greece. Most of the differences arise from the market power of the operators and the way that they are affected by national regulation. Given that half-circuits terminate in particular cities rather than on the beach, some variation in the cost of each end could be justified by differing costs in the land-based part of the connection. Prices in the US vary in this way.

Half-circuit pricing requires negotiation with both operators for a complete link. This involves transaction costs for users. On many international routes individual telephone calls have a higher percentage mark-up on cost than do leased lines, which creates arbitrage opportunities. Capacity on FLAG will break with tradition and be sold as complete links. This may precipitate the breakup of the half-circuit pricing system.

17.8.2 Restrictions on use

Users face restrictions on the use to which international private leased circuits (IPLCs) may be put, mainly to restrict arbitrage. The restrictions vary from country to country but they usually limit, and often prohibit, the use of IPLCs for the provision of service to third parties. Their main aim is to prevent market entry into the international switched voice market by unlicensed operators. There are fewer restrictions on the provision of data services, and on the use of leased lines to build private networks for the transmission of traffic between telephones within an individual company or other closed user group.

The definition of a closed user group has at times proved troublesome. It is intended to exclude the retailing of capacity to third parties in situations where simple resale is illegal. Large users of international services usually have several ways of dispatching their traffic, sometimes many. IPLCs are a viable alternative to the use of the PSTN where there is sufficient traffic volume. Closed user groups are sometimes permitted to use international resale facilities, depending on the local regulation. Restrictions on the connection of IPLCs to the PSTN are designed to limit competition between the PSTN and resold IPLCs, but they are difficult to enforce.

17.9 Forms of international competition

With the development of radiocommunications, three types of international infrastructure co-exist, based on terrestrial cable, terrestrial radio and satellites. Cable and satellite are for many purposes complementary technologies providing alternative means of transmitting calls on the public switched and leased circuit networks. Operators use one or the other depending on the cost and service quality offered. Satellites have been used by dominant operators and their competitors as the basis of private networks to take advantage of somewhat more liberal regulations governing the use of satellites as well as for their special operational advantages in some markets.

In Europe, the PSTN owners see the providers of private networks which accessed by VSATs as competitors, but VSATs are also a business opportunity for the them. Those with direct access to a satellite system can benefit by reselling VSAT capacity to users themselves.

Mobile networks serve a different market sector, where service can be offered at a premium to users without fixed terminal equipment. Where there is an overlap it is mainly among business users, some of which will also be customers for the global satellite-based mobile services being offered for the top few per cent of the international market. Some see the eventual merging of the mobile and fixed-link sectors, but so far the cost differences have been too great for this to happen at profitable and acceptable tariffs.

Network operators have attempted to manage competition by offering IVPNs (Section 9.4.2) to medium and large users. IVPN traffic is taken at a discount to normal PSTN rates. It permits market segmentation by using a multipart tariff which is only attractive for traffic above a certain level. IVPNs are a means by which operators can compete against each other abroad. An operator can use international circuits to offer IVPN services to companies in foreign countries, subject to whatever local regulation exists about concerning user groups and resale.

17.10 International resale

17.10.1 Description of resale arrangements

One of the quickest ways of breaking down the monopolistic structure of international telecommunications is to allow new entrants to lease international capacity and retail it at freely determined prices. There were numerous ways of preventing this in the 1980s, but the system then began to break down under the combined pressure of inventive entrepreneurs, new technology, complaints from customers and pressure from national governments.

International public telephone services can be offered over lines which are leased rather than owned by the operator where regulation permits. The lines can be leased from the operators which own them in the same way that lines are leased for the building of international private networks. The operation will be profitable if there is a sufficient margin between the cost of leased lines and the market price for telephone calls.

The operator needs to connect the international circuit to a customer at each end. The link to the customer can be made by connecting the circuit to the PSTN or to a domestic private network. A link to the PSTN at each end of the circuit enables a public service to be offered, whereas if the link is to a private network only users of that network can be reached.

Operations where unenhanced telephony is offered through a PSTN connection at both ends of the circuit are called international simple resale (ISR). In some configurations, designated international resale, they are connected to the PSTN at one end and a private network at the other. If there is a value-added element they are called international value-added network services (IVANS).

17.10.2 *Cost structure*

The cost structure of an international reseller has as its main elements:

- lease of the international circuits;
- interconnect costs with the PSTN;
- billing, administration and marketing;
- capital costs for switching, compression and other equipment.

To make an end-to-end connection, the reseller needs to lease international half-circuits on each end of the route. The capacity in the line is then split up, using compression and multiplexing equipment, to provide smaller private circuits for private networks and individual voice and data circuits for public use. Individual circuits are loaded with traffic up to a maximum determined mainly by service quality considerations. Any excess of traffic is turned away, usually by diverting it to the PSTN.

There are large bulk discounts in the retail IPLC tariff structure, shown in Figure 17.3. This is based on representative tariffs from the UK in 1996, assuming an average utilisation of ten per cent. They are bigger than those in call tariffs and a key factor in establishing the viability of leased circuit resale.

The cost of the capacity per minute to the reseller depends on the cost of the half-circuits, the degree of compression and multiplexing

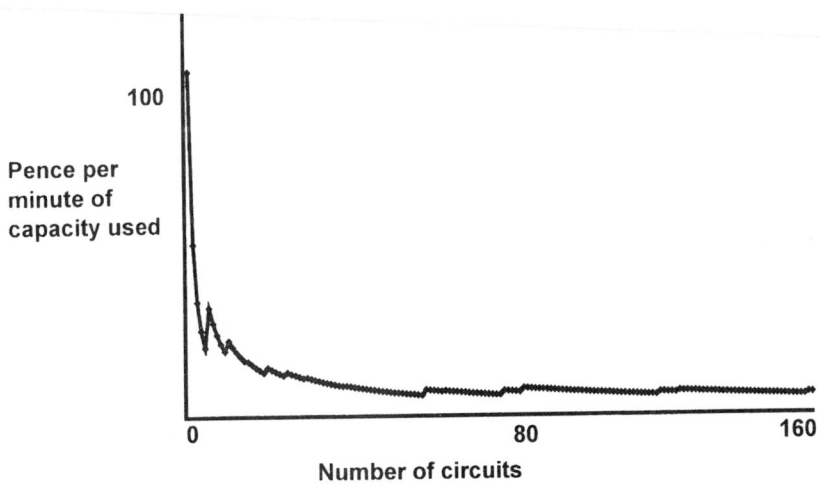

Figure 17.3 Bulk discounts in transatlantic IPLC tariffs

and the percentage occupancy of the circuits. There will also be capital costs. As an example of what may be achieved, the end-to-end cost of a 2 Mbit/s circuit between London and New York in 1996 at standard rates was about £48 100 ($72 700) per month or £578 000 per annum. The circuit is divided into 32 circuits of 64 kbit/s capacity. With 6:1 compression this can be increased to 192 voice-grade circuits with a total capacity of 101 million minutes in a year. With an average circuit occupancy or fill factor of ten per cent, this is equivalent to 10.1 million paid minutes of use. The cost in pence per minute is then $100 \times 0.578/10.1 = 5.7$ p (8.6 cents)

Public tariff rates offered by the facilities-based carriers on the route at that time were around 45p per minute, with discounts for bulk. The reseller has to offer a discount against the tariffs of the larger and better-known operators in order to get enough business. A typical reseller's price at the time was around 30p per minute, leaving 25p to pay for other costs and provide profit when competing with undiscounted public tariff rates.

All elements of this calculation are variable. The substantial volume discounts built into IPLC tariffs mean that the margin available from a single 64 kbit/s circuit is smaller than that from circuits with bigger capacities such as 34 Mbit/s, where they can be obtained. The degree of compression used may vary with traffic load and the type of traffic (voice, fax or data) for which it is used. The reseller may buy more capacity than is needed in order to take advantage of the volume discount, but then be unable to achieve a fill factor as high as ten per cent.

Discounted public tariff rates reduce the margin. In the long term, competition from resellers and other carriers reduces the general level of international tariffs. Public tariffs on the transatlantic route are subject to heavy competition, squeezing margins. Protected routes often have higher tariffs.

Although ISR may only be permitted over certain routes, there is often a more liberal regulatory regime for lines connected to the PSTN at one end only. The reseller may then connect customers to its IPLC by private circuit at one end and forward traffic through the PSTN at the other. For example, a bank in London may have a private circuit to a reseller with UK/US capacity and use it to make connections anywhere in the world if the reseller can hand non-US traffic over to a carrier in the US for onward delivery.

17.10.3 The international simple resale market

International simple resellers have typically taken medium and large business users connected by private circuit at one end of the line as their main target market. They have offered them 64 kbit/s private circuits and ordinary telephony discounted against public rates. Although this sector has been to some extent forced by regulatory limitations, a focus on the larger business users also reflects the greater profitability expected from the sector. In terms of tariffs, a high international traffic volume is needed to make direct use of an international reseller worthwhile, because of the fixed costs involved in setting up the private line and the two-part tariffs often employed.

The cost of accessing small business and residential customers may make their business unattractive to an international carrier with no local facilities. These groups may become customers of international resellers indirectly rather than directly, by taking service from a domestic carrier which has no international facilities and uses a reseller to handle international traffic. Cable TV operators that offer telephony are an example of this type of carrier.

17.10.4 Call-back operations

Where the cost of a call from country A to country B is higher than the cost of a call in the opposite direction, there is a chance for callers from country A to save money by getting the call initiated from country B. The simplest way of doing this is to make a brief call to country B and ask for it to be returned. The procedure can be turned into a business by having an agency in country B which can accept calls, return them and provide a connection to the required number in country B. The caller must have an account with the agency and is usually charged for the call at country B rates in country B currency.

This type of service is called 'call-back'. In its simplest form, an operator in country B can make the necessary connections. The procedure can be automated with relatively inexpensive equipment. Third country calling, sometimes called call re-origination, is a form of call-back where the caller reaches country C from country A via a call-back operator in country B. The caller pays the cost of a call from country B to country A and the cost of a call from country B to country C plus a small administrative charge. There are routes over which this is cheaper than paying for a call from country A to country C (see Section 15.5.4). In a common variant, offering bigger savings where the traffic level is large enough to support it, the caller reaches country B over a freephone line from country A.

The PSTN operator in country A may lose substantial amounts of revenue and profit from call-back operations if they become popular. The operations are difficult to block because they use ordinary telephone numbers in country B, which have to be identified by their unusual traffic volumes or by other means. Call-back operators may keep changing the numbers to make the operation more difficult to control.

17.10.5 The call-back market

Call-back had its origins in the needs of travellers from the US to countries such as Italy and Venezuela where outward international calls were greatly overpriced. Call-back services have become strongly developed in countries where international calls are very expensive. They can also be used from hotels to avoid expensive hotel surcharges, which may add 400 per cent or more to the cost of a call. Numerous companies offer the service. Customers tend to be private individuals, travelling business people or small companies. Commercial systems are not usually the cheapest for large volumes of business.

There are advantages for expatriates, since they no longer have to deal with an unfamiliar telephone system and can be billed in their own currency.

17.10.6 Card-based operations

Companies with their own international facilities can offer a similar operation to call-back, based on a calling card account. A customer is given a card, an account and an account number as with call-back. The card itself, known as a calling card, has no function apart from carrying information about the numbers to be dialled to use the service. The main operational difference lies in the way in which the call is routed. Originally, calls were often routed to country B over freephone lines, but it has become more common for them to be made to a node in country A, from which they are passed over an IPLC to country B.

Calling-card calls appear to be somewhat easier to block, especially in hotels. The user needs to know the number to dial for access to the system, which will vary from country to country. Card issuers provide a list of the codes required on a small piece of card or plastic. Although most cards are offered by facilities-based operators, there is no inherent reason why the operations cannot be run on the same basis as call-back.

Convenience plays a large part in the marketing of calling cards. By 1996 there were about 80 million in circulation, the large majority of them issued by US carriers. Most users have been expatriate Americans, e.g. servicemen stationed abroad, students, travelling business people and holidaymakers.

The cards have been used extensively for calls within the country in which they are used, for example Germany to Germany via the US using the dial-tone and operator services which are familiar at home, rather than those in Germany and with a bill in the home currency, US dollars. Price is frequently not the main factor determining use. Calling-card rates on some routes are higher than those on the PSTN, although hotel surcharges can more than restore the card's advantage.

17.11 Carrier refile

17.11.1 Transit bypass

With the development of competition on international routes, the facilities-based carriers themselves began to see opportunities for making money by bypassing the international settlements system. High accounting rates on direct routes can provide an incentive to use transit centres if the rates on the transit route add up to less than the rate on the direct route itself. The practice is known as transit bypass. It performs a useful economic function in bringing pressure to bear on monopoly profits made on the direct route, but is not encouraged by international telecommunications organisations.

Transit traffic is mixed with traffic originating in the transit country when forwarded to its destination, making bypass traffic difficult to identify.

17.11.2 Refile by IPLC

Facilities-based carriers can use IPLCs in the same way as do resellers, and the traffic carried will not be subject to accounting-rate settlements. This kind of routing is known as accounting rate bypass. A payment will be needed when the traffic is handed over to the PSTN for delivery. It might be at the ordinary inland tariff for traffic originating in the country, possibly qualifying for a bulk discount.

Carriers that own both ends of an IPLC themselves or through a foreign affiliate will find accounting-rate bypass particularly simple and it may be very profitable. Like transit bypass, the extent of refile by an IPLC is difficult to determine. The potential advantages are substantial on all routes with settlement rates in excess of local PSTN tariffs. An IPLC from New York to London can carry traffic for all european destinations and the accounting rates out of the UK are low.

Foreign-based carriers that use IPLCs for outward traffic to avoid accounting-rate liabilities but insist that inward traffic comes over the PSTN, thereby bringing accounting-rate credits, are practising what the FCC has called one-way bypass [8]. The FCC has said that it will permit a carrier to provide switched services over PLCs only if over half of the traffic on the route in question is being settled at or below the benchmark rates in Table 17.3. Sanctions included the enforcement of a 'best practice' settlement rate of only eight cents per minute.

17.12 UK call price and profit trends

International telegraph links from London to continental Europe were opened in the 1850s. They used undersea cables and had reached Germany by 1860, nearly twenty years before the launching of a domestic telephone service in the UK in 1879. The first international telephone service from the UK was the route to France, opened in 1891. Service to Belgium began in 1903 and to Switzerland in 1914. It did not reach Germany until 1926.

When the London–New York telephone service was opened in 1927, it linked the metropolitan areas of the two cities at a basic charge of £15 for a three-minute conversation. There were no undersea voice cables and transmission was by radio [13]. The average weekly earnings of a working man at the time were £4.06 [12]. The telephone service rate was reduced to £9 in 1928 and was down to four guineas (£4.20 in decimal currency) by 1936, with a cheaper night rate of £3. Service was suspended during the Second World War and reintroduced in 1945 at a basic rate of £3 for three minutes. The night rate was reintroduced in 1946 at £2.25. There was no further change in the basic rate until 1967, by which time average weekly male earnings had risen to £27.71. The retail price index rose by 229 per cent over the period 1927 to 1967.

Further changes, up and down, were made over the next twenty years until a period of more rapid reductions began after the introduction of competition in the mid 1980s. By 1996, the minimum three-minute charge had been replaced with one of a few seconds.

British Telecom's basic rate was down to 38p per minute (45p including VAT) and a wide range of discounted rates were available.

The long period of price stability in the post-war decades contrasts strongly with the rapid change of the 1980s and 90s. The stability cannot be attributed to unchanged technology. The first undersea transatlantic telephone cable was opened in 1956 (Section 7.4.2) and the first telecommunications satellite (TELSTAR) was launched in 1962. Nor was it due to sluggish traffic growth. The number of transatlantic calls rose from 49 000 (three times the pre-war level) in 1945 to 531,000 in 1965 [13]. The profit level may have been a reason. The Post Office report and accounts for the year to March 1966 showed a rate of return of 13.7 % on overseas telephone services, not greatly above the 8.2 % made by telecommunications as a whole. The absence of any competition on the route, coupled with administrative inertia, weakened any residual incentive for a change.

Unit costs fell strongly in the later years, permitting UK tariffs to fall, with an overall rate of return on all international services that fluctuated in the range 15-28 per cent from 1968–82. No further rate-of-return information was published until OFTEL started publication as part of the tariff consultation process. Nothing was published about the profitability of individual routes. The OFTEL figure for the overall rate of return in the year to March 1991 was 72.8 %, in spite of further tariff reductions. By 1994 this had fallen to 53.9 %. Although the methodology may not be fully consistent in the various rate-of- return calculations, a sharp increase in profitability and an even larger reduction in unit costs evidently occurred over the period. International service prices have continued to fall.

17.13 References

1   RICARDO, D. 'On the principles of political economy and taxation' (first published 1817, Cambridge University Press, 1951, edn.)
2   SAMUELSON, P.: 'Economics' (McGraw Hill, New York, 1976, 10th edn.)
3   CODDING, G. A. Jr, and RUTKOWSKI, A. M.: 'The International Telecommunication Union in a changing world' (Artech House, Inc., Dedham MA, 1982)
4   'World telecommunication development report 1994'. ITU, Geneva, 1994
5   'The devils and the deep blue sea', *Public Network Europe*, October 1996, pp. 43–46
6   STAPLE, G. (Ed.): 'Telegeography 1996/97' (Telegeography Inc., Washington D.C., 196)
7   'Statistics of the communications common-carriers' (FCC, Washington D.C., Annual)
8   'Commission adopts international settlement rate benchmarks' IB docket 96–261, FCC press release, Washington D.C., August 1997

9 JOHNSON, L. L.:'Competition pricing and regulatory policy in the international telephone industry' (Rand Corporation, Santa Monica, 1989)
10 JOHNSON, L. L.: 'Dealing with monopoly in the international telephone service: a US perspective', *Inf. Econ. Policy*, 1990, **4**, (3), pp. 225–247
11 ERGAS, H. and PATTERSON, P.: 'The joint provision of international telecommunications services: an economic analysis of alternative settlement arrangements'. Paper presented at ITS 8th biennial conference 1990, Venice.
12 'International telecommunications from an Australian perspective'. Bureau of Transport and Communications Economics report 82', Australian Government Publishing Service, Canberra, 1993
13 'Telecommunications statistics'. Annual until 1984 from the General Post Office and later from BT, London
14 London and Cambridge Economic Service: 'The British Economy, key statistics 1800–1970' (*Times* Newspapers, London)

Chapter 18
Strategic issues for suppliers and operators

18.1 The main issues

Business strategy is a large and complex subject, the full extent of which is outside the scope of this book. Strategies for growth or survival must, however, have secure economic foundations if they are to succeed. With the acceleration of technological change and the spread of deregulation, a new range of strategic issues has emerged. This section examines issues, old and new, now faced by equipment suppliers and operators. They include:

Traditional PTT issues
- Improvement of technology;
- Supply chain management; vertical integration versus out-sourcing for:
  equipment
  services
  research and development;
- Managing political relationships.

New issues
- coping with facilities-based competition in traditional monopoly areas:
  local loop
  long distance and international;
- competition based on new technology and deregulation:
  mobile communications
  internet telephony
  resale;

- holding existing customers and winning new ones:
  building brand loyalties
  tariff packages
  customer databases as strategic assets
  the separation of network ownership from operation;
- independent regulatory bodies;
- mergers, aquisitions and alliances;
- globalisation.

There has been extensive corporate and management restructuring, especially among operators, to handle these.

Commercially-driven organisations in service and equipment supply will try to develop strategies for the maximisation of the value of the business to its owners. This value is identified with discounted profits over time, share price or some related financial measure and will usually include expectations of business expansion. The long-term value is a function of the current level of profit and its expected rate of growth. Business planners think in dynamic terms and look at market growth, market share and diversification or demerger possibilities.

Expansion strategies usually require additional investment. Profits generated within the business may be paid out to shareholders or reinvested internally by the management. Retention of profits for reinvestment needs to generate extra profits at least as large as those which could be obtained by the shareholder from alternative investments if shareholder interests are not to be impaired.

Not every line of potential expansion can be expected to show a profit. In each direction there may be a marginal entrant, that is an entrepreneur who would go out of business if the market price were slightly lower. Profits that are as at least as high as the cost of capital will only be made if costs are lower than those of the marginal entrant or if a premium price can be charged. Two factors embrace the broad lines of most business strategies:

(i)   Building on comparative advantage, that is features of the business which give it lower costs or better market access than those of competitors.

(ii) Increasing market power, so that the price level can be maintained or raised.

A company which is able produce many types of output more efficiently than other companies can, might find that its best interests lay in specialising. The correct identification of special advantages that will give the enterprise an edge over competitors is essential for the protection of existing business and expansion into new areas of activity.

18.2 Sources of comparative advantage

18.2.1 Advantage at the level of the firm

Comparative advantage builds on the idea most commonly used in the theory of international trade and introduced in the previous Chapter. The principles are wide ranging. They explain why the labour of individuals is specialised, as well as being applicable to the economic activities of firms and nations. A firm with finite resources does best by deploying these on activities where it has the greatest comparative advantage. There are limiting conditions which may be important in telecommunications. The firm may monopolise the market and still have resources to spare. The challenge to find something better to do with spare resources than give them back to the shareholder, is a familiar one in boardrooms.

Porter [1] points to the ability of modern nations to create new factors of production, such as a skilled labour force or a scientific base, rather than having to live within the classical resource limits. He draws attention to the importance of successful innovation and the ability to upgrade productive ability. Without a relaxation of resource limits, a firm which is already optimally organised cannot expand production in one direction without reducing it in another. Among the stock of resources that are important in telecommunications are some which derive from special factors related to the monopolistic structure of the industry. For example, licensing and regulation are such a central feature of getting into the market at all that skills in getting required permissions from regulators are of high strategic and economic importance.

18.2.2 Identification of strengths and weaknesses

A common way of seeking to identify competitive advantage is through the so-called SWOT analysis of strengths and weaknesses, opportunities and threats. These can be fitted inside the Porter framework. Porter looks to the constituents of national competitive advantage in four areas:

(i)   The stock and quality of resources available for production
(ii)   The strength of the home demand base.
(iii)   The existence or otherwise of internationally-competitive related and supporting industries.
(iv)   Legal and cultural conditions governing company formation and behaviour.

These can be translated into corporate terms fairly readily although the conclusions, especially on international matters, are not all that clear cut. They need amplification, especially as regards telecommunications, to stress the role of knowledge and market power as sources of advantage. Economic pressures bring out a variety of other strategic issues, often involving the mixture of competitive and collaborative relationships between companies, which may compete with each other for the business of private and corporate users while trading equipment and network services amongst themselves. These conflicts of interest are becoming more difficult to accomodate.

Benchmarking (Section 15.6.2) is a means of trying to improve the competitive strength of a company by building on the best practice found in other companies.

18.3 Customer strategies

18.3.1 The home customer base

In all developed countries there is at least one operator with a large home-customer base. Such operators find it easy to co-operate with each other in exchanging international telephone traffic, but success in other overseas operations has been more difficult to achieve. The large home telephone service market may make them rich enough to afford substantial overseas ventures, but it contributes little to their likely success. Domestic scale operations translate less easily into lower costs for telecommunications service ventures abroad than is the case for using foreign manufacturing facilities.

18.3.2 The big customers

A relatively small number of large companies dominate the demand side of the market for business services nationally and internationally. The companies want to cut their telecommunications costs by simplified management and lower tariffs. The high rate of return on international services makes multinational companies an obvious marketing target.

The operators serving these customers act defensively, to protect their home markets and offensively, to gain market share abroad. Many of the moves are well publicised, and there is a summary in McCarthy [2]. Profitability is assisted by economies of scope, in supplying a variety of services to the same customer. Scale economies arise in billing, the bulk supply of leased lines and, to some extent, in the bulk handling of their switched traffic.

Operators can save money for their large customers by improving the terms of business. Examples are:

Simplified management

- availability of complete international circuits without the need to deal with foreign operators;
- a multinational company may want a single bill covering the whole of its world operations: others want bills sent to individual management units; customisation allows bills to be presented in a form and detail best suited to managerial control;
- accurate billing (not a requirement confined to the largest companies); many companies want bills presented in a way that makes them easy to check, and they may hire specialist firms to find errors in the bills;
- availability of customised reports on the cost and performance of private networks; for the business customer, the bill is management information which can help in network planning and the control of costs;
- a single point of contact for central and regional managers to use. Customisation of systems to offer local features such as language or character translation, for example into Chinese script, has been used in overseas ventures;
- system transparency, allowing the customer to monitor traffic flows, faults and performance;
- easy interconnection of PBXs and other customer equipment with and between facilities supplied by different operators.

Tariffs

- discount structures that are easy to understand and apply;
- discounts which run across service boundaries, rather than there being a separate scheme for each type of service.

Many operators have entered the market, where regulation permits, with offers of similar services and tariffs aimed at the same large companies and most focus on international services. By 1996, companies such as the Australia's Telstra, Japan's NTT and many others had opened for business in London, New York and other world trading centres.

The operators have, on the whole, a narrower base in overseas operations than the multinational companies whose business they are competing for, and, as discussed in Section 1.7, they have had less experience of competition. Their main overseas business has been in dealings with other carriers abroad in bilaterally-monopolistic markets.

The operators offer four kinds of service improvement to their large customers:

(i)   Greater availability of key services including leased lines, managed data network services, data services (such as ATM) which are capable of handling bursty traffic with wide variations in required bandwidth and facilities for outsourcing of private networks and PBX management.

(ii)   One-stop shopping services, where a single operator will handle all the arrangements needed to set up an international leased-line or private network, and provide a comprehensive package of other services at a competitive price.

(iii) Broadly-based tariff discount packages, such as VPN.

(iv)   New services to meet the requirements of the big companies and replace in-house telecommunications management.

They seek to differentiate their services from those of others and establish a reputation for quality and reliability.

18.3.3 Exploiting the customer database

Computerised billing systems can be designed in a way that allows marketing information to be extracted from them. They are used as valuable tool in modelling demand (Section 16.1.1).

Billing software may provide operators with detailed analyses of the expenditure patterns of individual customers which, when correlated with other data, will assist in the bundling of tariffs to maximise network usage and revenue. If the billing systems are flexible, they can then be used in discount structures that are aimed at segments of the market. An example was MCI's Friends and Family initiative in the early 1990s, widely credited with taking market share from AT&T.

Computer telephony integration (Section 3.2.6) makes for more productive relations with customers. Extranets (Section 3.8.4) are becoming important in the development of strategies for sales over the Internet.

18.4 Strategic alliances and acquisitions

18.4.1 Overseas diversification

Most large operating companies have made overseas expansion an important element in corporate strategy because it affords a means of expanding into a business area which is outside the reach of domestic regulation. As many of them have found, making money abroad, especially in competitive markets, is harder than making it in the domestic monopolies to which they had been accustomed (Section 1.7).

British Telecom began to develop an overseas strategy in the early 1980s, culminating in the purchase of Mitel in 1986. This was part of a diversification strategy based on terminal devices coupled with an interest in establishing a presence in the North American market. Several other small businesses were acquired, followed by a minority stake in the cellular operator McCaw in 1989. Most, if not all of these holdings had been disposed of by 1995 and none proved to be the basis of an enduring business success. A separate business division which had been established to take a strategic interest in the developing world was also closed down. Little interest was shown in the eastern Europe privatisations, which were thought to be unprofitable, although BT remained anxious to develop a substantial and profitable business outside the UK. BT then entered into an alliance with the MCI which, it was hoped, would give entry into the US business market and be the basis of a world trading operation based on private networks and VPNs. Later plans for a merger with MCI were stalled by higher bids from two US companies, both of which apparently had greater synergies with MCI, putting the future of a BT alliance in doubt. BT was at this stage left with a set of other alliances, mostly with european operators. The company had substantial resources for investment abroad but, after 15 years of effort, overseas operations were not making any contribution to profit.

The BT experience shows how hard it is to create a profitable business abroad, even for a company with adequate finance, an accommodating regulator and a well motivated board.

18.4.2 Acquisitions as an alternative to fixed asset construction

Telecommunications operating companies invest heavily in their own networks. Where they wish to expand into other territories or activities they often try to take stakes in existing enterprises if there are any, because this is quicker and the outcome more predictable. Investments in new services, such as mobile operations, require a licence and the construction of fixed assets. A strategic alliance with partners in other countries offers a way of expanding overseas with relatively low risk. It does not require the construction of physical assets or the complexity and finality of a merger.

18.4.3 Alliances

Operators entering into alliances seek three advantages:

(i)   The chance to develop better products, based on end-to-end service provision, matching the requirements of multinational customers.
(ii)   Access to foreign expertise and customer bases, and help with the provision of international leased lines.
(iii)   Greater market power, enabling the capture of traffic from the dominant local carrier.

In 1996 there were four main strategic alliances:

(i)   Concert, formed by British Telecom and the US company MCI in 1994.
(ii)   Global One (ATLAS), formed by Deutsche Telekom, France Telecom and the US company Sprint in January 1996.
(iii)   Unisource, which is a alliance between PTTs to provide private networks across Europe using a mixture of satellite and leased line facilities and offering international resale opportunities.
(iv)   WorldPartners, formed by companies including the US company AT&T, Japan's KDD, Singapore Telecom and the Unisource consortium in 1993.

The alliances have been examined by national and regional regulatory authorities. Kiessling and Johnson [3] have described the criteria applied in Brussels, where two main aspects, corresponding to the objectives of the alliance partners, were investigated:

(i)   Technical efficiency and interoperability.
(ii)   The effect of greater market power.

In assessing the economic logic of the alliances for the operators, Hewlett-Packard's alliance-making rules, as described by Wagstyl [4], fit well with observations in Chapter 11 on mergers and above on comparative advantage. They include these:

- be specific about what you want of the partner;
- start with a narrow alliance and broaden it;
- always look for early success;
- go for the market leaders, not the followers.

Telecommunications alliances have not always followed these precepts. For example, profitability seems to have been difficult to establish in the early years. The last rule has been breached on several occasions, e.g. BT and MCI. Operators say that they are optimistic about longer-term prospects and see the alliances as a strategic necessity to protect their own markets. The willingness of some operating companies to change partners suggests that alliance strategies may not be stable or durable. This has been the experience in other industries and may be of economic benefit, since the assets used in arrangements which break down can be redeployed more readily.

18.5 Cross-industry mergers

Some companies have seen economies of scope in mergers which link network operations in quite different commercial sectors, for example the supply of electricity and water. The 1990s saw many mergers of this type, although others had existed long before. The main advantages expected are:

(a)   Technical factors, such as the use of waterways, tunnels, ducts and electric power lines to carry telecommunications cables.

(b)   A common customer base for billing, if the service area is the same. If different areas are linked, the joint billing costs may be still reduced.

(c)   Services can be jointly marketed.

(d)   In regulated industries, regulatory expertise can be enhanced and shared.

(e)   Regulatory risk can be diversified.

(f)   Headquarters costs can be reduced.

The strength of these advantages, and their ability to offset the classic disadvantages of diversification into unrelated industries, will vary from case to case and will be most strongly tested in product markets which are competitive. Regulators in the UK have tended to apply conditions involving the reduction of tariffs to their approval of such mergers, diverting some of the potential cost savings to consumers at the outset. Examples of cross-industry mergers and diversifications planned or executed in the UK and elsewhere are given in Sections 3.2.2 and 3.9.1.

There has been extensive merger and alliance activity among cable TV, telecommunications operators, entertainment and film library companies in the US. Most of it relates to the provision of television and telephony services to the home. Internet applications are becoming more important.

In some countries the multi-utility company (usually a multi-monopoly) has long been an established form. Mergers may not always hold an advantage. Bell Canada, which bought pipeline interests to diversify regulatory risks, later sold them as part of a corporate restructuring.

18.6 Demerger strategies

Large, diverse companies with extensive vertically or horizontally-integrated interests are not necessarily efficient or easy to manage. Some analysts believe that they are inherently inefficient, because they present management with problems which are beyond the scope of central planning to solve and which are best resolved through competitive processes and the insights of managers at lower levels who are closer to the market. Without necessarily accepting this view, it provides a background to the strategies which some companies have developed of concentrating on the core activity and selling off the ancillary businesses or establishing them as subsidiaries.

In telecommunications, regulatory pressures have added to the complexity of the management problem. AT&T broke itself up on 31 December 1983, releasing the local operating companies (Regional Bell Operating Companies) and concentrating its activities on long-distance and international service provision and equipment manufacture. A further split involving the sale of the equipment manufacturing business under the name Lucent Technologies was carried out in 1995. The motive was said to be to overcome the reluctance of the RBOCs to buy equipment from AT&T, a company with which they were developing strong competitive contacts. AT&T

had continued to seek the business of large business customers through direct links to its network, after the 1984 divestiture.

18.7 Diversification and the value chain

Each stage of the production and distribution process uses additional resources and is a potential source of profit. Most boards see an expansion of the business as the best way of serving shareholder interests and will consider diversification into other products and services as a way of achieving this. Telecommunications forms part of a wider industry using networks to deliver information and communications services (Section 1.1). Operators position themselves in this wider market taking into account their own strengths, weaknesses and ambitions.

Diversification into less familiar fields is not without commercial risk and may unduly stretch the attentions of management talent which is itself a scarce resource. Market dominance is an advantage which plays a part in the strategies of large operators. Theory suggests that diversification into areas where a competitive advantage is not significant may waste business resources. Alternative strategies such as demerging and sticking to the core competence can draw strength from the theory of comparative advantage.

The total value added, which is the difference between the cost of non-labour inputs and the value of outputs, can be broken down into a chain of increments running from stage to stage, sometimes called the value chain. The proportionate value added and the return on capital employed may, in imperfect markets, vary between the stages, leading companies to examine the value chain for profit opportunities and activities of doubtful value to the shareholder.

The chain is most easily analysed in macroeconomic terms, picking out the relative profitability of related sectors of economic activity. A presumption that the chain points to profit opportunities for particular companies cannot be made without considerable qualification. The company must have adequate competitive advantage to exploit the opportunity. Vertical integration can be represented as movement both up and down the chain.

18.8 Product and technology strategies

18.8.1 Products

The evolution of new products is an essential in most industries. Migration strategies are then used to move existing customers to the new product rather than losing them to a similar product produced by a rival. Occasionally a new product quickly becomes so essential that it establishes a break point in market evolution. It then gives the producer a substantial market advantage, particularly if protected by a patent and is an example of the discovery process of Section 6.1.2. The search for products and applications of this kind is especially strong in the computer and software industries. It is also important in telecommunications, where it may concern the means of delivering an existing service, or the creation of new ones. ISDN combines features of both, as does the intelligent network (IN).

18.8.2 Examples concerning service delivery

Operators have a choice of technologies with which to address their target market. It has always been necessary to make these choices, from the earliest days of telegraphs to the present day. Small operators and those addressing a segment of the total market will choose the ones best suited to the task in hand from the range of switching, cable, radio, satellite and computer-based alternatives available. Larger operators and those addressing a broadly-based market will use a wider range of equipment. Customers may have their own technology preferences for equipment that they buy or lease, especially if they are business users with extensive private networks. All will take into account the comparative advantage that comes from using a technology which they understand well.

In some markets the choice between cable and satellite transmission can be made on quality and cost grounds and most large network operators use both. Mobile communication requires radio transmission from the handset to a call-collecting point on the network, using a variety of media from there to the hub. Satellite operators usually specialise in their own technology. INTELSAT, for example, does not use undersea cables.

Competition between fixed and mobile networks, and the desire of operators to enter each other's markets can lead to mergers or collaborative ventures between fixed and mobile operators, and to attempts by fixed-link operators to incorporate mobile features into their services. Integration of fixed and mobile networks can bring economies of scale and scope, mainly in billing and marketing.

Mobile operators can readily enter the fixed-link market by leasing capacity from an operator and reselling it. Whether they can make money from the resale operation depends on the arbitrage opportunities between PSTN and leased-line tariffs (likely to be eroded fairly quickly by competition where it is allowed, especially on domestic services). Domestic and international calls to and from fixed links can be exchanged with fixed-link operators through interconnect agreements. A mobile operator could build and own fixed links, but would then be likely to experience diseconomies of scale unless the operation were a very large one. Fixed-link operators can and do have stakes in mobile operators and, depending on local regulation, may control them and achieve economies through fuller network integration.

18.8.3 Mobility features in fixed networks

Fixed-network operators embody mobility features to enable them to reduce or avoid loss of business to mobile networks. Most of them add cost and are offered as value-added services, narrowing the price difference as well as the mobility difference. One example is the cordless handset for use in the home and garden. A second is its development into one which could also be used as a mobile telephone elsewhere. Personal numbering (follow-me-anywhere) services divert calls to any prenotified telephone, fixed or mobile.

Calls from other public or private telephones can be billed to the home address by the use of calling cards where facilities exist. Universal personal telecommunications (Chapter 4) extends the principle. In either case, there must be software in the network to route the bill for the call to the appropriate address. This is more complex for international applications, which require collaboration between operators in different countries.

18.9 Network intelligence strategy

18.9.1 Levels of intelligence

There are different levels of intelligence in telecommunications networks and, at each of them, there are struggles for competitive advantage in the location and function of computer power. Industry competition between networks and terminal equipment existed long before the evolution of intelligent network (IN) structures. Now that IN has arrived, it may change the balance of advantage between terminal and network equipment. Following Mansell [5], the production of network-intelligence systems requires inputs from at least four different industry groups:

(i)   The PTO for the physical network and its services.
(ii)   The switch supplier for network equipment.
(iii)   The computer supplier for transaction processing and database equipment.
(iv)   The applications supplier for software.

By extracting the network intelligence from the network itself, IN offers at least the notional possibility of separate ownership. The domains in which the four industry groups operate are not mutually exclusive. PTOs (e.g. AT&T) have manufactured equipment in the past, although most have now stopped doing so. Computer suppliers (e.g. IBM) have provided telecommunications services. The boundaries that these ventures seek to cross are strongly contested and the ventures themselves have a mixed track record.

It is not unusual to find that closely similar customer requirements can be met by different types of terminal equipment. A nontelecommunications example would be the tape, disk, radio, TV and video presentations of popular music. There are similar situations in telecommunications. The suppliers of switches, computers and applications can put intelligent features into terminal equipment and private networks as well as the networks of the PTOs. The use of network software to provide mobility features in terminals which are fixed was mentioned above. Some other examples follow.

18.9.2 Centrex versus PBX

The features of internal company switching can be offered in two ways:

(i)   Having a PBX physically on the premises of the customer.
(ii)   Using the intelligence in the network operator's local telephone exchange to mimic the functions of a PBX. This facility, known as Centrex, originated in the US.

To a considerable degree, centrex was a child of regulation, springing from the rules which prevented Bell Operating Companies from manufacturing customer premises equipment. It is a way in which network companies without manufacturing capabilities can compete for the business of PBX manufacturers. Centrex has a strong market position in North America, but elsewhere has been less successful. The two means of offering service are not complete substitutes. Centrex users do not own the equipment and hence have less control over system costs and features, but they do not have to commit capital for equipment purchase. There is a residual trade-off here between their own cost of capital and that implicit in the price paid to the centrex supplier.

18.9.3 Operator strategies for private networks

The advantages seen by large business users in having their own private networks were outlined in Chapter 3. From the point of view of the PTO, private networks enable large users to be offered cheaper service than they could get on the PSTN without having to offer similar reductions to low users. The same advantage accrues from volume-related call tariff schemes and both stem from the cross-subsidy from call charges to line rentals. In such circumstances, private network growth can be encouraged if leased lines are priced nearer to cost, offering discounts in a way that may be more convenient for the customer. Since the dedication of physical parts of the network to an individual user is likely to reduce network efficiency, the PTO has an incentive to keep the traffic on the network by other means. Large telecommunications operators are trying to get part of the business back by offering to build and operate private networks on behalf of corporate users. The aim is to maximise the amount of value which is added by the operator. There are four elements to the strategy:

(i)   Selling PBXs and other terminal equipment.
(ii)   Planning and building private networks.
(iii) Offering PSTN equivalents of physical private networks, examples being the centrex and virtual private networks cases discussed earlier.
(iv)   Providing network-management services, taking work from the company telecommunications manager.

Success or failure will depend on the comparative advantage of the operators, many of whom have not enjoyed the confidence of corporate users in the past. Several factors have tended to strengthen the hand of the operators, mostly stemming from the increasing computerisation and intelligence of their networks.

18.9.4 Networked computing

The stand-alone personal computer replaced and decentralised mainframe computing systems. Online access was replaced by processors on the desk. The growth of networked computing has re-opened the contest for comparative advantage between centralised and decentralised computer power. Desktop systems are expensive to maintain. Industry estimates in the region of $10 000 per annum, several times the cost of purchasing the actual equipment outright, are not uncommon. Much of the cost arises from wasted time as people struggle with systems which are not up to the job, or beyond their skill.

Networking allows users to access better information, software, computing power and storage from a server with a relatively simple terminal device. The domestic television set is seen as a key piece of apparatus in developing networked computing for the home. Time will tell where the advantage lies. The factors likely to be of most importance are information quality, ease of use and, especially in a domestic context, cost. Telephone companies have a lot to gain from networking, as have cable operators, because it creates traffic.

18.10 Internet and multimedia strategies

More reliable methods of packet switching for voice traffic have increased the potential for overlap between the PSTN and specialist data networks, including the Internet. The overlap is one of management more than of hardware, since Internet traffic is carried mainly over leased lines outside the ground covered by LANs. Presumably if packet switching proves to be a cheaper way of transmitting voice traffic with the reliability and quality of existing voice networks, then the operating companies will be among the first to adopt it for this purpose. Indeed, some are already doing so. One consequence is that, in the longer term, the value added by the conveyance of voice will be reduced as competition erodes prices down to the new cost base.

IN architecture will need continued evolution to accommodate future broadband and multimedia services. Videoconferencing, audio/visual information services and the linking of computers on different sites are among the multimedia applications most often used by multinational companies. Others existing or under development include distance learning, image processing and applications based on virtual-reality technology. Following Colombo and Garrone [6], the multimedia value chain has five main elements:

- service provision;
- the access network;
- the transmission network;
- terminals;
- applications (entertainment, information, etc.).

The achievement of market dominance in any one of these elements could be very profitable. This has provided an incentive to mergers of, for example, access and tranmission interests. Network operators can provide these facilities in an efficient way. Movement up or down the value chain would take them away from their core competence and seems unlikely to be successful. The information services industry could begin to control telecommunications through Internet access dominance rather than remaining as an important customer, but it may not have the competence. Technological convergence may not be enough to outweigh the management and marketing problems of combining the content and conveyance markets in a successful merger of interests. If structural change does occur it is likely to come from the eventual dominance of data traffic, which is set to replace voice as the major force driving network evolution.

18.11 Supply-chain management

18.11.1 Procurement policies

Operators have used three approaches to the long-term minimisation of procurement costs:

(i) Vertical Integration.
(ii) Sourcing from a restricted number of suppliers.
(iii) Open tendering.

Whichever form of procurement is chosen, the relationship between supplier and customer requires careful attention, through customer service agreements, and by other means, if it is to be efficient. Relevant areas include quality, stock levels and other matters about which there is a substantial management literature.

18.11.2 Vertical integration

Examples of vertical integration between any or all of the links of the network operation/equipment manufacture/equipment supply chain have appeared from time to time. At its period of maximum corporate extension, the US giant AT&T manufactured telephone exchanges, built them into the national network which it operated and manufactured telephones and other terminal equipment which it then provided to customers.

Vertical integration is commonly used to reduce market uncertainty by providing greater security of market to the producer and of supply to the user. It has existed in other industries with a large capital base, notably electricity and gas, and in others such as brewing, but is commonest in monopolistic utility markets. Proponents of vertical integration have argued that gains arise from the sophisticated nature of technology and consumer demand, which require free information flows between manufacturer and network operator that would be impeded if there were an ownership boundary to cross.

The economies of scale in the development and manufacture of telephone exchange equipment are greater than the economies of scope in network operation which were discussed in Chapter 6. The smaller networks may have low unit costs for the network traffic which they handle. They cannot present a large enough market for efficient in-house manufacture, however, unless they can sell to other operators or produce designs from large manufacturers under licence. Both patterns may be found. Bell Canada Enterprises' manufacturing division, Northern Telecom (Nortel), has built up a significant export market across the world. In 1995, Nortel sold 94 per cent of its products outside the BCE group. AT&T, Northern Telecom, Alcatel and others have in the past established small manufacturing subsidiaries with close links to the main local operator in many different countries.

Equipment manufacture in AT&T derived from the pioneering engineering work of Alexander Graham Bell and continued with the geographic extension of its monopoly under Theodore Vail. The large captive home market enabled a powerful research, development and manufacturing capability to be built up. Competition has changed the situation. As discussed in Chapter 9, vertical integration can produce conflicts of interest within companies competing with each other in one layer, such as network provision, but trading with each other at another.

Olley [7] describes the evolution of manufacturing and R&D in Canada, showing the importance of efficient information transfer. Bell Canada began manufacturing in the late nineteenth century, using Western Electric technology, after difficulties with imports and independent manufacturers. Technology transfer was effective but it left Bell Canada heavily dependent on Western Electric, and its operating practices were based on those of AT&T. By the mid 1950s Bell Canada and its manufacturing subsidiary, Northern, had begun their own R&D. Then, following a breakdown of arrangements for technology transfer from Western and AT&T, BCE chose to develop its own technology and retain manufacturing. AT&T had made its first electronic switch, the ESS1, by 1963. Bell Canada found the switch too large for much of the Canadian system and decided to create a smaller design of its own. An arms-length relationship with Northern was created, including a supply contract holding purchase prices at or below the level of other manufacturers, backed by joint control of R&D to match the product to Bell's needs. Northern established a large market outside Canada, enabling it to gain economies of scale, with competition to keep it efficient.

18.11.3 Sourcing from a restricted number of suppliers

Purchase from indigenous suppliers can bring long-term security of supply for equipment needed to construct new exchanges and extend old ones, as discussed in Section 10.4.2. It also avoids currency risk. These potential advantages can seem persuasive to a company anxious to maintain the integrity of its system, as they can to a government willing to support national champions. A stronger restriction is the use of a closed group of suppliers, usually all indigenous, practised in the UK, France, Germany and Japan to encourage domestic manufacture.

Most operators buy on the world market. Equipment suppliers may be able to use their market power to achieve good prices at home and sell on the margin abroad. This supports the theoretical expectation that any reduction in the number of potential suppliers brings other risks by reducing competition for supply contracts, a possibility sometimes countered by putting a proportion of the contracts out to open competition.

British Telecom, like the Post Office before it, initially bought the bulk of its network equipment from large independent suppliers, usually British based. The relationship with domestic suppliers has often been a troubled one. Harper [8] describes the problems from an inside perspective.

In the 1950s the UK Post Office had bulk-supply agreements under which 90 per cent of the tenders were put out to a closed group of suppliers. The group allocated market share between its members and they were all given the same price. The arrangements were modified in 1960 to permit 25 per cent of telephone apparatus to be procured by competitive tender from firms outside the group.

In 1962 the Post Office negotiated reservation clauses allowing it to place 25 per cent of telephone apparatus and ten per cent of exchange equipment out to open tender and obtained prices some 20 per cent lower than the agreement prices. This was thought to overstate the savings obtainable with competitive tendering, which was assessed at only eight per cent in the longer term [9]. The bulk supply agreements were terminated in 1969, but it was not until 1983 that switching tenders were put onto the world market. The Swedish company L. M. Ericsson was successful. A price ring which began with the end of the bulk supply agreements was discovered among the cable suppliers in 1975. It was eventually settled with an out of court compensation payment of £9m [8].

The development of Japanese procurement policy is described by Kawamata [10]. Japan was dependent on foreign switching suppliers before 1939, and the Ministry of Communications ran telecommunications and established joint ventures with foreign manufactures, to achieve technology transfer, and local manufacturing capacity. Two different systems were procured, to keep the market competitive. After World War II, the Ministry commenced restructuring and established Nippon Telegraph and Telephone (NTT) as a public utility in 1952. NTT's switching specifications made it difficult for foreign firms to break into the market, which was shared by a group of firms known as the NTT Family. Complaints made by the US about what were perceived as restrictive procurement practices by NTT [11], were similar to many made by other countries and manufacturing sectors in a wider context where cultural factors played an important part in Japanese buying behaviour. More recently, there has been a greater willingness to buy imported equipment although the proportion remains low.

In the 1980s AT&T in the US and Bell in Canada both had powerful and efficient manufacturing facilites and used their products extensively in their own networks. In these two cases there was pressure from regulators and shareholders for cost minimisation in procurement, reducing the risk of a barrier against cheaper foreign equipment.

18.11.4 Open tendering

The introduction of competition into final product markets puts pressure on the economics of vertical integration by making it more difficult to pass any excess costs on to consumers. OFTEL [12] discussed BT's change to open tendering, concluding that it was a reasonable commercial decision and that it need not destroy the basis for the success of the System X design, which was designed and built in the UK. The OFTEL report urged that the UK content of the Swedish System Y, which was to be purchased, should be maximised. Having selected the Swedish system, produced by Ericsson, BT had a second supplier, replacing a monopoly by a duopoly, and could play off one against the other. Further diversification of switch supply was limited by factors discussed earlier.

Fransmann [13] sees open tendering as providing the best supplier and inducing innovation. Disadvantages are seen as being uncertainty of supply and a limited incentive to invest in transaction-specific assets. BT's analysis at the time was similar. Open tendering is the norm for the US Independents and, since 1984, for the RBOCs. New entrants to the British market have acted similarly for transmission equipment. Switches have been chosen more for the facilities offered by their software, where the choice may be rather limited.

18.11.5 Outsourcing and the virtual company

By relinquishing the ownership of assets, outsourcing raises the relative importance of other, less tangible, factors of production, organisation and enterprise. Some management theorists have made this the basis of the justification of a new type of enterprise: the virtual company. This has minimal physical assets and human resources and uses information technology to manage its largely cerebral outputs. Such companies would represent an extreme case of supply specialisation, adding value through the generation of ideas, research, management and marketing, and leaving all physical processes to equally specialised suppliers, which might be based in developing countries. Virtual companies would be specialising in the supply of customer and market information and of intangible factors of production, such as enterprise, found at each stage of the process.

Outsourcing may reduce the capital base of the firm as well as the number of its employees, potentially reducing a source of added value and profit.

18.12 Strategies for research and development

18.12.1 R&D statistics

Equipment suppliers spend a higher proportion of their income on research and development, sometimes equalling or exceeding the really big spenders in pharmaceuticals, electronics and computing, see Table 18.2. A Bell Canada study found that the volume of R&D expenditure varied from 2.2 % (CIT–Alcatel) to 13.7 % (Plessey) of turnover among the 15 largest manufacturers in 1978 [14]. At this time, Plessey was heavily engaged in the development of the System X digital switching system.

Table 18.2 R&D as % of sales

Company	Year	%	Sector
Alcatel	1994	18.0	Equipment
Ericsson	1993	16.0	Equipment
Nortel	1995	14.6	Equipment
Italtel	1993	12.0	Equipment
Cabletron (equipment)	1993	11.0	Equipment
Siemens	1988	10.9	Equipment
Mitel	1993	10.0	Equipment
GTE (equipment sales)	1993	10.0	Equipment
Ascom	1993	9.0	Equipment
Phillips (Netherlands)	1988	8.1	Equipment
Motorola	1988	8.0	Equipment
Harris	1993	8.0	Equipment
NEC	1993	7.6	Equipment
AT&T Total	1988	7.3	Services and equipment
IBM	1993	7.1	Equipment
Cisco	1993	7.0	Equipment
Racal (equipment)	1993	7.0	Equipment
Nokia	1995	6.9	Equipment
NTT	1995	4.5	Services
AT&T Communications	1994	2.0	Services
BT	1997	1.9	Services
Stet	1988	1.9	Services
Telefonica Group	1995	1.0	Services
RBOCs in USA	1994	0.3	Services
Belgacom	1995	0.0	Services

(Source: company annual reports, Mosbacher [15], Rao [16], Communications International, January 1994)

Research and development makes up a large part of the cost of bringing a new switching system to the market. Software development is a major part of the cost many new services. For network operators it represents a diversification away from the core business, justifiable to the extent that it can be built on comparative advantage. The advantage may not be sufficient to overcome extra costs in other directions, including:

- scope and scale economies available to others serving a larger market;
- lack of the necessary competence.

Bell Canada's R&D strategy has been discussed above in the context of vertical integration. The company decided to end its dependence on AT&T and Western Electric technology 1956 and an independent research facility was created. Central laboratories were set up as a separate corporate unit in 1957. Bell Northern Research was established in 1971, taking over research functions from the operating and manufacturing units and serving them both; contract work was also carried out for third parties. With the manufacturing arm forced to make sales on the world market to generate sufficient economies of scale, the strategy managed to create a world-class research and product capability. In Table 18.2, all R&D in the BCE group is charged to NORTEL. There is none shown against the Bell Canada operating companies in the BCE accounts.

Elsewhere, only the largest network operators now carry out a significant amount of R&D. They use it to be able to specify the most up-to-date systems, to negotiate for their purchase and to assist in the planning of networks and markets. Their R&D expenditure is subjected to severe market tests. NT&T stands out among the others as the highest spender, reflecting its continuing role in steering the direction of technical development among the family of suppliers from which it buys most of its equipment. NTT took control of the government R&D when it was created in 1952 and established an independent technology which promoted domestic industry. The company continues to be closely involved with industry R&D, underpinned by statutory obligations to carry it out and disseminate the results. Smaller companies spend a very small proportion of their income on R&D and most of them do not consider it worth identifying the amount in their annual accounts.

8.12.2 R&D strategies in the UK Post Office and British Telecom

The UK Post Office had a substantial R&D function in the 1960s, with some 3000 staff and an expenditure which was nearly two per cent of the total telecommunications operating cost. Most of the work was used to support the design of switching and transmission systems. Equipment was at the time purchased from a closed group of firms under the bulk supply agreement referred to above. The control exercised at the time by the Post Office over the suppliers in the matter of design was tight [9]. Control developed further in the evolution of the System X digital switching system.

In the years up to 1975, R&D expenditure was typically under two per cent of operating cost and about three per cent of total capital expenditure, and switching R&D was about three per cent of capital expenditure on telephone exchanges. There was then a sharp increase focused mainly on digitalisation. By 1982, the last year for which detailed figures were published, switching R&D had risen from £9 m to £87 m and represented 13.6 % of the value of exchanges purchased, see Table 18.3.

Total R&D spending moved to over ten per cent of total capital spending and was still at that level in 1995. The increase was not accompanied by a proportionate increase in the number of research staff. Much of it was in the form of research contracts put out to suppliers, together with intellectual property rights. MMC [17] reported that £60 m of the £182 m total in 1984–85 consisted of contracts with industry and a small amount of research financed at universities. NTT also funded substantial research on behalf of its suppliers, as did the French PTT.

Research and development strategy was re-appraised by British Telecom in the years up to the 1984 privatisation. Internal and external sponsorship was made more rigorous. Much of the effort was carried out under contract to operating divisions. Intellectual property rights were more carefully guarded.

British Telecom attempted to find an outlet for its own research and development capability through external ventures. A joint venture, BT&D Technologies, with Dupont was set up to exploit BT's expertise in opto-electronics. The purchase of a majority holding in the Canadian PBX manufacturer MITEL in 1986 gave BT experience in manufacturing without getting into the cost of manufacturing equipment as large as that needed for the PSTN. Mitel had a significant share of the North American market and interests in many other countries.

Table 18.3 Research and development expenditure by UK operators, 1969–95

Year ending March	Total R&D (£m)	Total operating costs (£m)	Total capital expend. (£m)	Switching R&D (£m)	Switching capital expend. (£m)	Total R&D % of operating	Total R&D % of total capital expend.	Switching R&D % of switching capital expend.
1969	8.3	518	330	2.9	100.4	1.6	2.5	2.9
1970	11.8	591	376	5.5	111.2	2.0	3.1	4.9
1971	11.9	692	432	3.0	127.0	1.7	2.7	2.3
1972	17.1	826	518	5.2	162.1	2.1	3.3	3.2
1973	19.0	1 012	627	5.5	209.2	1.9	3.0	2.6
1974	20.7	1 222	712	6.5	266.4	1.7	2.9	2.4
1975	27.7	1 583	775	9.4	293.5	1.8	3.6	3.2
1976	39.3	2 012	882	15.1	306.6	2.0	4.5	4.9
1977	50.7	2 293	840	23.3	292.8	2.2	6.0	8.0
1978	55.8	2 597	860	24.7	275.2	2.1	6.5	9.0
1979	73.4	2 908	1 002	42.4	360.3	2.5	7.3	11.8
1980	80.9	3 430	1 224	54.3	450.6	2.4	6.6	12.1
1981	123.4	4 374	1 574	75.7	594.3	2.8	7.8	12.7
1982	157.6	4 701	1 387	87.2	641.0	3.4	11.4	13.6
1983	172.0	4 843	1 500	N/A	662.0	3.6	11.5	N/A
1984	179.0	5 342	1 513	N/A	586.0	3.4	11.8	N/A
1985	182.0	5 778	1 822	N/A	643.0	3.1	10.0	N/A
1986	161.0	6 269	1 945	N/A	698.0	2.6	8.3	N/A
1987	190.0	6 990	2 056	N/A	727.0	2.7	9.2	N/A
1988	195.0	7 576	2 379	N/A	712.0	2.6	8.2	N/A
1989	214.0	8 264	2 947	N/A	852.0	2.6	7.3	N/A
1990	228.0	9 105	3 115	N/A	952.0	2.5	7.3	N/A
1991	243.0	9 623	2 758	N/A	799.0	2.5	8.8	N/A
1992	240.0	9 922	2 446	N/A	722.0	2.4	9.8	N/A
1993	233.0	10 806	2 155	N/A	545.0	2.2	10.8	N/A
1994	265.0	10 660	2 171	N/A	493.0	2.5	12.2	N/A
1995	271.0	11 200	2 638	N/A	605.0	2.4	10.3	N/A

Percentages calculated from unrounded data

(Sources: Post Office and British Telecom Annual Reports. Figures relate to these only)

The MMC allowed the merger to go ahead but laid down stringent arms-length conditions because of BT's dominance in the supply of small and medium-sized PBX's in the UK:

- BT could not use Mitel equipment in its own network or sell it to end users;
- Mitel could not use a BT logo on its products;
- BT and Mitel could not make joint use of any resources.

These largely destroyed the scope for any synergy between the two operations. Even without the conditions, integration of the two companies may not have yielded enough economies to overcome differences in the product markets and corporate cultures. BT gained valuable experience of acquisition hazards at an affordable price, and the holding was disposed of in 1992.

18.13 References

1   PORTER, M. E.: 'The competitive advantage of nations', *Harv. Bus. Rev.*, March–April 1990
2   McCarthy, C.: 'Battle of the giants', *Commun. Int.*, June 1996, pp. 63–64
3   KIESSLING, T., and JOHNSON, G.: 'Strategic alliances in telecommunications – an economic analysis of recent European Commission Decisions'. Paper presented at ITS 11th biennial conferece, June 1996, Seville
4   WAGSTYL, S.: 'When even a rival can be a best friend', *Financial Times*, 22 October 1997
5   MANSELL, R.: 'The new telecommunications' (Sage Publications, London, 1993)
6   COLOMBO, M., and GARRONE, P.: 'Common carriers' entry into multimedia services'. Paper presented at ITS 11th biennial conference, June 1996, Seville
7   OLLEY, R. E.: 'The process of technology transfer and application in telecommunications: a case study', in SAHAL, D. (Ed.): 'The transfer and utilisation of technical knowledge' (Lexington Books, 1984)
8   HARPER, J. M.: 'Monopoly and competition in British telecommunications' (Pinter, London, 1997)
9   'Comptrollers and auditor general's report, Post Office report and accounts 1963/63'. General Post Office, London, 1964
10   KAWAMATA, T.: 'R&D investment and spillover of R&R in Japanese I&CT industry: perspective of the NTT family R&D system in the era of multimedia'. Paper presented at ITS 11th biennial conference, June 1996, Seville
11   'World trade news digest, US appeal on Japan telecoms', *Financial Times*, 7 August 1997
12   'British Telecom's procurement of digital switches' OFTEL, London, 1985
13   FRANSMANN, M.: 'AT&T, BT and NTT: vision, strategy, corporate competence, path-dependence and the role of R&D' in POGOREL, G. (Ed.): 'Global telecommunications strategies and technological changes' (North-Holland, Amsterdam, 1994)
14   'On the need for large size and vertical integration in the research and development and manufacturing of telecommunications products'. Bell Canada Special Task Force, paper presented in Paris, June 1980
15   MOSBACHER, R. A.: 'US telecommunications in a global economy: competitiveness at a crossroad'. Report of Secretary of Commerce', Washington D.C., 1990
16   RAO, P. M.: 'R&D and innovation in US telecommunications: recent structural changes and their implications'. Paper presented at ITS 11th biennial conferece, June 1996, Seville
17   Monopolies and Mergers Commission: 'British Telecommunications PLC and Mitel Corporation Cmnd. 9715' (HMSO, London, 1986)

Chapter 19

Diversification into less developed countries

19.1 Commercial background

Liberalisation and privatisation trends around the world have opened up many opportunities for operators to invest in the development of foreign networks. This type of operation carries a relatively high risk, but it has the advantage of being free of domestic regulatory constraints. The long-standing agreements which, for example, C&W has with governments in Commonwealth countries are not representative of the political and operating conditions likely to be encountered in more turbulent parts of the world.

Developed and underdeveloped countries have different perspectives on their problems, but the nature of the business is the same. There is common ground on:

(*a*)   The fundamentals of telephony economics and technology
(*b*)   Conversion from analogue to digital technology.
(*c*)   Economies of scale.
(*d*)   The need for heavy capital investment.
(*e*)   Long time horizons.
(*f*)   The value of telecommunications to the community.
(*g*)   Exchanging traffic with other operators and, in some cases at least, competing with them.
(*h*)   The frequent existence of local monopoly, with only one operator providing the local loop.
(*i*)   Even monopolists face competition in two areas:
    1   The capital markets. Operators have to compete with other borrowers for funds, from government or in capital markets.

Their competence and financial soundness will affect the cost of borrowing, and perhaps make it impossible if lenders think they are bad risks.

2   In regulatory bodies, where their performance and efficiency may be compared with those of monopolies in other areas. A poor position in the comparisons may create serious difficulties for management.

Telecommunications services are potentially profitable at all stages of economic development. It takes a wilful act of politics to drive them into loss, although some governments have managed to achieve it through tough price controls.

As with all, or at least most, material things, society can have too much telecommunications as well as too little. The underpricing of service stimulates demand among customers not able or willing to pay the full economic cost. Some underdeveloped countries show an abnormally high penetration rate for their national income, which can be shown to derive from artificially low prices. The role of the entrepreneur is to find the markets where demand can be met with tariffs which fully cover the cost of the resources employed and within these to find the cases where business prospects look best.

19.2 Technical background

19.2.1 Modernisation

After the collapse of their governments, the networks in the old command economies were often underdeveloped, having been designed to prevent, rather than facilitate, effective communication. Decades of market failure under communist regimes, combined with the COCOM (Coordination Committee for Multilateral Export Controls) rules of the US kept modern equipment and services off the market. Sometimes the networks were overextended, with demand encouraged by low tariffs. Either way, there was a large suppressed demand for higher quality and for modern services with better international links. The networks themselves were sometimes in good condition, maintained by careful engineers but overloaded and without the facilities needed to provide modern services.

New investment may be for system growth, quality enhancement or the replacement of old plant. The existing equipment in the old command-economy networks usually needs complete modernisation and replacement. Migration from old plant to new requires careful planning if it is to be carried out in an affordable way. This is a difficult task even in developed economies, where the digitalisation of local exchanges may proceed at a slower rate for residential customers than for businesses. Older exchanges can sometimes be made to simulate the services available from digital equipment by the addition of extra pieces of electronic equipment. The preference, however, is likely to be for new digital systems. Modernisation will stimulate calling rates, perhaps dramatically so, by reducing congestion and introducing direct dialling.

Governments will sometimes formulate network modernisation targets as part of a regulatory system or when specifying conditions for the privatisation of a substandard network. Various objectives may be set in relation to network modernisation, including timetables for the full deployment of such elements as digital transmission, digital trunk and local switching, ISDN availability, subscriber trunk dialling and international direct dialling. The roll-out targets that are initially presented by local managements may be too optimistic and unfinance-able from likely demand. They may be a poorer investment, at the margin, than other forms of development. Modern equipment bought in hard currencies on the world market can be expensive compared with the book cost of the obsolete equipment which it is to replace.

One exmember country of the USSR publicised plans for a billion dollar investment package to come from a foreign partner, but it eventually settled for less than a fifth of that figure. What it is obtaining looks like being good value. It will get a modern business and trunk network and the eventual elimination of gross overloading on the local network by investing in better exchanges and the introduction of local call charging. The public-service potential of payphones can be exploited to put basic telephone services within the reach of everyone without big development costs.

19.2.2 Radiocommunications

There are various roles for radio fixed and mobile communications, distinct from each other and sometimes from western countries:

- to bypass the local network, where service is bad or rapid roll-out is required;
- to secure privacy – an odd claim to those in Britain but one which has been made in India;
- to improve international communications, in small countries, assisted by GSM or by handover to an international operator for transmission over special lines at premium rates.

Tariffs may be set at western levels even though costs are eastern. Terrestrial radio may be used in the local loop and for trunk and international transmission. Its effectiveness for long-distance transmission was demonstrated by MCI, which used it as a quick and economic way of providing trunk capacity when competing with AT&T in the US. It has been used by dominant operators and monopolists as a long-term way of providing service in remote areas but it does not usually prove to be economical against cable on denser routes unless the terrain is especially difficult for land lines.

There is growing interest in the use of radio in the local loop, either by the adaptation of mobile technology to use fixed terminals, or by the use of equipment which has been specifically designed for fixed-link communications. Costs have fallen to a level where they are fully competitive with cable over certain types of terrain and further falls are expected. Telefonica used radio to extend service to remote Spanish villages (Section 7.3.7). This was in the early 1990s, when under pressure to extend universal service provision. There are many systems in developing countries.

19.3 The legal framework

Even in developed western countries, such as the UK before the privatisation of telecommunications, the legal framework for operations can be vague and arbitrary. In 1980 Britain, for example:

- there was no explicit statement of universal service objectives;
- no case law on unfair telecommunications trading practices existed;
- customers did not have a service contract with the operator;
- many of its operations were protected from legal challenge by Crown immunity.

Uncertainty about the legal framework, particularly on such matters as tariff regulation and the right to repatriate funds, increases the undiversifiable market risk for a potential investor and hence the rate of return required from any investment made. The perceived risk may be so high that, in the imperfect state of capital markets which usually exists, no investor will be willing to commit funds.

Although not usually a problem associated with developed countries in the West, it may sometimes be encountered there. The *Financial Times* of 3 June 1996 carried a report that the chief executive of Tractabel, a multinational Belgian utility, had told the newspaper that the UK had become too unpredictable for his company to invest in the energy sector because of instability and political involvement in the regulatory systems.

The minimum legal framework that is needed to bring the risks down to levels which will draw investment finance into a developing country includes:

- a telecommunications law;
- a licensing system specifying the rights and obligations of operators;
- property rights concerning who owns what, and who is authorised to sell it;
- clarity about how tariffs are fixed;
- an arbitration procedure for disputes.

Locally there are mixed views about the benefits of network competition. There may be a national monopoly, as in Latvia, or there may be regional monopolies, as in Hungary.

The terms of reference of regulators are rather variable. Tariffs in the USSR were usually set by independent local committees. The distinction between setting and monitoring tariffs is not always well understood in the new regulatory bodies, where they exist. Some Russian operating companies have complained that these committees are damaging liquidity by preventing them from disconnecting customers who do not pay their bills. The situation varies from committee to committee.

19.4 Costs and tariffs

19.4.1 Financing the service

The prime requirement for tariffs is that they are high enough to fund service. A firm commitment to this principle will be wanted by the investor. Western companies are likely to look at both the cost of capital and the rate of return in real terms when appraising an investment. They may choose to do this through accounts expressed in US dollars at current exchange rates. The chance to make extra large profits in the short term can be important for the attraction of foreign capital to local investments.

Practical implementation may have its difficulties. Cross-subsidies may be rife and need rethinking in the context of a commercial network. When one of the states in the old Yugoslavia negotiated development finance, it was necessary for the government to prepare a tariff policy at high speed. The job was done and it covered such matters as the frequency of tariff revisions, links to the consumer price index, a medium-term commitment to aligning tariffs towards cost, a rate of return high enough to meet the cost of capital and non-discriminatory pricing. The initial commitment of management was obtained in an intensive seminar where ideas where solicited, presented and discussed. An action plan for implementation included milestones against which to measure progress.

Tariffs in the Russian Republics and most East European countries bore little relation to cost until recent reforms. A 1994 survey of 21 local Russian operating companies showed monthly rentals of 1000–2500 roubles (£0.33–£0.83) for residential users and 11 250–39 000 (£3.75–£13.00) for business. In Latvia business users used to pay more for their lines than residential users. 1993 tariffs were 4.44 Lats (£5.55) per month as against 0.6 Lats (£0.90) for residential users but they have since converged.

Russian economic theory of the Soviet period said that users should pay according to the benefit which they obtain (value-of-service pricing) and that this was eight times higher for business users than residential users. Even western economists might see some sense in this. Current theory on socially and commercially optimum tariffs points to demand-sensitive (Ramsey) prices. The literature of price discrimination points to different prices in different markets as enhancing welfare.

Schenk, Kruse and Müller [1] mention problems encountered in the Czech Republic, Hungary and Romania over tariff levels, which failed to move adequately towards cost or to reflect inflation in the early 1990s, prejudicing the profitability of inward investment. Since inflation arises from an attempt by the government and the citizens to spend the same money, this is an indication that modernisation plans may be too ambitious for private values. Another indication lies in the waiting lists, which were high in the Czech Republic, Hungary and Poland from 1990–94 despite substantial system growth [2]. Table 19.1 shows that the waiting lists were probably stimulated by low tariffs.

Table 19.1 Rentals, revenues and waiting lists in 1993

Country	Waiting list as % of systems size	Residential rental ($ per month)	Revenue per line ($ per annum)
Czech Republic	29.2	1.8	297
Hungary	51.5	4.6	393
Poland	51.4	3.7	332

(Source: ITU [3])

Czech and Polish revenues are not out of line with the average for their income level but Hungary is low. None of them looks adequate to sustain the capital charges on new network investments of perhaps $1500 per extra line.

19.4.2 The control of prices and profits

Little of the data used to regulate tariffs in a developed country may be present in one with poor statistics, a chequered political history and rampant inflation. Although a foreign investor will want an adequate return on capital from the investment made, there may be little perception, among the local managers, of the true cost of capital, or the need to spend it where it yields most benefit at the margin. Basic understanding may be hindered by the use of accounting systems inherited from the old regime.

It is difficult to make accurate estimates of the cost of capital to a foreign investor and excessive downward pressure on it may make the investment look less attractive. Practical implementation of rate-of-return controls is likely to have to be in terms of a range, preferable agreed in advance with the operator:

- a minimum, related to the cost of capital as far as can be estimated;
- a maximum, beyond which the profit rate looks unreasonably high by comparison with similar enterprises;

However arrived at, the method must take risk and inflation into account. There seems little alternative for the first few years but to work on these rather subjective lines, but it may be the only chance to retain the confidence of the operator, the user and the regulator. There should be an incentive to improve operating efficiency.

19.4.3 Cost structures

Central and eastern european operators face sharply different cost structures. In 1993 Latvia, for example, average earnings were 48 Lats (£60) per month. Capital assets per line were approximately £66 and income £50. This compared with assets of £585 and income of £513 at British Telecom.

New exchange equipment might cost at least £100 per line which, if depreciated over ten years and financed by borrowing at 20 per cent per annum, could generate an annual charge of around £30, or four per cent of average earnings.

Even a few staff at western rates can have dramatic effects on the pay bill and required tariff, because of low local pay rates.

There may be ambitious plans for developing the service, but what are they worth at the margin? Will the next available $100 be best spent on:

- a new exchange line;
- a better electricity network;
- better TV.

If market intervention is attempted a cost-benefit analysis will be needed to identify the way in which maximum utility can be obtained. The analysis might identify social and third-party benefits, as the Maitland Commission did for countries with substantial low-income rural populations [4].

19.4.4 Quality of local staff

There are plenty of able people and they can be quick learners. This is most obvious in the more modern networks, especially mobile. Some of the new entrepreneurs have made a great deal of money quickly, and the better ones know how to act quickly when things go wrong. Investors may find more problems with older networks where staff may be lacking in basic management skills:

- management accounting;
- financial accounting;
- investment appraisal;
- marketing and sales management;
- efficiency measurement;
- econometrics.

Often they will be looking back to a world of certainties which has vanished even in the West. For success with investment projects, management expertise is needed as well as cash. This may be hard for the existing managements to accept. It is not uncommon to hear that they can do a perfectly adequate job given the funds. This may lack credibility with the investor without a good business plan and a plausible way of carrying it through.

19.5 Partnership, trust and control

19.5.1 Control and management

An essential problem with overseas investments and joint ventures is the matter of control. In most industries, a 20 per cent stake would be regarded as inadequate for control but too big to be a trade investment. The track record of joint ventures in other industries suggests that a majority holding is needed to get effective results for the investor.

Governments are rarely happy about letting their major telecommunications operator fall to the control of a foreign company and may protect them from takeover without explicit consent (Section 17.2.3). The transfer of outright control to US and european investors has been frequent in South America (Table 10.1) but is resisted as a form of imperialism in many African countries. This is a problem of political economy (Section 1.1.1). Control is therefore a critical element in overseas investment and the issue needs to be faced squarely. In 1998, Cable & Wireless began to pull out of foreign investments which it did not control. There may be compromises, of which these are examples:

- majority ownership can remain with the national government, safeguarding the national interest
- a management contract can give operational control to the investor, safeguarding the investor.

Whatever is agreed, there must be a clear basis in telecommunications and commercial law.

At the end of the day the absence of a track record means that there must be a large element of trust in the deal between investor and government. The investor must obtain the best contractual basis which can be achieved. The national government must avoid excessive suspicion about the bona fides of the investor. Both must seek the common ground where the investor makes an adequate return and the government gets the services that it wants.

Small things can make a big difference. Publication of commercial accounts in the local language as well as the language of the investor is an elementary courtesy. It is not unusual for there to be false starts on the route to foreign involvement. The partial privatisation of the Greek operator OTE was advanced and delayed several times before eight per cent of its equity was sold in 1996. The allocation of mobile licences in Poland has suffered similar oscillations.

19.5.2 Stakeholders

The telecommunications operator usually has a widely distributed customer base, so it is essential to satisfy the reasonable ambitions of the all the stakeholders, the requirements of which were discussed in Chapter 13.

19.5.3 Taxation

The local tax regime needs careful investigation. Social security taxes, for example, may be a high percentage on the pay bill. Rates of 37 per cent or more may be encountered. Investors in Russia have complained of steadily rising corporate taxes in a climate where anti-Americanism appears to be rising.

19.5.4 Local equipment manufacture

Local manufacture of telecommunications equipment is a feasible and popular option in some of the larger countries and may assist with transition problems. Many countries are too small to support a viable independent operation. Also, there may be a problem with attempts to use existing facilities, where products are now obsolete and management not ready for change. Latvia was a major supplier of PSTN switching equipment to the USSR before political changes and currency realignment made the market inaccessible. The home market is not large enough to support a significant operation, but prospects could change if there were better political relationships with neighbouring countries and export prospects improved.

19.5.5 Use of available statistical sources

Any quantitative study of an investment opportunity needs to be based on the best available source material. There may be surprisingly good local data on traffic, staff and other operational statistics – better than can be found in many western countries. Good demographic and social statistics may also exist and a check should certainly be made with the government statistics office. A £10 investment in the latest yearbook can be money rather well spent. Data on interest rates and inflation should be available from one of the major banks.

Interviews with major users can give a useful perspective on local services and operating conditions. Local representatives of foreign banks and airlines may be particularly useful. Local opinion formers should also be consulted, especially if they are likely to have anything to do with present or future regulatory systems. There may be a good deal of information outside the country. ITU statistics [3], for example, provide country information about many aspects of local services and a 1994 publication included projections of future investment needs. The figures are updated at intervals. Although these projections may have been derived locally and may be hard to substantiate analytically, they are a useful reference point against which to judge investment proposals.

Gwartney, Lawson and Block [5] report on studies of the relationship between economic performance, measured by GDP and economic freedom, measured with a composite index combining a group of indicators covering law, regulation, tax, international exchange and market-entry restrictions. They find a significant positive correlation between freedom and economic growth. The study covered countries at all stages of development and is consistent with other studies of the difference between, for example, parts of the Far East and Africa.

19.6 Criteria for investment

19.6.1 General considerations

There are general investment criteria:

- risk assessment;
- cost of capital;
- comparative advantage;
- market identification;
- priorities;
- raising finance.

Investors are well able to take care of these. Some have been discussed earlier and they will not be elaborated here. There are, however, a few other matters.

19.6.2 Affordability

Telephone bills have to be at an affordable level if the investor is to avoid antagonising customers, regulators and the government. Typical western telephone bills are in the region of one to three per cent of family income. Telephone revenues as a percentage of GDP may be a useful indicator of the realism of tariff proposals. It is rare for this to exceed around three per cent, although it may be more in poorer countries. Figures range up to six per cent in developing countries, e.g. Belize, Guyana and several others in 1991. Rentals are a bigger percentage of earnings in low-income countries, where the penetration rate is lowest. Figure 19.1 shows the relationship.

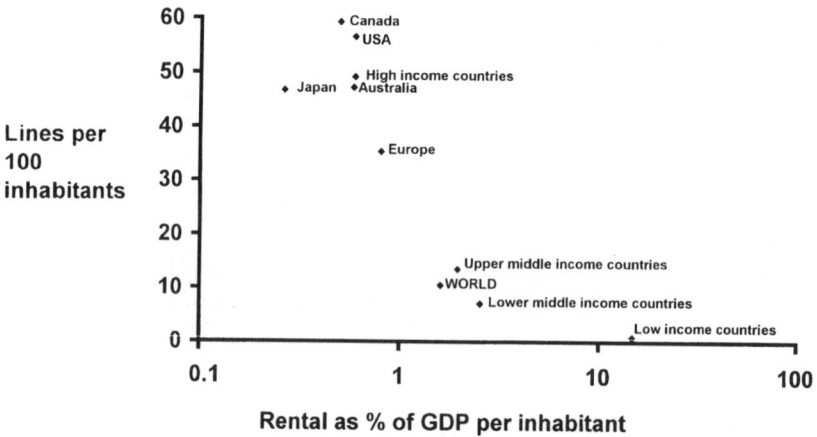

Figure 19.1 Rental, income and penetration, 1992

All telephone companies have to meet social obligations, although the exact form of these varies from country to country. There needs to be a clear distinction between meeting a local universal-service requirement by external government subsidy and by internal subsidy from other customers.

A common problem is political or social pressure to provide cheap basic access for people with relatively low incomes. If this cheap basic access becomes the norm for all customers, it will prevent adequate revenue from being raised and there may be serious consequences for better-off users:

(a)   The operator may be starved of finance and unable to provide the service which they want and can afford.

(b)   Costs may be recovered by excessively high call charges.

(c)   In general, there will be economic inefficiency which reduces the overall value of the services to customers.

As discussed in Chapter 12, dominant operators enjoy a variety of advantages from their size and market power. These provide a margin from which an element of cross-subsidy towards residential services can be afforded. Competition and privatisation have led British and American operators to develop tariff alternatives which avoid this. The general principle is that a basic low-rental service is made available to anyone who wants it. The cost per call is higher than standard, but it will suit poor people who expect to make few calls. Above this, there is a series of options offering cheaper calls in exchange for higher rentals.

An economic model is essential in exploring the full ramifications of tariff structures. The core of the model needs to contain the elements used for a financial appraisal, with projections of cost, income, profit and cash flow and an estimation of the internal rate of return on investment. Additional requirements are:

- productivity, efficiency and unit-cost calculations needed to satisfy the regulator,
- affordability measures to gauge the likely political acceptability of any proposed scheme of tariffs.

The model needs to be run under various assumptions to find the common ground (if it exists) between an acceptable set of tariffs and a rate of return which will induce the foreign investor to part with funds.

There may be a problem in getting reliable figures for local average earnings. In one of the Baltic states, for example, official figures for average earnings were hard to reconcile with other information about the economy. Of the three measures of GDP used in western countries only two, output and expenditure, had been developed by 1993. No estimates were available for the third, based on incomes. The three should, of course, all be the same in expectation. Wages and salaries are typically about two thirds of GDP in the UK, the bulk of the remainder going into company profits. When average earnings were multiplied by official figures for the number of people employed in the Baltic state concerned, the resultant figure came to not much more than a third of GDP, as estimated from the other two measures. The reason for the difference could not be identified at the time. It might have been due to double jobbing or the black economy. Whatever the cause, the affordability of tariffs could not reliably be judged.

Another practical difficulty lies in traffic measurement. Old equipment may not have the ability to count or time local calls. This may come about where calls are free and it brings to mind North American studies into the economics of charging for local calls, which is not necessarily worth doing if the equipment is expensive. Local calling rates can be extremely high where calls are free, even in relatively poor countries. It may be possible to get sample statistics of calling rates carried out by the engineers and extrapolate to a charging situation using price elasticities derived elsewhere. Digital exchanges can meter and bill local calls without difficulty, but there may be a real problem in knowing beforehand how much money the introduction of charging for local calls will raise after they are installed. This can complicate studies of the affordability of the tariff package as a whole.

Price elasticities for long-distance and international services may be obtainable from local data, which can be good in terms of traffic volume even if shaky on inflation rates. A real price elasticity of -0.1 was obtained from the analysis of monthly data in a country where general prices increased by a factor of seven and tariffs were raised twice during the year. Although it is common practice for visiting consultants to bring elasticities derived from other countries in their travel bags as proxies, it is satisfying to be able to derive them from local data and help to persuade local managements that the relationship exists in their own country and not just in the remoter countries of the West.

Although this discussion of economic models is in the context of investor interests, the host country has an equal interest and may wish to equip its regulatory body with this kind of tool. Successful economic modelling depends on the availability of trained local staff who can use and develop it. Such staff may be hard for the regulator to come by at the levels of pay prevailing from the public sector. The people with these skills are the ones most likely to be offered jobs in the commercial sectors of the growing new industries.

Implementation is a critical part of the task. The modelling has to be presented to local officials in terms which they can understand and which will make them willing and able to use it. This calls for a variety of presentational skills and also requires that the model should:

- use data which they will be able to find without undue difficulty;
- produce results that are fully relevant to management and other problems as they are perceived locally;
- run effectively on equipment which is locally available;
- be fully documented in a language which is locally understood.

19.6.3 Country risk

The country may not have a credit rating with the standard rating agencies. In some countries, such as Latvia, annual price inflation has reached 1000 per cent. Most of the East European and ex-USSR governments are new, with short track records from which to judge future political or financial stability. The legal framework may be vague or untested. The presence or absence of telecommunications, commercial, property and competition law will affect the risk. Investors need to determine whether their contacts have negotiating authority, and to investigate the enforceability of contracts and the right to repatriate revenue.

Private investors are likely to regard all investment in eastern Europe, Russia and the developing countries in Africa as being high risk and appraise new projects with a relatively high discount rate, often 25–30 per cent. The future may be hard to predict in countries with strong monarchies, dictatorships or idiosyncratic forms of government, another factor adding to country risk.

19.7 Choosing a partner: choices for the host country

19.7.1 Adjudication

Countries seeking a partner to help develop the domestic telephone network may find many applicants queuing at their door. Adjudication procedures can be complex, with detailed evaluation systems giving points to each aspect of the performance requirements in the contract. There are, however, a few more general considerations which may guide the decision, based on the performance of the investor in other markets, as discussed next.

19.7.2 Home network quality

Some operators have modernised their home networks more quickly than others. Positive features here would include: commercial priorities, getting modernisation quickly to those who need it most, the absence of wasteful investment, a good record for innovative commercial services and good quality-of-service statistics.

19.7.3 Operating efficiency

Operators with high costs at home are less likely to deliver low costs abroad. Points to look for in the home network are: low unit costs, high labour productivity and cost-efficient procurement.

19.7.4 Tariffs

Operators with high or inefficient tariff structures at home could well repeat these features abroad. Points to look for are: low tariffs when ranked in international comparisons, tariffs reasonably related to cost, the absence of undue price discrimination, an ability to meet social goals within acceptable tariffs, an absence of significant public complaint and no reported problems with interconnect.

19.7.5 Experience outside home territory

A relevant matter to look at is the operator's experience in other countries outside the home market. An absence of such experience could be a warning sign. And, if the operator does provide service abroad, are the local people happy with it? If quality-of-service undertakings were given, have they been met? Does the local operation appear to be efficient?

19.7.6 Procurement practices

Procurement practices are worth looking at. Is the operator a cost minimiser on procurement or does it have a favoured, high-cost supplier? What has been done about local manufacture?

19.7.7 Type of operator

The most significant factor of all may be the type of operator involved. Some are privately-owned commercial companies, others are state monopolies. There are likely to be considerable differences in management style, often favouring the private company. State under-takings, on the other hand, may have freer access to cash. If they have empire-building rather than profit-maximising objectives, this may bring significant subsidies to the host.

19.8 Other points for the host country

19.8.1 Investor perpectives needing attention

The host country should pay attention to investor perspectives on key issues, especially those with a critical affect on cash flows. Although there is no set list of factors, four are likely to be of special importance:

(i)   Tariffs and how they are set.
(ii)   The track record and credibility of local management.
(iii) The commercial realism of network modernisation targets in the context of tariffs and likely demand.
(iv) The financial return expected in relation to the cost of capital.

The probable approach of investors on these matters has been described above.

19.8.2 Currency risk

There can be significant exchange-rate risks when development in the host country is financed by loans taken out in foreign currency. Western companies usually cover these by back-to-back or other financial methods to reduce or cap the risk. This is not so easy to do at reasonable cost in countries where financial markets are poorly developed or currencies are unstable, and these are the ones where the risk to the borrower is largest.

Countries with currencies tied to a strong unit, such as the dollar or the French Franc, are exposed to risks which may become critical where local political, inflation and trading conditions differ widely from those in the country where the currency is based. Just before the 50 per cent CFA devaluation of 1994 (Section 12.5.4), one of the countries had arranged a loan in a mixed bag of foreign currencies. This effectively doubled the capital-repayment costs on which the investment projects were based and stimulated an internal discussion on who should carry the losses: the government (through taxation), the PTO (through tariffs) or the lenders (by renegotiation of terms). The episode provided a useful, but potentially expensive, lesson in risk and its management.

Investors wanting to reduce the risk in their investment may also seek guarantees, particularly on the level of tariffs. There may be pressure to have them denominated in US dollars, translated to local currency through the appropriate exchange rate. This transfers the risk back to the host country and may result in tariffs which are politically unsustainable.

19.9 Examples of international investment financing

19.9.1 Hungary

The Hungarian network is provided by MATAV, there are 54 local monopoly concessions, 39 of which are held by MATAV. Network modernisation required more funding than could be raised in Hungary, and MATAV was privatised by a tendering process, which began in 1993 and was won by a consortium of Deutsche Bundespost Telekom, Cable & Wireless and Ameritech International. Cable & Wireless resigned and the other two partners paid $875 m for a 75 per cent share of a 25-year concession which can be extended by half of this period. Adjudication criteria were capital contributed, new technology offered and managerial and operational expertise. A higher bid from STET of Italy was not accepted. Modernisation targets were set, including an annual 15.5 % growth in lines, if there is a demand for it (details from Schmideg [6]). Foreign companies with stakes in the 15 nonMATAV local concessions and in mobile communications came from the US, Canada, France and Israel.

19.9.2 Indonesia

The President of Perumtel, the monopoly operator, said in 1991 that the funding of investment was a serious problem and that private sector finance was needed. Perumtel could manage 25 per cent self-financing but would need external assistance for an accelerated investment plan. A revenue-sharing plan had been introduced and 425 000 lines were being provided by private companies on this basis. About 75 per cent of the revenue would go to the private operator for six to 15 years, after which the title would be transferred to Perumtel. The revenue-sharing arrangement would provide an incentive to keep tariffs on a realistic basis.

19.9.3 Latvia

A proposal to allow Swedish interests into Latvia was aborted after a vigorous political campaign for a broader system of international tendering. The government then launched an international tender for part privatisation and modernisation of the network. Two bids were eventually obtained and the winner was Tilts, a consortium of C&W (70 per cent) and Telecom Finland (30 per cent) which was allocated 49 per cent of the shares. Tariffs were to rise, and there were undertakings about investment levels and service standards. Guidelines on the lines of a limit of four per cent of average earnings for residential phone bills were discussed. C&W announced its intention to pull out in 1998, because it did not have control.

19.10 An example of overseas investment strategy: Deutsche Telekom

Telecommunications were supplied under a joint monopoly with postal services held by the Deutsche Bundespost in the early 1990s, when steps to separate the two services and privatise telecommunications were initiated. The telecommunications arm was set up as Deutsche Telekom and strategic changes soon became apparent, notably in procurement policy where the traditional practice of sourcing most switching equipment from a single supplier, Siemens, was replaced by a more competitive tendering system.

German reunification in 1990 after the collapse of the East German communist regime was followed by a government-led crash programme to modernise the infrastructure of the eastern states. Deutsche Telekom took over the telecommunications services and by 1997 had spent DM50 bn (£21 bn) in upgrading the network [7]. Although the prime purpose of this was political, and the expansion was within the national boundaries, the investment opportunity was similar to the strategies for international geographical extension pursued by other operators. The main difference was in the priority given to speed of provision over cost. Thus, mobile technology was used in the initial phase. Deutsche Telekom became the monopoly supplier of services to 17 million as part of the arrangement.

As late as 1994 Deutsche Telekom was forbidden to own telecommunications businesses outside Germany. However, in 1997 the operator had mobile and paging ventures in Austria, the Czech Republic, Indonesia, Kazakhstan, the Netherlands, the Philippines, Poland, Russia, Switzerland and Thailand and fixed-network ventures in Hungary, Malaysia, the Philippines, Kazakhstan and Thailand. Investment shares varied from 21 to 67 per cent (*Financial Times*, 4 June 1996, Deutsche Telekom). The strong mobile focus reflected experience in eastern Germany. A bid to provide fixed-network services in Latvia had been lost to Tilts (Section 19.9.3). Deutsche Telekom was also a member of the Global One alliance in the international business-services market, which incurred start-up losses of DM 390 m (£130 m) in 1997 (Deutsche Telekom).

The motives behind this rapid diversification were partly political and partly driven by the belief that overseas diversification was likely to be profitable, and even essential to establish a reputation as a global operator. A belief in competitive advantage is reflected in the mobile focus. Some of it has been forced by circumstance or by the belief that the best opportunities abroad would be lost to other operators if no immediate action was taken. The long-term success of the moves remains to be tested. At home, the investment per household is above average and needs a high level of tariff to sustain the level of profitability unless business levels can be raised. From the experience of other operators the profits from investments in developing countries, like those in global alliances, may be elusive, especially in the early years. Deutsche Telekom was partially privatised in 1996 but is currently (1998) still under the control of the government. The strategies, therefore, are best understood as being those of the state.

19.11 Looking ahead

Investment in telecommunications is a long-term business. An investor facing a negative cash flow for the first six years of an investment might, even with a UK project, have some misgivings. The world can change radically in ten years. The establishment of local stability is therefore important if long-term inward investment is to be attracted.

Some of the changes ahead are broadly predictable and positive. International services, for example, are expanding and profitable. Of the 60 billion or so international outward minutes paid for in 1995, only a small fraction came from the countries of central and eastern Europe. Tokyo has more telephones than the whole of Africa. Other changes break new ground. Mobile communications have a growth path far less dependent on the exact level of national income than the fixed networks. Radio and satellite communications have the potential to bypass stages of technological development.

There is a need for both imagination and realism. Some markets have potential for rapid expansion. Budapest once planned to be a Hong Kong to western Europe. Riga could establish itself as an entrepot centre for trade with the Russian republics. In some countries there may be oil and other maverick factors in the economy. In others, the existing network may be overwhelmed with demand stimulated by trivial charges for access and calls. Commercial tariffing may be unpopular, and the political infrastructure may be poor. Underdeveloped countries need western markets but many in the European Union and developed countries elsewhere want to keep them out. Their products are a threat to the high-cost industries of the Rhine and Rhone, whatever their benefit to consumers.

The dynamism of the world's telecommunications markets creates opportunities for a new range of entrepreneurial skills. There seems little doubt that in all countries men and women will readily be found to take them up, practical people as well as those looking to far horizons. Explorers and people of vision have an essential role but, on a company board, their advice has to be accepted within the limits of what is financially prudent. Realism requires a proper appreciation of what is difficult.

19.12 References

1   SCHENK, K–E., KRUSE, J. and MÜLLER, J. (Eds.): 'Telecommunications take-off in transition countries' (Avebury, Aldershot, 1997)
2   KONTKIEWICZ–CHACHULSKA, H. and PHAN, D.: 'From path-dependent processes of structural change to a diversity of market models in central european countries' telecommunications'. Paper presented at ITS 11th biennial conference June 1996, Seville
3   'World telecommunication development report 1994'. ITU, Geneva, 1994
4   GWARTNEY, LAWSON, and BLOCK: 'Economic freedom of the world 1975-1995' (Frazer Institute, Vancouver, 1996)
5   SCHMIDEG, I.: 'Hungary: problems of over- and under–capacity and demand barriers' *in* SCHENK, K–E., KRUSE, J. and MÜLLER, J. (Eds.): 'Telecommunications take-off in transition countries' (Avebury, Aldershot, 1997)
6   'Bonn ruling on charges hits telekom', *Financial Times* 13th September 1997

Index

transmission 44–5, 301
transmission quality required 44,
78, 397
via satellites 74–9
volume measurement 60
Data-envelope analysis 364–6
Data network service
tariffs 220–4
Delaware effect 316
Demand schedule 89
Demand side 129, 299, 420
Depreciation 75, 77, 116–7, 168,
172–5, 180, 182, 184, 253, 258,
315, 349, 357
backlog 175
economic 172, 392
rate 174
reducing balance 173
straight line 172–3
supplementary 175
Derived services network 48
Deutsche Telekom 19, 71, 244, 424,
460–2
Dialling parity 143
Digital compression and multiplexing
equipment 35
Digital european cordless
telecommunications 44
Digital subscriber loop 36–7, 47, 160,
277
Digital transmission 33–4, 45, 159, 445
Diminishing returns 309
Direct broadcasting by satellite 38, 245
Directory enquiries 25, 62, 103, 142–3,
164–5, 192–3, 236, 291
Discounted cash flow 8, 194, 260–1
Discovery process 130–1
Discrimination
price 215, 290, 292–4, 326–7, 392,
448
undue, unreasonable 293–4, 320,
458
Diseconomies of scale and scope 117,
120, 122–3, 169, 366, 429
Dividend policy 185
Divisia index 355
Dominant supplier 137, 139
Downlink 38
Drop-off 43
Dumping 394
Duopoly 137, 139, 150–1, 437
Durbin–Watson test 380

E-commerce 81–2
Econometric models 323, 378–87
Economic depreciation 172, 349, 352
Economic geography 93–7
Economic legislation and treaties 278
Economic value of an asset 39, 172,
175, 273
Economies of scale 68–9, 108, 117–21,
131, 135, 141, 148, 160, 164–6,
171, 180, 220–1, 224, 241, 250,
257, 298, 330, 356, 366, 396,
429, 434–5
Economies of scope 82, 121–3, 143–8,
155, 168, 170–1, 249, 275–6,
298, 330, 421, 425, 429, 439
Efficiency
allocative 137, 194–5, 219, 225,
235, 287–8, 315
distributional 137
dynamic 144–7, 287
economic 137, 148, 189, 197, 213,
236, 270, 333, 340–1, 402
frontier 114–6, 144, 364–6
market 290, 370, 402
network 45, 221, 431
operational 287, 458
productive 136, 197, 287
technical 365–6, 424
Elasticity
GDP against infrastructure
income 90, 106
market 90
own price 90, 382
price 89–90, 101, 103–6, 151–2,
385–6, 405
price and Ramsey pricing 200
spending 206
supply 175–6
Electronic document interchange 61
Electronic funds transfer 61
Electronic mail 80–2
Entropy index 151
Equal access 143
Equipment trade 68
Equity premium 183
European Telecommunications
Standards Institute 301
Exchange connections 9–10, 91, 100,
279
Exit costs 148
Exogenous
variables 373
Externalities 190–1, 195, 198, 203,